Energy Efficiency and Energy Saving in Buildings

Energy Efficiency and Energy Saving in Buildings

Guest Editors

José Luis Molina Félix

Teresa Rocío Palomo Amores

MDPI

Basel • Beijing • Wuhan • Barcelona • Belgrade • Novi Sad • Cluj • Manchester

Guest Editors
José Luis Molina Félix
Department of Energy Engineering
University of Seville
Sevilla
Spain

Teresa Rocío Palomo Amores
Department of Energy Engineering
University of Seville
Sevilla
Spain

Editorial Office
MDPI AG
Grosspeteranlage 5
4052 Basel, Switzerland

This is a reprint of the Special Issue, published open access by the journal *Energies* (ISSN 1996-1073), freely accessible at: https://www.mdpi.com/journal/energies/special_issues/ZP5DU736P6.

For citation purposes, cite each article independently as indicated on the article page online and as indicated below:

Lastname, A.A.; Lastname, B.B. Article Title. *Journal Name* **Year**, *Volume Number*, Page Range.

ISBN 978-3-7258-7474-3 (Hbk)
ISBN 978-3-7258-7475-0 (PDF)
https://doi.org/10.3390/books978-3-7258-7475-0

Contents

About the Editors

José Luis Molina Félix

José Luis Molina Félix is a researcher within the Department of Energy Engineering of the University of Seville (Sevilla, Spain), where his academic and scientific activity focuses on the analysis of and improvement in building energy performance through advanced thermal modeling and simulation. His research interests include the dynamic simulation of buildings, the study of building elements and HVAC systems and components, the optimization of thermal systems, and the development of object-oriented programming approaches for thermal models. He has contributed to the advancement of computational tools that support the design and evaluation of energy-efficient buildings, enabling more accurate prediction of thermal behavior and system performance. Throughout his career, he has participated in research projects related to energy efficiency, sustainable building design, and the integration of innovative thermal solutions in the built environment. His work combines methodological rigor with practical applicability, supporting the development of optimized strategies to reduce energy consumption while maintaining indoor comfort. As Guest Editor of this Reprint, he contributes his expertise in modeling, simulation, and system optimization to foster high-quality research on energy efficiency in buildings and sustainable energy solutions.

Teresa Rocío Palomo Amores

Teresa Rocío Palomo Amores is a researcher within the Department of Energy Engineering of the University of Seville (Sevilla, Spain), with a research profile centered on energy efficiency in buildings and the environmental quality of indoor and outdoor spaces. Her main research interests include outdoor thermal comfort, the characterization and integration of passive heating and cooling techniques, and the development of strategies to enhance energy performance in buildings under different climatic conditions. She has been involved in research initiatives addressing sustainable design, passive conditioning solutions, and climate-responsive approaches aimed at reducing energy demand while improving comfort and resilience in urban and building environments. Her academic contributions emphasize the relationship between energy efficiency, environmental conditions, and user comfort, particularly in warm climates where passive strategies play a key role. Through her role as Guest Editor of this Reprint, she supports the dissemination of interdisciplinary research focused on practical, scientifically robust, and context-sensitive solutions for achieving more sustainable and energy-efficient buildings.

energies

MDPI

Article

Global Solutions for Sustainable Heating, Ventilation, Air Conditioning, and Refrigeration Systems and Their Suitability to the New Zealand Market

Nicholas Andrew Harvey and Eziaku Onyeizu Rasheed *

School of Built Environment, Massey University, Auckland 0632, New Zealand; nicharvs@outlook.com
* Correspondence: e.o.rasheed@massey.ac.nz

Abstract

This paper attempts to find alternative ways in which heating, ventilation, air conditioning and refrigeration systems can be made more energy efficient and sustainable at a global level. Eight technologies or solutions that either passively or supplementarily reduce the heating or cooling load required by a structure are detailed. These technologies or solutions were then presented to heating, ventilation, air conditioning and refrigeration industry professionals in New Zealand to determine their viability and further establish market readiness towards integrating new, innovative, and sustainable solutions in New Zealand. A literature review was conducted to establish the performance of the selected solutions and understand their operational principles and the efficiency they provided. Qualitative research and data collected via semi-structured interviews provided the data for assessing the viability of the selected technologies in the New Zealand market. Following a thematic and hybrid-thematic analysis of the data, the technologies were ranked, and suggestions were made to help improve innovation and energy efficiency in the heating, ventilation, air conditioning, and refrigeration industry in New Zealand. Of the technologies selected, airtightness, heat recovery ventilation retrofits, materials and design principles, and photovoltaic hot water heating were identified as the most viable. The New Zealand market was deemed not to be in a good position to adopt new or alternative solutions. The main issues affecting New Zealand's market readiness to assimilate innovative and energy-efficient solutions are a lack of new technologies, poor standards of education throughout the industry, a lack of regulation, and a lack of government incentives.

Keywords: heating; ventilation; air conditioning; HVAC; New Zealand; energy; efficiency; sustainability; innovation

1. Introduction

The World Bank predicts that nearly 70% of the global population will live in cities by 2050 [1]; in New Zealand, this is already at 87% [2]. The energy demand from New Zealand's residential sector accounts for 33.8% of total demand. If commercial usage is included, that number increases to 57.9%. Residential energy usage related to the heating, ventilation, air conditioning, and refrigeration (HVAC&R) trade in New Zealand is 74%, or 25% of the country's total production [3].

With more people living and working indoors, and with the temperature and weather associated with global warming, the need for indoor environment control for either the inhabitation or the storage of goods is expected to increase. There are two potential

 https://doi.org/10.3390/en18092190

solutions to this problem: either produce cheaper, more accessible systems that increase energy demand, or explore integrated solutions that reduce the heating and cooling demand of our buildings, so they use less energy and are more efficient. This research details eight technologies that fit into the categories of geothermal, building envelope, design, and alternative energy sources and assesses them for viability in the New Zealand market.

Interior temperature control in buildings already accounts for 32% to 40% [4] of the total international electrical energy generation. With significant populations around the world and large minority groups rising out of poverty, the option for indoor climate control will become attainable. An expected increase in demand for heating or cooling leads to either uncomfortable living conditions or power/energy poverty in low socio-economic communities [5]. When combined with expected temperature increases associated with global warming, the requirement for electrical energy to power these systems will dramatically increase.

With the expected increase in demand for indoor temperature control, mass-producing cheap HVAC&R systems will significantly increase the electricity demand of buildings. An alternative approach exists to reduce the energy demand of traditional systems or supplement their performance with passive solutions to improve efficiency. The integration of passive design and the availability of more efficient technologies will reduce energy demand and improve access to better quality, more economical, and healthier systems for lower-income families in emerging markets.

It is plausible that the motivation that drives the customer's decision to choose a product based on the principles of its sustainability or efficiency is not definitive. The industry should strive for efficiency and provide healthy, affordable, and comfortable environments for the people who occupy and inhabit the spaces we build.

This work aims to find alternative ways for HVAC&R systems to be made more energy efficient and sustainable at a global level, and, further, to establish market readiness towards integrating new, innovative, and sustainable solutions in New Zealand.

There is a need to understand the challenges the New Zealand HVAC&R market faces when implementing innovation, identify any underlying barriers that restrict improving the efficiency of the systems, and uncover attitudes towards new or alternate techniques or technologies.

2. Background

A joint agreement is that building energy requirements make up a significant portion of global energy consumption. Although there is no exact number agreed upon across the literature reviewed, the consensus is that it is a considerable proportion, likely between 20% and 50% globally; 22% of the residential sector in the United States of America [6], 50% of global energy usage worldwide [7], and 20.1% of the total delivered energy worldwide [8]. This demand is likely to increase in the future with a global move towards a high-density, urban population whose inhabitants' time is spent primarily indoors [9]. The literature shows that refrigeration and air conditioning (excluding heating and ventilation) are responsible for about 15% of the total electricity consumption around the world [10].

New Zealand's climate, as described by the Köppen–Geiger climate classification system, is temperate, with no dry season, warm summer (Cfb) [11]. Whereas other climate types exist, Cfb is predominant and describes all areas of high-density population. This work focuses on identifying HVAC&R systems that are implementable in New Zealand's climate. As such, the coefficient of performance (COP), a common thermodynamic methodological measure that represents the efficiency of a system, is used to compare the systems' useful heating or cooling produced (Q) to the work required (W). This comparison is conducted based on a widely accepted framework to engineer passive cooling systems consisting of three steps: prevention of heat gains, modulation of heat gains, and heat dissipation [10].

This process was used as a starting point to identify possible solutions. The identified solutions are discussed under the following categories:

(1) Geothermal solution,
(2) Envelope and Building Design solution, and
(3) Alternative energy sources.

2.1. Geothermal Solution

The geothermal principle is a renewable energy application. Outside air temperatures fluctuate with the yearly seasonal cycle; however, the temperature of the soil below the surface remains relatively constant. Caused by the Earth's temperature, this temperature stays at a constant just below the outside air mean temperature [10].

Geothermal heating and cooling uses the subterranean soil as an energy transfer medium. At a depth of up to 10 m below ground level, soil temperatures are affected by the daily and seasonal weather cycles; below 10 m depth, temperatures become relatively constant, followed by a minor increase in temperature with depth at a rate of 2.5 °C/100 m [12].

Soil recharge is a drawback of all geothermal applications. As heat is transferred or absorbed by a heat exchanger, the temperature of the soil changes, and the differential between the soil temperature and the outdoor air temperature reduces, reducing the potential of the system. A recharge period is required in all geothermal applications to allow the soil temperature to return to its natural value.

2.1.1. Geothermal Heat Pump (GEP), Ground Source Heat Pump (GSHP)

A Geothermal Heat Pump (GHP), also known as a Ground Source Heat Pump (GSHP), can be either installed horizontally (in shallow trenches 1 to 2 m deep) or vertically down boreholes up to 250 m in depth. Heat is extracted or rejected to the ground by means of buried pipes, through which a heat transfer fluid usually circulates, either water or a water/antifreeze solution [13]. Despite having higher installation costs, vertical installations are more common.

GHP/GSHP limitations to implementation can be described as non-standardised system designs, significant capital cost compared to other systems, limited skilled individuals knowledgeable in the installation process, and government policy limitations [14]. Upfront capital cost for installation and groundwork is a disadvantage of the Geothermal Heat Pump solution [14].

2.1.2. Foundation/Slab Heat Exchanger (FHX), Geothermal Heat Exchanger Pile (GEP)

Geothermal Energy Piles (GEP) are systems that are not exposed to the groundwork cost associated with Geothermal Heat Pumps. This is because the technique mitigates the requirement for additional excavation and groundwork by utilising the excavation or pile boring process during construction. The heat exchanger components (absorber pipes) are installed in the excavation voids before the pile/foundation material is poured; this transforms the pile into a heat exchanger. Reinforced concrete piles have been found to be advantageous in thermal energy transfer due to the material's high thermal storage capacity and enhanced heat transfer capabilities [15].

The system can be used for both heating and cooling demands with the seasonal cycle. In warmer seasons, when occupants require cooling, the system will use the reinforced concrete pile and soil material as a heat sink. In the colder seasons, this energy is recovered from the ground, which is used as a heat source for the building [16]. GEPs cannot completely offset the energy usage, and electrical energy input will still be required to drive a heat pump [17].

A Foundation Heat Exchanger (FHX) is a concept for a Horizontal Ground Heat Exchanger (HGHX) and latent heat thermal energy storage (LHTES) system [18]. An FHX system uses absorber pipes embedded in the foundation (rather than piles) during the construction phase. FHX installations have properties like GEPs, and such will be considered a similar concept/technology.

GEPs and FHXs do not have the excavation costs of GHXs and are shown to be economically beneficial with time; they, too, experience low maintenance costs as most of the material is embedded in concrete. As it is a new technology, there will be a lack of experienced professionals with experience in these installs; universal guidelines are not fully developed and will likely be required to ensure consistency. GEPs with closed-loop systems offer a COP of between 3.0 and 5.0 [16].

2.1.3. Earth Air Heat Exchanger (EAHE)/Earth Air Tube Heat Exchanger (EATHE)

A building structure can either be in direct or indirect ground communication [19]. Direct ground communication occurs when the building envelope directly interacts with the earth, either partially or fully embedded, for example, in subterranean or underground basements. Indirect communication occurs with systems such as Earth, Air Tunnel Heat Exchangers (EATHE), Earth Air Heat Exchangers (EAHE), or Geothermal Heat Pumps, where heat is transferred through a medium to the building. In the examples of EATHE and EAHE, the air is passed through pipes buried underground, the temperature of the soil is transferred to the air via the pipe walls, and the air is then passed into a structure that is not completely embedded in the terrain.

Ahmed et al. [20] identify two different categories of EAHE, closed loop and open loop. The closed-loop system does not exchange the air in the system with the outdoor air. The air inside the room is cycled through the EAHE and back into the room, heating or cooling depending on the temperature difference between the soil and air. In the open-loop system, fresh air is passed through the EAHE before being distributed to the room and later vented to the atmosphere via an outlet. Where there is the benefit of improved air quality for the open system, the closed loop is considered more energy efficient as the continual cycling reduces the size of the required exchanger tube.

Reported performance of Earth Air Heat Exchanger systems is varied. Literature shows the system's performance in dry climates where the technology originates, where the system was able to reduce incoming air temperature from a mean of 30.9 °C by 7.5 °C; however, as the heat was transferred to the soil, the effectiveness reduced [21]. These types of applications may be viable in the summer hot, dry regions of New Zealand, such as Canterbury or Hawkes Bay.

2.2. *Envelope and Building Design Solution*

The literature for locations under the Köppen–Geiger climate classification temperate (Cfb) states that interior thermal performance and comfort for occupation should begin with an airtight envelope, supplemented with energy-efficient passive heat sources [22].

A combination of a Trombe wall, shading, ventilation, and insulation is the most effective and low-cost passive option for this climate type [9]. Many literature sources agree with the requirement for insulation and suggest reducing the building envelope's exposure to the outside environment [8,23,24]. The following categories, namely, Trombe wall, roof insulation, materials, reflective surfaces, insulation, and phase change materials, will be considered together as envelope and building design principles.

2.2.1. Trombe Wall

A Trombe wall is an indirect gain passive solar system; its purpose is to supply heating and ventilation in buildings by means of solar radiation [25]. Typical construction of this feature consists of a thick slab wall with a heat-absorbing material exposed to a solar source. A transparent glazing, typically glass, is placed with an air gap 2–10 cm away from the wall [26].

Literature is inconclusive regarding the optimal thickness of the slab wall; however, 30 to 40 cm is around the commonly stated dimension [27], 15 cm is very sensitive to changes in solar radiation, while 30 cm thickness was enough to dampen these changes to an acceptable degree for occupation. While a thickness of 30 to 40 cm is recommended [26,28], and 35 cm is the mean [29], 45 cm (thickness) should be used when energy is desired to store energy for nocturnal warming [29]; however, if warming during the day is the priority, a thickness of 20 cm is the best choice. Ideal materials for construction vary, with solid brick having the highest cumulative heat gains in summer and lowest in winter [29], whereas others recommend a concrete-constructed wall [26,28]. A common agreement for the performance of the system is highly dependent on design details, such as size, glazing type, absorber, air layer thickness/channel width, and vents [29]. Divergence from an effective 201 design would limit peak efficiency and energy reductions.

Literature shows that the application of a Trombe wall in Ancona, Italy (climate classification Cfb) contributed to a reduction of heating demand of around 28% [29]. Furthermore, recommending tailored shading devices and vents with daytime operating schedules and a night ventilation strategy for similar regional locations can result in a reduction of cooling loads by 35% or more. All materials and techniques described in the literature would be accessible and implementable in the New Zealand market.

2.2.2. Roof Insulation

In the United States, 14% of total heat gain for single-story residential buildings comes from the roof structure; this figure is lower for multi-story structures [30]. Literature shows research conducted on the possibility of heating and cooling with air recovered from the roof space in typical existing New Zealand residential buildings, and recommends that roof space air should not be used for this purpose, as the benefits were not significant [31]. It was found that a lack of airtightness, lack of insulation, and high thermal conductivity of materials typically used in New Zealand result in significant reductions in the temperature of the roof space when incident solar radiation is no longer available, be it seasonal, meteorological or diurnal [31].

Another research study explored different roof configurations and material performance under tropical conditions and found that the temperature of common flat roofs is 10 °C above the ambient temperature, rising to 15 °C on hot days [30]. They compared the thermal performance of different materials and found that clay tile-clad roofs had better performance over steel roofing, with the latter having an additional maximum ceiling temperature rise of 6 °C along with higher daily fluctuations. When comparing roof configurations, flat roofs, compared to roofs of different geometries, changing to a gabled or hip roof design results in a reduction of temperature by up to 5 °C [30].

By using better available cladding materials, ensuring airtightness, and using favourable roof configurations, the stability of roof space temperature can be controlled. This minimises the losses of energy to and from the building through the roof space, stabilising the internal temperature and resulting in less energy requirement to meet a comfortable standard.

2.2.3. Materials, Reflective Surfaces and Insulation

It is widely accepted that to conserve energy, the thermal performance of buildings needs to be upgraded [32]. Additionally, modern homes are easy to heat and difficult to cool. Several studies have shown the potential of using cheaply attainable materials to control the amount of energy transmitted into and attained by a building envelope. By comparing several building materials (hollow block, red, brick, and solid block) treated with different layers of plaster, insulation material, and reflective coatings, and found that the treatment of two layers of plaster insulation and a reflective surface showed a reduction of thermal load by up to near 49% to a non-treated configuration [7]. When compared to plaster and painting against exterior shade and natural ventilation, the combination of the latter was found to be more effective as a cooling technique [33].

A standard agreement is the unavailability of inferior environmental conditioning for low socioeconomic households. Where this occurs in hot, dry, and equatorial climates, natural ventilation should be implemented to reduce the energy demand required by active environmental controls [5,7,24,34]. Furthermore, natural ventilation [5], as a means for controlling thermal comfort in low socio-economic households under hot conditions, only caused large periods of thermal discomfort, so a mixed mode solution of air-conditioning during the middle of the day, with natural ventilation, was required during dawn and evening. For these climates, the agreement is made for the benefit of reflection, shading, and insulation as an affordable method for reducing energy consumption [7–9,21,24,32,34].

2.2.4. Phase Change Materials

Phase Change Materials (PCM) are a Latent Heat Thermal Energy Storage (LHTES) technology. PCMs work by either storing or releasing latent heat during a phase change process within the material. With this application, PCMs make fantastic insulation materials that reduce heat loss in winter and slow heat gain in summer.

Considerable energy performance can be gained by incorporating PCMs into building envelopes, such as walls, floors, and roofs; however, several factors affect the performance that can be achieved by considering the seasonal effects on temperature, climate, orientation, thickness, phase-change temperature and position of the PCM placement [35]. By retrofitting building envelopes with LHTES, building energy consumption can be reduced by 20 to 50% [36].

PCMs as insulation material are recommended to be used as a supplement to traditional methods, not to replace them. Newer PCM products are easy to install, nontoxic, and can be produced from organic matter. As they are considered new technology that has not been fully adopted by the market, they are still considered expensive. PCM materials work best in temperate climates, where there is a significant difference between outside air temperatures during a diurnal cycle, for recharging purposes. For this same reason, they should not be recommended to tropical coastal installations where the temperature fluctuation is low. Phase Change Materials in Trombe wall designs have been proven to better store energy than conventional materials [37].

2.2.5. Airtightness, Solar, Heat Recovery Ventilation Retrofits

Solar heating, via direct methods (Trombe wall, equatorial facing facades) or transfer medium (Solar/Photovoltaic water heat exchanger), integrated with a Heat Recovery Ventilation (HRV) system, is a current and recurring methodology that exists globally for the climate type Cfb [22]. These methods reflect the theories of the Zero Energy Building (ZEB) methodology [8].

Heat Recovery Ventilation (HRV) or air-to-air energy recovery systems are defined as a device that is able to remove, extract, recover, or salvage heat from one airstream and transfer it to another airstream using a mechanical approach [38]. These provide energy efficiency by reducing the energy consumption required by either preheating or cooling air coming in from outside by transferring or absorbing thermal energy from the exhaust airstream. This effectively minimises the heating or cooling load required by a supplementary traditional air conditioning system. Based on climatic conditions, energy savings in the range of 20 to 35% for annual energy consumption can be achieved by employing air-to-air energy recovery systems [39].

However, heat recovery ventilation systems are not always efficient compared to standard ventilation systems, especially in low-energy buildings with a primary energy consumption of about 50 kWh/m^2 per year [40].

Retrofitting can be difficult due to the requirement for installing ducts and vents in the post-design phase or construction stage. This can be a compromise that risks lowering the efficiency of the system. While the plant can be designed well, the installation requires adequate design for the particular building, as well as proper installation and commissioning [41]. Without it, the system's benefits will be degraded.

2.2.6. Thermal Buoyancy Principles

○ Solar Chimneys.

This passive ventilation system operates by a density gradient that stimulates thermal buoyancy. By using the natural convection of air to ventilate a building, solar chimneys normally consist of a vertical shaft (chimney) vented at the top and open at the bottom. An external source heats the shaft, typically solar radiation. Through conduction, the air molecules' temperature increases and begins to rise, creating an updraft. The updraft lowers the pressure at the base of the chimney; this, in turn, draws in air from the interior of the building. The reduced pressure of the air inside the room then stimulates natural ventilation in the form of fresh air being drawn in from the outside.

Solar chimneys are a form of solar-induced ventilation that supplies natural ventilation to a building [42,43]. The feature has the benefit of lowering the indoor temperature of the building, improving the ventilation airflow rate, and improving the fire rating of the building through smoke ventilation.

Solar chimneys need to be designed properly to be effective. Tall, narrow structures are best suited; however, adverse flow reversal effects can occur if the ratio of the height to the diameter is less than 16 [44,45]. Research on the effects of the design of the roof structure on the performance of a solar chimney shows that buildings with solar chimneys and sloping roofs induced better ventilation rates over buildings with solar chimneys with flat roofs [44]. The best performance was attained when the roof structure sloped upward toward the chimney from all sides because of Bernoulli's principle.

These structures can be utilised for energy generation with the installation of turbines in the shaft [46], but this application is moving away from the scope of reducing the energy required by traditional HVAC&R systems.

Solar chimneys are suitable for hot and arid climates [44], so they may be implementable for summer conditions on the east coast of the North and South Islands of New Zealand [47]. Other applications are possible when coupled with an evaporative cooling system; the thermal buoyancy induced by a solar chimney, drawing air past an evaporative cooler, was able to reduce the peak cooling load of the evaporative cooling system in an office in Japan by 10% [10].

○ Domed Ceilings with Cupolas.

In the United States of America (and, to an extent, New Zealand), roof design is selected based on cost and aesthetic aspects [48]. This overlooking of the thermal performance of roof configuration leads to thermal discomfort and the use of fans and air conditioning systems as a corrective activity.

Solar energy collected by any roof configuration causes an increase in roof temperature. The characteristic of a domed roof or vaulted roof is the auto-shading effect. With the daytime passage of the sun, one side of the roof structure will be subject to direct sunlight whilst the other remains in the shade (with the sole exception of midday). This difference in transmission results in a warmer and cooler surface transmitting to the interior. This surface temperature differential promotes internal air movement as the warming of one side of the structure causes the internal air in contact with the surface to rise, with the opposite happening on the opposing side, where the cooler surface causes the air to subside. The hemispherical exterior shape of the structure has a unique effect via Bernoulli's principle: as air moves over the external surface, it accelerates over the crest, creating an area of low pressure. If a cupola or vent is placed at the peak of the dome, the negative pressure caused by air movement will cause suction and promote natural ventilation and extraction for the internal space. Under normal circumstances, this vent will expel warm air, thus lowering the interior temperature.

The literature provides ample information about the performance of the domed roof solution when compared to flat roofs. Further literature shows that domed/vaulted roofs provide lower internal temperatures than flat-roofed air-conditioned buildings in hot dry climates [49,50]. Solar energy absorption levels are undecided as Gómez-Muñoz et al. [51] found that hemispherical vaulted roofs received approximately 35% less solar energy than a flat roofs; however, this is inconsistent with the findings of Sedighi et al. [52] who found that heat transmitted through a domed roof is 70% more than a flat roof configuration for the same base area. Further, multiple domes did not show heat transfer improvement over flat roofs; however, heat transferability can be improved if cupolas are integrated into the design.

This design principle would be best suited to areas of New Zealand where there are high sunlight hours, as the system is designed around solar incidence and the requirement for ventilation and cooling.

2.3. *Alternate Energy Sources*

2.3.1. Integrated Solar/Photovoltaic Thermal (PV/T) and Hot Water Heating

Due to continuous efforts to reduce the heating demand of buildings, the percentage of energy used for hot water supply is predicted to increase decisively during the upcoming years [53], with the relative weight of energy consumption for domestic hot water having already notably increased [54].

Photovoltaic panels for electricity generation are a solution that supplements a building's energy requirements and can be used to provide energy for hot water heating. When comparing a photovoltaic loop heat pipe/solar-assisted heat pump (PV-LHT) water heating system to an air-source heat pump, it was shown that the lifecycle costs of the new system could be reduced by almost 30% over a standard air-source heat pump [55].

As for photoelectric efficiency, the highest value appears in winter [8,55], which is due to the effect temperature has on the efficiency of PV panels. It is well known that electrical efficiency in PC systems is reduced with increases in panel temperature [56]. As the temperature of the panel increases, the efficiency of electricity production decreases. This becomes problematic when peak electricity generation is typically available in summer months when hot water demand is generally at its lowest; the opposite is the case for winter, where generation is lowest and demand is highest, resulting in electrical generation being insufficient to meet the required heating load.

Photovoltaic Thermal (PV/T) is a solution that uses either air or liquid, acting as a cooling medium for the PV panel. Heat is transferred away from the back of the panel, increasing the panel's efficiency, and the heated air or liquid is then either moved into the home or elsewhere where it may be needed. A reverse cycle is possible where air or liquid can be cooled during the night. Relatively high initial investment costs and low overall efficiency of photovoltaic panels limit their viability [56].

2.3.2. Wastewater Source Heat Pump (WWSHP) and Heat Exchangers (WWHX)

As mentioned above, the relative weight of energy consumption for domestic hot water is expected to increase notably. Wastewater Source Heat Pumps (WWSHP) and Wastewater Heat Exchangers (WWHX) work by using the thermal energy that is stored in grey water and reducing it through heat exchangers. Heating is the primary cycle; however, cooling can be achieved in a reverse cycle, with heat being rejected into the wastewater stream.

There are two distinct applications: local systems, which exchange energy as the greywater leaves a premises, and district systems, where an exchanger integrated into the wastewater infrastructure benefits multiple buildings within its locality. Local systems are regularly used for hot water heating, achieved by recovering the thermal energy stored in water as it leaves the building and transferring it to cold water that acts as a heat transfer medium; this is cycled into the facets to increase the temperature of the water before reheating, this solution has been proven to reach energy savings between 30 to 50% in energy required to heat water for a shower [54].

Whereas local systems provide energy savings, which should not be ignored, district systems that recover heat from the sewer system are more relevant to the scope of the research. The constant flow and availability of wastewater, especially in urban and metropolitan areas, have led to its being deemed a reliable and renewable energy source. Heat recovery from sewer systems occurs in one of three ways: (1) sewer-integrated heat exchangers placed directly into the sewer system, (2) sieved wastewater, where particulates are removed to the water being passed through the heat exchanger, and (3) collection shafts, where the sewerage is stored in a collection chamber while energy is recovered, then returned to the main line [57,58]. The temperature of the sewer content is shallow in daily fluctuation [59]; however, it yearly fluctuates in a wider range, between 10 °C and 20 °C, with the lower being aligned with the warmer seasons and the higher with colder seasons [60]. This fluctuation aligns with the thermal demand from occupants of a building. Estimates show that over 500 district-level WWSHPs operate worldwide [53]. The feasibility of the district WWSHP depends on (a) the distance of the district to the WWSHP, with the shorter being beneficial, and (b) the volatility of the cost of electricity as a fuel for alternative heating applications [59]. District heating is expected to become more favourable as the cost of energy is generally expected to increase in the future.

The literature highlights the efficiency of the system with a COP range of 4.0 to 4.6 [53], when a constant flow rate of 5 L/s of sewerage is provided, a WWSHP can provide heating to an office building with a heating load of 60 W/m^2 [61]; this flow rate is considered replicable in urban areas. The efficiency of the system in economic terms is providing heating at €0.12 per kilowatt/hour (kWh) [57].

Issues experienced when implementing the system around the world include regulation modifications that were required to modernise and regularise wastewater laws to encourage private sector investment in Japan and China [61].

3. Method

The type of research conducted was qualitative. It began with a background study that utilised secondary research techniques and concluded with qualitative data collected through purposive sampling supplemented by snowball sampling.

The first stage of the research was to conduct research focusing on secondary research methods, such as online research, literature research, reviews, and case studies. Secondary research is a process that uses data that already exist [62]. Expanding on an in-depth literature review, the exploratory research was conducted using a systematic process of investigation. This investigative process analysed existing data to determine a solution's suitability to be assessed further by assessing its viability in the New Zealand Market.

3.1. Sampling and Interviews

Data were collected through semi-structured expert interviews. A purposive sampling method was implemented. This selection method was used as the expert opinion was deemed to be essential to attaining data that would accurately answer the research questions. As of January 2025, approximately 247 HVAC&R contractors are operating in New Zealand, with the majority being small, single-owned businesses [63]. A sample of eleven participants was selected for the study. While the sample size may appear small, expert interviews are a recognised qualitative method where depth of insight is prioritised over quantity. According to Gill et al. [64], data saturation in expert interviews is often achieved with fewer participants when the group is homogenous and highly knowledgeable. Hennink and Kaiser [65] noted that 9–17 interviews or 4–8 focus group discussions reached saturation in empirical studies.

The participants were selected based on two different group criteria. Group A participants were the initial participant set and consisted of professionals who met a minimum experience threshold. The experience requirement was fifteen years of construction industry experience, of which ten years' experience is required to be relevant to the HVAC&R trade. Group B participants were the secondary sample set and were selected based on recommendations from participants in Group A. Group B participants did not meet the minimum experience requirements of Group A participants; however, they were deemed competent by way of a professional association. Participants from both Group A and Group B were required to have New Zealand industry employment experience. Participant demographics are described in Table 1. The geodemographic information of participants is from Auckland, New Zealand (9) and Christchurch, New Zealand (2). All participants gave their time voluntarily.

Table 1. Participant Title and Group Composition.

Qualification/Position	Number of Participants	Group A/B
Engineer/Designer	4	3/1
Company Directors	2	2/0
Business Development Manager	1	1/0
Sustainability Manager	1	0/1
Market Manager	1	0/1
Project Manager	1	1/0
Quantity Surveyor	1	1/0

Participants were asked the following peer-reviewed questions to collect data that would be used to address the research questions:

(1) What is the current state of HVAC&R integrated solutions in New Zealand? Is New Zealand a leader or a follower on the global stage?

(2) The following technologies have been identified and are used in other regions of the world. Based on your experience, please rank from most viable to least viable in New Zealand. Please elaborate on any strengths or weaknesses of these identified technologies and any factors that may affect this ranking:
- Geothermal Heat Pump (GHP), Ground Source Heat Pump (GSHP).
- Foundation/Slab Heat Exchanger (FHX), Geothermal Energy Pile (GEP).
- Earth–Air Heat Exchanger (EAHE)/Earth–Air Tunnel Heat Exchangers (EATHE).
- Material and Design (Trombe Wall/Roof Insulation/Materials/Reflective Surfaces/Insulation and Phase Change Materials).
- Airtightness, Solar, and Heat Recovery Ventilation retrofits.
- Thermal buoyancy principles (Solar Chimneys, Domed Ceilings with Cupolas, etc.).
- Photovoltaic Thermal (PV/T), Hot Water Heating.
- Wastewater Heat Recovery (WWSHP), Wastewater Heat Exchangers (WWHX).

(3) Are there any key requirements, legislation, technologies, or infrastructure that New Zealand lacks that would support improving HVAC&R system efficiency?

(4) Please give me your opinion on New Zealand's readiness to assimilate new technologies.

When all questions had been answered, the participants were given the opportunity to raise or discuss any subjects, points of view, or issues that they personally felt were of concern to the industry that had not been expressly identified or covered, which would benefit the themes of the research project.

These questions were revised and refined, and ethics approval was granted on 20 June 2024, Ethics Notification Number 4000029195 for the collection of required data.

3.2. Analysis

An inductive thematic analysis was conducted on interview questions 1, 3, and 4, while an inductive/deductive hybrid thematic analysis was applied to question 2. NVivo was used as a coding mechanism, and Microsoft Excel was the software used to analyse the data. Data were manually coded, and themes were identified by the process outlined [66]. In qualitative research, few guidelines exist to justify an adequate sample size [63]. Common practice is to justify the sample size when the principle of data saturation has been reached. This occurs when no new information is obtained from interviews and/or observations. Saturation was met with a sample size of eleven.

4. Results

The results are detailed as the themes that were developed from the inductive thematic analysis of the data retrieved from the interview process. A short description detailing the common ideas that emerged from each theme has been presented below under the title of each interview question. Where an inductive/deductive hybrid thematic has been completed on interview question two, the data have been made available in Table 2.

- Interview Question One: What is the Current state of HVAC&R integrated solutions in New Zealand? Is New Zealand a leader or a follower on the global stage?

  - Leader or a Follower?

Most participants agreed that, due to New Zealand's relatively small population and market, the New Zealand HVAC&R industry is forced to be a follower on the global stage. Innovation in HVAC&R predominantly originates from the European market. Regulations tend to follow Australia's direction, and materials manufacturing is predominantly sourced from Asia. The geographical relationship of Australia also influences the New Zealand market by affecting the product lines that are made available to the market by the most prominent manufacturers.

- The state of the HVAC&R industry in New Zealand.

Participants described the current state of the industry as neutral, neither good nor bad. More than half (6) of participants had a neutral opinion on the situation, with more than a quarter (4) believing the industry performance was poor, stating development and products being sourced overseas as reasons for this point of view.

- Government & Industry.

Government policy and regulation have been identified as having a prominent influence on the position of the industry in New Zealand. Poor education, low skill set, integration between the trades, regulation, and legislation of the trade limit the market from establishing practices and hold back the industry from adopting new technologies. Other factors that influenced the position of the industry were the lack of government incentives to innovate, geographic location, and market size.

- Motivation.

Most participants agreed that a lack of funding and a lack of incentive to innovate were the main issues preventing clients from exploring more energy-efficient and alternative solutions. Most participants noted the temperate environment, which much of the New Zealand population occupies, with temperatures rarely venturing into extremes, as a limiting factor to exploring alternative solutions. In addition, some participants mentioned a cultural change underway presently where houses are unoccupied for long periods of the day and require different and adaptive solutions.

- Areas in which New Zealand is well-positioned.

A participant believed that the New Zealand market was in a good position. This is attributed to the understanding that New Zealand is on par with the rest of the world when it comes to single-split technology. Additionally, New Zealand has an established and reliable electrical grid, which bodes well for future electrification. Electrification was identified as the future direction that the world was heading towards.

- Interview Question Two: Of the identified technologies used in other regions of the world, with regard to your experience, please rank from most viable to least viable in New Zealand. Please elaborate on any strengths and weaknesses of these identified technologies and any factors that may affect this ranking.

The participants first identified if they were aware of the technologies and the advantages and disadvantages of each application; they then provided insight into viability in New Zealand with a yes or no answer before ranking the options from most viable to least viable (Table 2).

Table 2. Participants' rankings of the selected technologies from most viable (1) to least viable (8).

HVAC&R Technology	Participants Ranking											Average
	P1	P2	P3	P4	P5	P6	P7	P8	P9	P10	P11	
Earth–Air Heat Exchangers (EAHE)/Earth–Air Tunnel Heat Exchanger (EATHE).	8	6	7	6	8	8	8	2	8	5	3	6.27
Thermal Buoyancy Principles (Solar Chimneys, Domed Ceiling with Cupo- las, etc.).	3	2	8	5	7	5	4	8	3	7	6	5.27
Geothermal Heat Pump (GHP)/Ground Source Heat Pump (GSHP).	6	8	5	4	6	4	3	6	6	2	8	5.27
Foundation/Slab Heat Exchangers (FHX), Geothermal Energy Pile (GEP).	7	5	3	1	3	7	5	7	5	3	7	4.82
Wastewater Heat Recovery (WWSHP)/Wastewater Heat Exchangers (WWHX).	2	3	4	2	5	2	6	4	7	8	5	4.36
Photovoltaic Thermal (PV/T), Hot Water Heating	5	7	2	8	4	3	2	1	4	6	2	4.00
Material and Design	4	4	6	3	1	6	7	5	1	1	4	3.82
Airtightness, Solar, and Heat Recovery Ventilation retrofits.	1	1	1	7	2	1	1	3	2	4	1	2.18

After the ranking, the participants were then given the opportunity to describe their reasons for this ranking. The awareness and viability of each technology are depicted in Figure 1.

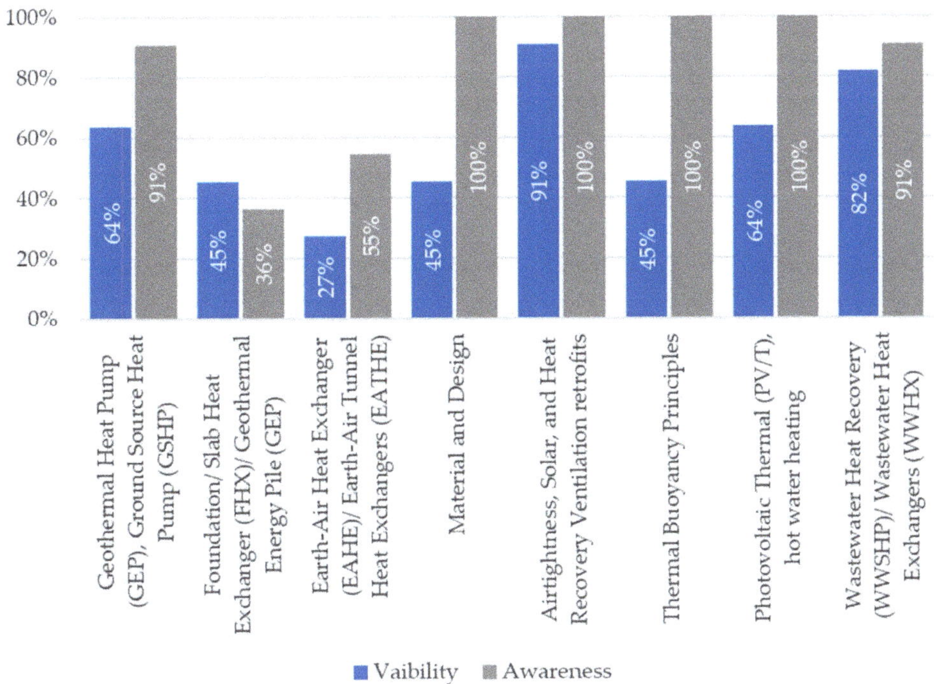

Figure 1. Viability and awareness of technologies by participants.

A summary of the themes derived from their reasons is described below.

○ Geothermal Heat Pump (GHP)/Ground Source Heat Pump (GSHP):

It has a higher COP than a normal air-to-air heat pump and is widely scalable, so huge efficiencies can be found there. The participants stated that the requirement for land, the cost of earthworks, the cost of consent, and compliance would limit technology in New Zealand. The demand for this in New Zealand is unlikely to be high, and this is unlikely to drive the price down into the realm of the mass market. Temperature differential in New Zealand was also stated as being too low in the coastal areas, with more temperate climates; however, there is a possibility that conditions may exist in areas such as the Central Plateau and in the Southern Alps.

- Foundation/Slab Heat Exchangers (FHX), Geothermal Energy Pile (GEP).

Likely viable in areas south of Wellington, New Zealand. Issues stated included seismic concerns, high risk, limited available energy, and the cost to implement. This solution had the lowest awareness rating of all technologies. However, this was not reflected in the perceived viability when the concept was described.

- Earth–Air Heat Exchangers (EAHE)/Earth–Air Tunnel Heat Exchanger (EATHE).

Whereas aquifer water could potentially be used to recharge the system, the system would come with some setbacks, including low ΔT, high cost associated with groundwork, consent, operation, and a high water table.

- Material and Design.

Participants felt that the principles were viable. However, the lack of access to proper materials is limited or incurs an inflated price. The principles covered would be more suited to the alpine or colder regions of New Zealand, such as the Central Plateau and the Southern Alps. Poor insulation in the existing housing stock would require attention prior to any further thought.

- Airtightness, Solar, and Heat Recovery Ventilation retrofits.

Ranked the most viable in New Zealand by Yes/No, 91%, and by scored ranking. The strengths mentioned include that it would be effective in reducing humidity in temperate climates and address the need for better ventilation. Weaknesses to implement included poor insulation in the existing housing stock, high cost, low efficiency, and the requirement for specialist design to be an effective proposition for the unique design characteristics of the existing New Zealand housing market.

- Thermal Buoyancy Principles (Solar Chimneys, Domed Ceiling with Cupolas, etc.)

This could be a good option if the heating of the roof structure could be utilised more. However, the cost to mass implementation is too large, cultural design principles tend away from this solution, and a lack of familiarity, education and experience to get the system to operate as intended drives up the installation costs.

- Photovoltaic Thermal (PV/T) and Hot Water Heating.

Photovoltaic panels have succeeded solar panels in the New Zealand market. Substantial momentum existed in technology through the mid-2010s. However, poor insulation led to inefficiency, coupled with high maintenance costs, and the payback period became inflated. Photovoltaic panels are now a more viable option for this application; however, storage and synchronisation of high-demand peaks and available energy production limit their feasibility. Thermal integration of panel cooling was not recognised.

- Wastewater Heat Recovery (WWSHP)/Wastewater Heat Exchangers (WWHX).

This technology ranked the second most viable to the interview participants when asked if it was viable or not viable, with 82% stating viable and placed fourth in the scored ranking. All participants deemed the technology to be an economically viable option; however, they stated issues with implementation when considering that the wastewater network is governed by local councils, which may include legal concerns regarding public asset use.

- Interview Question Three: Are there any key requirements, legislation, technologies, or infrastructure that New Zealand lacks that would support improving HVAC&R system efficiency?

○ Government, Regulation, and Authority

Many participants identified legislation as an area that needs to be improved to support improving system efficiency. The main concern raised regarding legislation is the lack of regulation in the HVAC&R trade. Two things that were identified as should be provided are standard guidelines for installation, which are currently not available to the market, and more government incentives.

An idea that some participants recommended as implementable in New Zealand was mandatory Operational Efficiency Labels on buildings. Described as a "Carrot and stick" approach that would generate incentives for people to invest in improving building energy performance rather than adopting a prescriptive approach. It was mentioned that this is common in Organisation for Economic Co-operation and Development (OECD) countries, Europe, and Australia; however, New Zealand only has voluntary schemes.

○ Industry Standard, Education, and Competency.

The participants felt that the professional environment in the HVAC&R industry is lacking. The reason for this belief is that the industry is unregulated, and with this, there is no requirement to upskill or continue education beyond the minimal acceptable standards. Some participants felt that this further results in below-satisfactory levels of installation and sub-optimised, energy-inefficient installations and solutions.

Availability and accessibility to technology are further issues. Despite its willingness, the industry in New Zealand does not have access to the latest global technology due to market size and demand, so it faces the challenge of being unable to train a suitably experienced workforce to deliver innovative technology to the market successfully. Furthermore, a lack of regulation makes the prospect of introducing new innovations risky to a manufacturer, so there is hesitancy.

When asked why these issues are present, the reply was that the industry body was either not big enough or strong enough to implement change. This is the belief that there is an underlying attitude that encourages the industry to solve its own problems, rather than adopt processes or solutions that have been successful in similar climates and situations.

○ Motivation and Market factors.

The lack of manufacturing or infrastructure places a reliance on products and technology being imported into the country. This factor makes innovation unaffordable. Innovation, at the scale of the New Zealand market, is a risky and unprofitable venture, driving a low incentive to innovate.

○ Areas for Improvement.

Areas in the industry mentioned by participants that were in urgent need of improvement are ventilation and dehumidification, air quality in buildings, and the adoption of more green building trends.

- Interview Question Four: Please give me your opinion on New Zealand's readiness to assimilate new technologies.

○ Attitude toward innovation.

Despite a quarter of participants (3) stating that the industry was willing to explore new or innovative solutions, particularly in the high-end market, the consensus was that the New Zealand industry was not ready. Financial constraints, government policies, and lack of access to new technology are all attributed as reasons that the market is not ready.

○ Industry Governance.

Training and education were identified as the biggest issues when it comes to trying new things. There is an agreement that there is a skills gap in the industry, from the bottom all the way to the top. This shortcoming is not exclusive to trade but includes the client aspect, as the consumer is neither educated nor aware of the alternative or how these may affect their building, operational costs, or indoor air quality. Participants mentioned that there is a gap in knowledge and experience when it comes to writing and implementing effective legislation, initiatives, and building standards. Some stated that this is exacerbated by New Zealand's relatively short government term, resulting in potentially volatile building policies.

○ Motivation and a need to change.

Motivation is predominantly determined by financial flexibility to explore alternative solutions. In the current market, and for much of the population, there is hesitancy and a lack of finance to explore alternatives and step away from tried and tested solutions. Historical data are unavailable to depict return on investment, incentives to innovate, and the hierarchy of fund allocation to different trades in the building process, which affect the availability of capital to be directed to explore alternative options.

○ Direction of the Market.

A small number of participants believed that the New Zealand industry was in a good position for the adoption of new innovations and technologies. With the energy grid in New Zealand being heavily electrified, New Zealand stands well to adopt new technologies as the global trend heads toward renewable energy electrification.

○ Current situation.

There is no agreement that new technologies are the solution; however, there is agreement that access to technology is limited. Technology is not the only thing that is limited, from basic building materials and insulation through to materials specific to industry, such as low Global Warming Potential (GWP) refrigerants. "90% of the required housing stock in New Zealand is already built", so there lies a problem that a solution is required from the approach of improving the quality of air in these buildings rather than focusing purely on new construction projects.

5. Discussion

This work aimed to explore global alternative energy-efficient HVAC&R solutions for New Zealand and establish market readiness towards integrating new, innovative, and sustainable solutions.

The first part of the research identified alternative HVAC&R solutions that reduce traditional installations' energy demand and establish their viability in the New Zealand Market. The second part of the research looked at what underlying factors influenced the readiness of the market towards integrating new, innovative, and sustainable solutions in New Zealand. The findings are discussed below.

Viable Solutions

Airtightness, solar, and heat recovery ventilation retrofits were the most viable solutions, according to the participants interviewed. It is interesting to note that the strengths of the system, being that it reduces indoor humidity in temperate climates and provides better standards of ventilation to buildings, are in line with some of the identified areas that some participants stated as being urgent areas for improvement. It is essential to realise that despite this solution being noted as the most viable, the weaknesses identified by participants and in the literature are in line with some of the issues the industry currently faces in New Zealand [67]. These are expensive to design and implement, and poorly insulated homes with complex designs remain challenges that will affect the solution's viability in the market. Regardless of this, the system was still identified as the most viable, both in terms of ranking and by a yes or no answer.

Materials and Design, as a category, did not rank highly on viability when asked if it was a viable strategy or not. However, it was the second most feasible option scored. The category included a wide variety of material and design principles, of which some of the participants had negative viewpoints or were questioned on whether they were implementable, namely, phase change materials, stating access to materials, cost of the product, and unavailable performance data [68]. Trombe walls were mentioned as expensive propositions due to the costs associated with the design and the lack of skilled professionals to implement this type of system successfully. These themes align with other themes identified in other interview questions, as discussed later. Other principles, such as using better material, reflective surfaces, insulation, and roofing materials, were highly praised, with one participant noting that the solutions "align with the Zero-Energy Building (ZEB) methodology, which should be integrated more into the New Zealand construction industry". This mismatch of opinions in the category may explain the high viability ranking, although further research should be pursued regarding specific methods and technologies.

Photovoltaic, thermal, and hot water heating, despite participants being vocal about the failures of solar thermal technology in New Zealand, remains a viable option through photovoltaic technology as a substitute. Access to material remains a limiting factor, as with most new technologies in New Zealand, which increases the cost and limits accessibility to a large percentage of the population. Peak electrical generation in summer, energy storage requirements, and low sun hours during winter were all disadvantages identified that aligned with the literature. Although the application of hot water heating was deemed viable, further integration with panel cooling, heat diversion, and subsequent indoor heating is an unfamiliar application for most participants; however, it would likely not negate the identified disadvantages.

Wastewater Heat Recovery (WWSHP) and Wastewater Heat Exchangers (WWHX) were the second most viable options by yes or no opinion; however, they scored fourth when ranked against others. The participants identified issues regarding legislation and implementation of these systems as potential issues in New Zealand; this is in line with limitations identified in the literature. Despite not all being aware of the technology, most participants express positivity towards this solution, yet not one ranked it first.

Geothermal applications were deemed among the least viable of the technologies identified for this research. Although one of the cleanest and most stable alternative energy sources was identified, the results show that the participants do not favour the technology. Cost of groundwork, access to land, council consent, and seismic concerns were all identified as deterring factors towards geothermal in general. Where the cost of groundwork was identified in the literature as being a disadvantage, other issues mentioned were not and could be specific to the New Zealand context. Participants were aware of the technology; however, most associated geothermal with hydrothermal principles associated with the New Zealand Taupō Volcanic Zone, or Kawerau region.

Foundation/Slab Heat Exchanger (FHX) and Geothermal Energy Piles (GEP). Despite having the lowest awareness of all technologies, its viability by ranking and yes or no was neutral, seismic concerns were raised, and statements about favouring locations south of

Wellington were made. These statements do not align with the limitations identified in the literature, so further research into these concerns will be required.

Geothermal Heat Pumps (GHP) or Ground Source Heat Pumps (GSHP) have certain general shortcomings. It has been suggested that the temperature differences typical of temperate climates may not be ideal for these systems, and that they are better suited for colder regions such as the Southern Alps and Central Plateau. While this perspective is not widely documented in the literature, other shortcomings are noted. Themes that emerged from subsequent interviews included limited experience and skills in implementation, high costs of the technology, and a lack of coordinated efforts among different trades during installation.

The participants stated that thermal buoyancy principles are widely used in many applications in New Zealand, particularly warehouse structures. Although there are examples of this theory in non-warehouse installations, the temperate climate with lower sunlight hours and high humidity levels is not conducive to an effective solution. This finding reflects the specific geographic locations of the literature; further research into the application of the principle in temperate climates in non-warehouse buildings should be conducted.

Where thermal buoyancy principles were related to design, such as domed ceilings with cupolas and solar chimneys, participants stated that pre-existing biases towards established design principles tended to dictate the direction of structural design and construction techniques in that regard. This was not considered or reflected in the literature; in addition, unfamiliar practices lack skilled individuals and so incur higher installation costs.

The Earth–Air Heat Exchanger (EAHE)/Earth–Air Tube Heat Exchanger (EATHE) was the least viable option, both by opinion and ranking, for the reasons associated with geothermal as a category and issues identified with temperate climate conditions, particularly humidity, groundwater, and bacterial spores developing in the system. Solutions to overcome these issues were not identified in the literature, and further investigation would be required to validate these statements.

Despite the literature showing that all of the selected technologies are capable of reducing the energy loading required by traditional HVAC&R systems, their viability was very specific to New Zealand's climate, industry, legislation, and culture. There is an implication when assessing how to implement an alternative solution to consider these factors in the feasibility and subsequent payback period assessment.

- Assessing the New Zealand HVAC&R market's readiness towards improving efficiency and innovation.

Some prominent themes emerged when trying to establish market readiness toward integrating new, innovative, and sustainable solutions in New Zealand. The government's influence, education, finance, access to technology, and underlying cultural themes were recurring.

The first thing to consider is New Zealand's unique geographic situation. The country is relatively isolated from most of the world, with its nearest trading partner being Australia. The country has a small population, and despite being a developed country, the HVAC&R industry is small, and so is the market. Due to these factors, manufacturing is almost non-existent, the capital available in the market, and the demand for research and development of product lines to meet the country's environmental and market conditions is not a scalable endeavour and would not likely be able to compete with global technology and manufacturing.

Because of this, the HVAC&R industry in New Zealand is almost forced to purchase and import materials and technologies from overseas. A secondary effect of this is that the industry is forced to follow the direction of Australia for product lines and Europe for legislation and technology direction.

New Zealand has a temperate climate, no dry season, and a warm summer climate (Cbf). This climate does not experience extreme temperatures or significant variations across the year; however, due to the maritime influence, high humidity does have a considerable impact on the nation's requirements.

Education in industry has a large impact on the ability to implement new or innovative technologies into the market. Some of the reasons behind the lack of industry skill set can be attributed to the restricted availability or access to new technology. Because of the difficulty faced by the industry in acquiring technology and materials, there is not enough demand to push education and skill development forward to a point where there is an adequately skilled workforce to complete installations to an appropriate level.

There are no industry-enforced standards or regulations for working in the HVAC&R trade in New Zealand. Unlike other trades, electricians, plumbers, gas fitters, etc., where there is a requirement to be licensed and registered, the HVAC&R industry does not require a recognised qualification. There is a very limited professional environment in the HVAC&R trade, and with this, there is no requirement to upskill or continue education beyond the minimum acceptable level. This promotes the minimum standards to be practised. This results in the New Zealand market being a higher-risk environment for manufacturers to operate in, further leading to possible hesitancy in introducing new products into the market. Where industry participants repeatedly stated that the industry should require better professional standards, further suggestions were made for using standard guidelines for installation.

Lack of education is not limited to industry professionals. The opinion of some participants is that clients, building owners, building operators, landlords, and people associated with selecting the mechanical services to be installed or designed for a structure are typically not aware of the performance or efficiencies available by alternate solutions. Furthermore, the cost of consent associated with alternative and innovative solutions deters the appetite to innovate. The participants recommended (1) more government incentives for projects aimed at improving energy efficiency and (2) building operational efficiency labels to promote public awareness and generate investment in energy efficiency.

While there seems to be an attitude and willingness toward innovation and energy efficiency in New Zealand, the results show that the market's priority needs regarding efficient HVAC&R systems do not reflect this. Where motivation can be spurred by the methods suggested above, a significant proportion of participants agreed that there were some areas of concern in the New Zealand industry, particularly around industry standards for dehumidification and ventilation. Where this would require further investigation, it is beyond the purpose of the research.

The implications of these findings are that the industry has some challenges that will need to be addressed if improvements to energy efficiency are to be prioritised; these challenges include overcoming poor education, lack of regulation, lack of initiatives, and lack of access to new technologies.

6. Conclusions

New Zealand's temperate climate was a common feature when discussing the viability of all technologies/solutions. With a large percentage of the population occupying coastal and temperate climates, humidity, relatively shallow differences in high and low temperatures, and low ΔT negate the requirement for integrated solutions in most circumstances. Due to geographical isolation and small market size, access to technology is somewhat limited, coupled with low demand, which results in installation costs being high. Legislation, consent, and a lack of skilled workforce further led to increased risk and projected costs.

The New Zealand market is not in a good position to adopt new or alternative solutions. There is a willingness toward innovation; however, the ability to explore new technologies is reserved for organisations or individuals with financial means. Lack of manufacturing, relative market size, and industry organisation have resulted in high installation costs, supply issues, and a position of followers on the global stage. Key issues that put the industry in this position are a lack of industry regulation, professional qualifications, and inadequate standards.

Recommended and most viable technologies or strategies by industry opinion to improve the energy efficiency of HVAC&R systems were as follows:

- Airtightness and Heat Recovery Ventilation Retrofits.
- Material and Design (Roof Insulation/Materials/Reflective Surfaces/Insulation).
- Photovoltaic Thermal (PV/T) and hot water heating.
- Wastewater Source Heat Pump (WWSHP) and Heat Exchangers (WWHX).

Where other innovative technologies exist (not considered in this research), recommendations should be made to assess the specific characteristics of the solution to New Zealand's climate, industry, legislation, and culture and to consider the industry's ability to deliver successful installation. Further recommendations can be made to evaluate the efficiencies provided by the selected technologies of this research, where examples exist within New Zealand, if available.

Before the implementation of any incentives, further study should be conducted into the success of other countries' government initiative programs and their effectiveness in improving the energy efficiency of HVAC&R systems.

Further areas of study would include an investigation into effective industry regulation and policy, along with the level of qualifications to improve the quality of installations before any recommendations could be made in this regard.

Also, studies on other countries would provide a comparative analysis of barriers to the implementation of innovative solutions and the development of specific strategies to overcome the identified barriers. The authors note that a considerable proportion of participants were based in Auckland, New Zealand, and Christchurch, New Zealand. It was mentioned that the different climatic conditions of the Southern Alps and Central Plateau regions of New Zealand may be more suited to some technologies identified in this research. Involving participants with expertise in these regions would have provided a deeper insight into the situation. Throughout the research process, information became available about some technologies that had recently been implemented in some construction projects around New Zealand; although data was requested, it was either non-existent or not available at the time of publication. This data would have been beneficial to the project. Further limitations would be the input of the customer or client regarding attitudes toward innovation, energy efficiency, and motivations.

Author Contributions: Conceptualization, N.A.H.; Methodology, N.A.H.; Formal analysis, N.A.H.; Writing—original draft, N.A.H.; Writing—review & editing, E.O.R.; Supervision, E.O.R. All authors have read and agreed to the published version of the manuscript.

Funding: This research received no external funding.

Data Availability Statement: The original contributions presented in the study are included in the article, further inquiries can be directed to the corresponding author.

Acknowledgments: The author wishes to acknowledge Thermal Solutions Ltd., which supported this work. The author declares that all work is their own, and, to their knowledge, has referenced all material that they did not produce.

Conflicts of Interest: The author declares that they have no known competing financial interests or personal relationships that could have appeared to influence the work reported in this paper.

References

1. Zhang, M. Urban Development. 2023. World Bank Group. Available online: https://www.worldbank.org/en/topic/urbandevelopment (accessed on 15 November 2024).
2. Urban Population (% of Total Population)—New Zealand. 2018. World Bank Group. Available online: https://data.worldbank.org/indicator/SP.URB.TOTL.IN.ZS?locations=NZ (accessed on 24 March 2025).
3. Ministry of Business, Innovation & Employment. Electricity Statistics. 12 September 2024. Available online: https://www.mbie.govt.nz/building-and-energy/energy-and-natural-resources/energy-statistics-and-modelling/energy-statistics/electricity-statistics (accessed on 15 November 2024).
4. Singh, R.; Sawhney, R.; Lazarus, I.; Kishore, V. Recent advancements in earth air tunnel heat exchanger (EATHE) system for indoor thermal comfort application: A review. *Renew. Sustain. Energy Rev.* **2018**, *82*, 2162–2185. [CrossRef]
5. Bienvenido-Huertas, D.; Sánchez-García, D.; Rubio-Bellido, C. Analysing natural ventilation to reduce the cooling energy consumption and the fuel poverty of social dwellings in coastal zones. *Appl. Energy* **2020**, *279*, 115845. [CrossRef] [PubMed]
6. Gluesenkamp, K.R.; Chugh, D.; Abdelaziz, O.; Moghaddam, S. Efficiency analysis of semi-open sorption heat pump systems. *Renew. Energy* **2017**, *110*, 95–104. [CrossRef]
7. Uriarte-Flores, J.; Xamán, J.; Chávez, Y.; Hernández-López, I.; Moraga, N.O.; Aguilar, J. Thermal performance of walls with passive cooling techniques using traditional materials available in the Mexican market. *Appl. Therm. Eng.* **2019**, *149*, 1154–1169. [CrossRef]
8. Sun, X.; Gou, Z.; Lau, S.S.-Y. Cost-effectiveness of active and passive design strategies for existing building retrofits in tropical climate: Case study of a zero energy building. *J. Clean. Prod.* **2018**, *183*, 35–45. [CrossRef]
9. Tejero-González, A.; Andrés-Chicote, M.; García-Ibáñez, P.; Velasco-Gómez, E.; Rey-Martínez, F.J. Assessing the applicability of passive cooling and heating techniques through climate factors: An overview. *Renew. Sustain. Energy Rev.* **2016**, *65*, 727–742. [CrossRef]
10. Santamouris, M.; Kolokotsa, D. Passive cooling dissipation techniques for buildings and other structures: The state of the art. *Energy Build.* **2013**, *57*, 74–94. [CrossRef]
11. Beck, H.E.; Zimmermann, N.E.; McVicar, T.R.; Vergopolan, N.; Berg, A.; Wood, E.F. Present and future Köppen-Geiger climate classification maps at 1-km resolution. *Sci. Data* **2018**, *5*, 180214. [CrossRef]
12. Sani, A.K.; Singh, R.M.; Amis, T.; Cavarretta, I. A review on the performance of geothermal energy pile foundation, its design process and applications. *Renew. Sustain. Energy Rev.* **2019**, *106*, 54–78. [CrossRef]
13. Minea, V. *Heating and Cooling with Ground-Source Heat Pumps in Cold and Moderate Climates: Design Principles, Potential Applications and Case Studies*; CRC Press: Boca Raton, FL, USA, 2022.
14. Self, S.J.; Reddy, B.V.; Rosen, M.A. Geothermal heat pump systems: Status review and comparison with other heating options. *Appl. Energy* **2013**, *101*, 341–348. [CrossRef]
15. Gao, J.; Zhang, X.; Liu, J.; Li, K.; Yang, J. Numerical and experimental assessment of thermal performance of vertical energy piles: An application. *Appl. Energy* **2008**, *85*, 901–910. [CrossRef]
16. Tarnawski, V.; Leong, W.; Momose, T.; Hamada, Y. Analysis of ground source heat pumps with horizontal ground heat exchangers for northern Japan. *Renew. Energy* **2009**, *34*, 127–134. [CrossRef]
17. De Moel, M.; Bach, P.M.; Bouazza, A.; Singh, R.M.; Sun, J.O. Technological advances and applications of geothermal energy pile foundations and their feasibility in Australia. *Renew. Sustain. Energy Rev.* **2010**, *14*, 2683–2696. [CrossRef]
18. Hughes, P.; Im, P. *Foundation Heat Exchanger Final Report: Demonstration, Measured Performance, and Validated Model and Design Tool*; Oak Ridge National Laboratory (ORNL): Oak Ridge, TN, USA, 2012.
19. Fazlikhani, F.; Goudarzi, H.; Solgi, E. Numerical analysis of the efficiency of earth to air heat exchange systems in cold and hot-arid climates. *Energy Convers. Manag.* **2017**, *148*, 78–89. [CrossRef]
20. Ahmed, S.F.; Saha, S.C.; Debnath, J.; Liu, G.; Mofijur, M.; Baniyounes, A.; Chowdhury, S.; Vo, D.-V.N. Data-driven modelling techniques for earth-air heat exchangers to reduce energy consumption in buildings: A review. *Environ. Chem. Lett.* **2021**, *19*, 4191–4210. [CrossRef]
21. Bulut, H.; Demirtaş, Y.; Karadağ, R.; Hilali, İ. Experimental analysis of an earth tube ventilation system under hot and dry climatic conditions. In Proceedings of the 8th Mediterranean Congress of Heating Ventilation and Air-Conditioning-Climamed, Nice, France, 10–11 September 2015.

22. James, M.; Bill, J. *Passive House in Different Climates: The Path to Net Zero*; Routledge: London, UK, 2016.
23. Gustafsson, M.; Dermentzis, G.; Myhren, J.A.; Bales, C.; Ochs, F.; Holmberg, S.; Feist, W. Energy performance comparison of three innovative HVAC&R systems for renovation through dynamic simulation. *Energy Build.* **2014**, *82*, 512–519.
24. Lee, S.; Park, S. Zero-energy building integrated planning methodology for office building considering passive and active environmental control method. *Appl. Sci.* **2021**, *11*, 3686. [CrossRef]
25. Hernández-López, I.; Xamán, J.; Chávez, Y.; Hernández-Pérez, I.; Alvarado-Juárez, R. Thermal energy storage and losses in a room-Trombe wall system located in Mexico. *Energy* **2016**, *109*, 512–524. [CrossRef]
26. Jovanovic, J.; Sun, X.; Stevovic, S.; Chen, J. Energy-efficiency gain by combination of PV modules and Trombe wall in the low-energy building design. *Energy Build.* **2017**, *152*, 568–576. [CrossRef]
27. Balcomb, J.D.; Hedstrom, J.C.; McFarland, R.D. Simulation analysis of passive solar heated buildings—Preliminary results. *Sol. Energy* **1977**, *19*, 277–282. [CrossRef]
28. Agrawal, B.; Tiwari, G.N. *Building Integrated Photovoltaic Thermal Systems: For Sustainable Developments*; Royal Society of Chemistry: London, UK, 2011; Volume 4.
29. Simões, N.; Manaia, M.; Simões, I. Energy performance of solar and Trombe walls in Mediterranean climates. *Energy* **2021**, *234*, 121197. [CrossRef]
30. Sirimanna, M.; Attalage, R. A model for analyzing the thermal performance of roof configurations with flat inclined surfaces. *Energy Build.* **2016**, *116*, 122–132. [CrossRef]
31. Fitzgerald, W.B.; Fahmy, M.; Smith, I.J.; Carruthers, M.A.; Carson, B.R.; Sun, Z.; Bassett, M.R. An assessment of roof space solar gains in a temperate maritime climate. *Energy Build.* **2011**, *43*, 1580–1588. [CrossRef]
32. Ponmurugan, M.; Ravikumar, M.; Selvendran, R.; Medona, C.M.; Arunraja, K. A review on energy conserving materials for passive cooling in buildings. *Mater. Today Proc.* **2022**, *64*, 1689–1693. [CrossRef]
33. Samani, P.; Leal, V.; Mendes, A.; Correia, N. Comparison of passive cooling techniques in improving thermal comfort of occupants of a pre-fabricated building. *Energy Build.* **2016**, *120*, 30–44. [CrossRef]
34. Bisoniya, T.S.; Kumar, A.; Baredar, P. Energy metrics of earth–air heat exchanger system for hot and dry climatic conditions of India. *Energy Build.* **2015**, *86*, 214–221. [CrossRef]
35. Li, C.; Wen, X.; Cai, W.; Wu, J.; Shao, J.; Yang, Y.; Yu, H.; Liu, D.; Wang, M. Energy performance of buildings with composite phase-change material wallboards in different climatic zones of China. *Energy Build.* **2022**, *273*, 112398. [CrossRef]
36. Harish, V.; Kumar, A. A review on modeling and simulation of building energy systems. *Renew. Sustain. Energy Rev.* **2016**, *56*, 1272–1292. [CrossRef]
37. Rabani, M.; Kalantar, V.; Dehghan, A.A.; Faghih, A.K. Experimental study of the heating performance of a Trombe wall with a new design. *Sol. Energy* **2015**, *118*, 359–374. [CrossRef]
38. Ahmad, M.I.; Mansur, F.Z.; Riffat, S. Applications of air-to-air energy recovery in various climatic conditions: Towards reducing energy consumption in buildings. In *Renewable Energy and Sustainable Technologies for Building and Environmental Applications: Options A Greener Future*; Springer: Berlin/Heidelberg, Germany, 2016; pp. 107–116.
39. Rasouli, M.; Ge, G.; Simonson, C.J.; Besant, R.W. Optimal Control of Energy Recovery Ventilators During Cooling Season. *ASHRAE Trans.* **2014**, *120*, 386–396.
40. El Fouih, Y.; Stabat, P.; Rivière, P.; Hoang, P.; Archambault, V. Adequacy of air-to-air heat recovery ventilation system applied in low energy buildings. *Energy Build.* **2012**, *54*, 29–39. [CrossRef]
41. White, J.; Gillott, M.; Wood, C.; Loveday, D.L.; Vadodaria, K. Performance evaluation of a mechanically ventilated heat recovery (MVHR) system as part of a series of UK residential energy retrofit measures. *Energy Build.* **2016**, *110*, 220–228. [CrossRef]
42. Monghasemi, N.; Vadiee, A. A review of solar chimney integrated systems for space heating and cooling application. *Renew. Sustain. Energy Rev.* **2018**, *81*, 14–27. [CrossRef]
43. Maghrabie, H.M.; Abdelkareem, M.A.; Elsaid, K.; Sayed, E.T.; Radwan, A.; Rezk, H.; Wilberforce, T.; Abo-Khalil, A.G.; Olabi, A. A review of solar chimney for natural ventilation of residential and non-residential buildings. *Sustain. Energy Technol. Assess.* **2022**, *52*, 102082. [CrossRef]
44. Awad, H.H.; Desouki, M. Integrating physical experiments with computational fluid dynamics to transform mosque minarets into efficient solar chimneys. *Sci. Rep.* **2024**, *14*, 9721. [CrossRef] [PubMed]
45. Kouravand, H.; Omidvar, A.; Brambilla, A. Evaluation of the Buoyancy-Induced Natural Ventilation Performance of PVC Tube-Type Solar Chimneys: Numerical Study and Experimental Validation. *Iran. J. Sci. Technol. Trans. Mech. Eng.* **2024**, *49*, 361–372. [CrossRef]
46. Afsari, H.; Yousefi, M.R.; Sajjadi, M.; Shirvani, M. Experimental and numerical study on effects of additional air height and existing soil layer under the collector of solar chimney. *Energy Syst.* **2024**, 1–17. [CrossRef]
47. Sturman, A.P.; Tapper, N.J. *The Weather and Climate of Australia and New Zealand*; Oxford University Press: New York, NY, USA, 1996.

48. Huang, J.; Hanford, J.; Yang, F. *Residential Heating and Cooling Loads Component Analysis*; Building Technologies Department, Environmental Energy Technologies Division, Lawrence Berkeley National Laboratory, University of California: Berkeley, CA, USA, 1999.
49. Tang, R.; Meir, I.; Etzion, Y. Thermal behavior of buildings with curved roofs as compared with flat roofs. *Sol. Energy* **2003**, *74*, 273–286. [CrossRef]
50. Pearlmutter, D. Roof geometry as a determinant of thermal behaviour: A comparative study of vaulted and flat surfaces in a hot-arid zone. *Archit. Sci. Rev.* **1993**, *36*, 75–86. [CrossRef]
51. Gómez-Muñoz, V.M.; Porta-Gándara, M.Á.; Heard, C. Solar performance of hemispherical vault roofs. *Build. Environ.* **2003**, *38*, 1431–1438. [CrossRef]
52. Sedighi, E.; Yaghoubi, M.; Mousavi, S.; Siahpour, S. Thermal study of domed roofs in a traditional bazaar (the case of old Ganj-Alikhan bazaar in Kerman, Iran). *Energy Sustain. Dev.* **2017**, *39*, 67–81. [CrossRef]
53. Shen, C.; Lei, Z.; Wang, Y.; Zhang, C.; Yao, Y. A review on the current research and application of wastewater source heat pumps in China. *Therm. Sci. Eng. Prog.* **2018**, *6*, 140–156. [CrossRef]
54. Postrioti, L.; Baldinelli, G.; Bianchi, F.; Buitoni, G.; Di Maria, F.; Asdrubali, F. An experimental setup for the analysis of an energy recovery system from wastewater for heat pumps in civil buildings. *Appl. Therm. Eng.* **2016**, *102*, 961–971. [CrossRef]
55. Li, H.; Sun, Y. Operational performance study on a photovoltaic loop heat pipe/solar assisted heat pump water heating system. *Energy Build.* **2018**, *158*, 861–872. [CrossRef]
56. Nižetić, S.; Čoko, D.; Yadav, A.; Grubišić-Čabo, F. Water spray cooling technique applied on a photovoltaic panel: The performance response. *Energy Convers. Manag.* **2016**, *108*, 287–296. [CrossRef]
57. Erhorn, H.; Görres, J.; Illner, M.; Bruhn, J.-P.; Bergmann, A. "NeckarPark Stuttgart": District heat from wastewater. *Energy Procedia* **2018**, *149*, 465–472. [CrossRef]
58. Sun, Y.; Zhai, P.; Ning, J.; Zheng, C.; Zhang, S.; Liu, C.; Yan, W. An integrated system of sinusoidal-corrugated plate heat exchanger and thermoelectric modules for wastewater heat recovery. *Energy Convers. Manag.* **2022**, *267*, 115885. [CrossRef]
59. Somogyi, V.; Sebestyén, V.; Domokos, E. Assessment of wastewater heat potential for district heating in Hungary. *Energy* **2018**, *163*, 712–721. [CrossRef]
60. Guven, H.; Ersahin, M.E.; Ozgun, H. Energy self-sufficiency in wastewater treatment plants: Perspectives, challenges, and opportunities. *Clean Energy Resour. Recovery* **2022**, *2*, 105–122.
61. Gualotuña-Gualoto, D.; Martínez-Pérez, I.; Laera, R.; Pereda, L.d. The Thermal Potential of Wastewater for Heating and Cooling Buildings: A Case Study of a Low Exergy Building in Madrid. *Buildings* **2023**, *13*, 2057. [CrossRef]
62. Largan, C.; Morris, T.M. *Qualitative Secondary Research: A Step-By-Step Guide*; SAGE Publications: Los Angeles, CA, USA, 2019.
63. SmartScrapers. 2025. List of HVAC&R Contractors in New Zealand. Available online: https://rentechdigital.com/smartscraper/business-report-details/list-of-hvac-contractors-in-new-zealand?utm_source=chatgpt.com (accessed on 20 April 2025).
64. Gill, S.L. Qualitative sampling methods. *J. Hum. Lact.* **2020**, *36*, 579–581. [CrossRef] [PubMed]
65. Hennink, M.; Kaiser, B.N. Sample sizes for saturation in qualitative research: A systematic review of empirical tests. *Soc. Sci. Med.* **2021**, *292*, 114523. [CrossRef] [PubMed]
66. Clarke, V.; Braun, V. Thematic analysis. *J. Posit. Psychol.* **2017**, *12*, 297–298. [CrossRef]
67. Byrd, H.; Matthewman, S.; Rasheed, E. Air-conditioning in New Zealand: Power and policy. *Build. Cities* **2022**, *3*, 1–9. [CrossRef]
68. Panchalingam, K.; Rasheed, E.O.; Rotimi, J.O. Cost-Related Drivers and Barriers of Passivhaus: A Systematic Literature Review. *Sustainability* **2023**, *16*, 6510. [CrossRef]

energies

MDPI

Article

Utilizing Fuel and Energy Sector Waste as Thermal Insulation Materials for Technical Buildings

Artem Pavlychenko [1], Dariusz Sala [2], Michal Pyzalski [2,*], Serhii Dybrin [1], Olena Antoniuk [3] and Roman Dychkovskyi [1,2]

1 Dnipro University of Technology, D. Iavornytskoho Ave. 19, 49-027 Dnipro, Ukraine; pavlichenko.a.v@nmu.one (A.P.); dybrin.s.v@nmu.one (S.D.); dychkovskyi@agh.edu.pl (R.D.)
2 AGH University of Krakow, A. Mickiewicza Ave. 30, 30-059 Kraków, Poland; sala@agh.edu.pl
3 Ukrainian State University of Science and Technologies, 4 Gagarina Ave., 49-010 Dnipro, Ukraine; 101sprava@gmail.com
* Correspondence: michal.pyzalski@agh.edu.pl; Tel.: +48-12-617-25-15

Abstract

The growing demand for sustainable construction materials has prompted intensive research into the potential reuse of waste from the fuel and energy sector as effective thermal insulation materials. This study examines the feasibility of utilizing ash–slag mixtures, fly ash, and aluminosilicates as insulation materials for technical buildings. These materials were selected due to their availability and potential to improve energy efficiency in construction. Practical tests were carried out to determine the thermal conductivity coefficients of various samples, which were produced using different cement mixtures as binders to ensure adequate structural strength. The results demonstrated that the use of industrial waste-derived materials not only provides satisfactory thermal insulation properties but also contributes to environmental sustainability by reducing the challenges associated with the disposal of industrial by-products. The study highlights the crucial role of cement as a binder, enhancing the mechanical strength and durability of the insulation samples. The integration of ash–slag mixtures, fly ash, and aluminosilicates into the construction sector may foster the adoption of more environmentally friendly building practices, thereby supporting a circular economy and mitigating the environmental impact of construction activities. The study showed that the lowest thermal conductivity coefficient (0.24 W/m·K) was achieved for mixtures containing fly ash and cement, while the highest value (0.30 W/m·K) was recorded in samples incorporating aluminosilicates. The obtained results confirm the effectiveness of fly ash as a cost-efficient additive that improves the thermal insulation properties of the material.

Keywords: sustainable technologies in buildings; thermal insulation materials; fuel and energy sector waste; circular economy; energy efficiency in buildings; construction process management; zero energy buildings; building energy

1. Introduction

As Ukraine strives to enhance its energy independence, there is growing interest in the implementation of materials that not only provide adequate thermal insulation properties but also support the country's broader sustainable development goals [1]. The use of waste from the fuel and energy sector, such as ash–slag mixtures, fly ash, and aluminosilicates as insulation materials in construction, represents a strategic approach to achieving these objectives. These materials offer dual benefits, contributing to the development of a circular economy while addressing the urgent need for cost-effective and environmentally friendly insulation solutions [2].

The integration of these materials into the construction sector supports both short-term and long-term goals by reducing dependence on conventional insulation materials and limiting the environmental impact associated with the operation of new and existing buildings [2,3].

Thermal insulation materials play a key role in reducing energy consumption in buildings by limiting heat transfer between interior and exterior environments. They help maintain comfortable indoor temperatures, lower energy costs, and enhance the overall energy efficiency of buildings [4]. In particular, the use of waste from the fuel and energy sector, such as ash–slag mixtures, fly ash, and aluminosilicates, is emerging as an innovative insulation approach [4,5].

The choice of thermal insulation materials often depends on the type of building and its specific requirements. In residential buildings, the priority is comfort and cost-effectiveness, while commercial buildings focus on reducing operating costs and meeting energy certification standards [1,6]. Industrial and technical facilities, such as factories, laboratories, and utility buildings, require materials that combine high thermal performance with the durability necessary to withstand harsh operating conditions [7].

In residential construction, traditional insulation materials such as glass wool, mineral wool, and foam boards are commonly used. However, recycled materials such as fly ash and ash–slag mixtures demonstrate promising thermal insulation properties, making them a viable alternative [6].

Traditional materials, such as mineral wool, glass wool, or synthetic foams, are characterized by very good insulation properties, but they are produced from primary raw materials and often require energy-intensive manufacturing processes. In contrast, newly developed composites containing fly ash, ash–slag mixtures, and aluminosilicates achieve thermal conductivity within an acceptable range for technical insulation and represent a significant step toward sustainable construction.

The comparison of these two groups of materials not only helps to understand the current capabilities of available insulators but also highlights the contribution of this study to the development of knowledge, particularly regarding the use of industrial mineral waste as functional, inexpensive, and durable insulation components. The adopted methodology allows for a practical comparison of the insulation properties of new mixtures and their potential in the design of materials dedicated to technical facilities.

In terms of technical parameters, traditional insulation materials such as mineral wool and polystyrene are characterized by very low thermal conductivity values, typically ranging from 0.030 to 0.040 W/m·K. For comparison, the materials described in this study achieved values ranging from 0.24 to 0.30 W/m·K, which, although indicating lower insulation effectiveness than wool and polystyrene, are comparable to the thermal conductivity of ceramic hollow blocks or perforated bricks. This makes them suitable for application in the insulation of technical and industrial facilities. Nevertheless, their use offers significant ecological and economic benefits, including the reduction in industrial waste, lowering CO_2 emissions, and support for a circular economy. Therefore, these materials can represent a valuable solution in the context of insulating technical buildings, where mechanical strength and structural durability are also of key importance. Such parameters are considered promising, especially given that the tested materials were derived from industrial waste, are characterized by low production costs, and have a positive environmental impact. The thermal conductivity coefficient λ of different insulating and structural materials is summarized in Table 1.

Table 1. Comparison of the thermal conductivity coefficient λ of various insulating and structural materials.

Material	The Thermal Conductivity Coefficient λ, W/m·K	Reference
Sample: Cement + Fly Ash	0.240	[8]
Sample: Cement + Ash–Slag Mixture	0.249	[8]
Sample: Cement + Aluminosilicate	0.299	[8]
Structural Concrete (Reinforced Concrete)	≈1.7	[9]
Cement (Dry)	0.7–1.1	[10]
Mineral Wool	0.035–0.045	[11]
Polystyrene Foam (EPS/XPS)	0.030–0.040	[11,12]

In commercial buildings, thermal insulation plays a crucial role in reducing operational costs and meeting stringent energy efficiency standards. When fly ash and aluminosilicates are combined with cement as a binder, durable insulation materials with high thermal resistance and strength are produced. These composites can be applied to walls, floors, and roofs, enhancing the thermal envelope of commercial buildings [6,13]. By improving the thermal resistance of building partitions, these materials help maintain stable indoor temperatures and reduce long-term energy consumption. Moreover, their use supports sustainability efforts by reusing industrial waste that would otherwise require disposal.

Technical buildings such as data centers, control rooms, or research laboratories generate significant amounts of heat due to the presence of high-energy equipment. For these types of facilities, insulation materials must demonstrate exceptional thermal resistance and long-term durability. Ash–slag mixtures and aluminosilicates, combined with high-strength cement, provide insulation that meets these requirements while also offering fire resistance and sound absorption properties [7,14].

Cement is commonly used as a binder to enhance the structural integrity of insulation materials. The inclusion of cement in mixtures containing fly ash, ash–slag mixtures, or aluminosilicates imparts adequate compressive strength and resistance to mechanical loads to the resulting composites [15]. This is especially important in technical buildings, where insulation panels are exposed to more demanding operational and environmental conditions.

Modifying the stability characteristics of materials through thermochemical degradation is a particularly relevant approach in the context of sustainable construction [16]. This process involves the use of surface or underground generators to gasify coal or biomass, resulting in valuable by-products that can be reused [16,17]. During gasification, an artificial shell forms around the material, enhancing its structural integrity. By selectively melting a portion of the surface to a controlled depth, a protective layer is formed that serves as a barrier against environmental factors [18]. This layer prevents degradation caused by moisture, temperature fluctuations, and other external influences. As a result, the stability and durability of thermal insulation materials are significantly improved, allowing them to perform effectively over extended periods [16,19]. This innovative approach enhances both the performance and lifespan of insulation systems used in modern construction.

The use of waste from the fuel and energy sector as insulation materials not only reduces the amount of waste sent to landfills but also lowers the carbon footprint of the construction industry [19,20]. Fly ash and slag, for example, are by-products of coal combustion that can effectively replace virgin raw materials in insulation applications. This practice supports the principles of a circular economy and contributes to the achievement of sustainable development goals across various sectors [21]. Thermal conductivity is a key parameter in evaluating the effectiveness of insulation materials. Studies have shown that ash–slag and aluminosilicate mixtures exhibit low thermal conductivity, making them suitable for retaining heat during winter months and protecting interiors from overheating in the summer [22]. This reduces the demand for heating and cooling, ultimately lowering energy costs for building owners.

The use of fuel and energy sector waste as insulation materials is gaining importance not only in the context of technical buildings but also in residential, commercial, and industrial construction. As research progresses, the development of advanced insulation materials based on recycled content is likely to become a cornerstone of sustainable building practices. With the growing demand for environmentally friendly solutions, these materials offer an opportunity to improve energy efficiency while addressing large-scale waste management challenges.

The innovation of this study lies in the direct comparison of the thermal insulation properties of various materials derived from fuel and energy sector waste, conducted under standardized experimental conditions. Particular attention was given to the phase structure, particle morphology, and specific surface area of the analyzed materials. Additionally, the study combines the evaluation of thermal efficiency with a practical approach to the production technologies of materials intended for technical buildings. This integrated approach offers a new perspective on optimizing the use of waste in construction, supporting both energy efficiency and the implementation of circular economy principles.

2. Reference Background of the Thermal Insulation Materials for Buildings

Thermal insulation materials are essential components in construction, playing a key role in energy conservation and ensuring indoor comfort. Their function is to reduce the rate of heat transfer between the interior of a building and its external environment [23]. Insulation materials can be organic, such as wool or cellulose, or inorganic, such as fiberglass or mineral wool, and each type is characterized by different properties. The selection of an appropriate insulation material depends on factors such as the type of building, the climate, and the desired level of energy efficiency. The use of thermal insulation in construction dates back centuries, with early forms such as straw and clay being used in primitive structures. Modern insulation materials began to appear during the Industrial Revolution, with mineral wool and fiberglass introduced in the early twentieth century [1,23,24]. Since then, these materials have been refined to improve their insulating performance, fire resistance, and ease of installation, making them widely used in today's construction industry.

The most commonly used thermal insulation materials in modern construction include fiberglass, mineral wool, expanded polystyrene (EPS), extruded polystyrene (XPS), and polyurethane foam. Each of these materials has unique characteristics that influence its effectiveness in reducing heat loss. They differ in aspects such as R-value, which measures thermal resistance, as well as in moisture resistance, durability, and ease of application [23,25]. Fiberglass and mineral wool are especially valued for their affordability, non-combustibility, and excellent fire resistance properties, making them a safe choice for various building types. On the other hand, foam-based insulations such as EPS and XPS are highly appreciated for their versatility, low weight, and high R-values, which make them extremely effective thermal insulators [23,26]. Together, these materials offer a wide range of options for construction projects aiming to improve energy efficiency and meet specific insulation requirements [23,27].

In recent years, there has been a growing interest in sustainable insulation materials derived from natural and recycled sources. These include materials such as hemp, sheep's wool, and cellulose (recycled paper) [23,28]. Additionally, advancements in aerogel insulation, vacuum insulation panels (VIPs), and phase-change materials (PCMs) have improved thermal efficiency and expanded the range of environments in which insulation can be effectively applied. Such innovations contribute to the overall reduction in energy consumption in buildings and support the achievement of global sustainability goals.

The use of waste from the fuel and energy sector, such as fly ash, ash–slag mixtures, and aluminosilicates, as thermal insulation materials has gained significance due to the dual benefits of waste reduction and improved energy efficiency [29]. These materials are by-products of coal combustion and other industrial processes, they are readily available and possess favorable insulating properties [30]. After appropriate processing and combination with binders such as cement, they can form stable and durable insulation materials suitable for construction applications [31].

The ash–slag mixture is a material composed of ash and slag formed during coal combustion in power plants or boilers. It can be used in construction, especially in the production of cement or concrete, due to its binding properties. Fly ash is a material consisting of fine ash particles carried by flue gases during coal combustion. It is often collected using electrostatic precipitators and is also used in construction, particularly as an additive in cement and concrete mixtures. Aluminosilicate is a mineral composed of aluminum, silicon, and oxygen. It is commonly found in the ash remaining after coal combustion, especially in the case of lignite. Aluminosilicates are used in the production of various materials such as fiberglass, ceramics, and other construction products.

These materials are by-products not only from the extraction and processing of coal but also from the development of ore deposits and the thermochemical processing of biomass [19,32]. During biomass processing, coal is often added to increase the calorific value of the fuel, resulting in higher energy output. This process generates valuable by-products, including ash–slag mixtures and other mineral residues, which can be repurposed in various industrial applications [26,33]. These by-products possess beneficial properties that make them suitable for use in thermal insulation materials [34]. Their production aligns with sustainable practices by repurposing industrial waste, thereby reducing the environmental impact of mining activities and fuel production [33,35]. As a result, these materials contribute to more sustainable energy solutions while offering practical benefits for building insulation and other applications.

Traditional insulation materials, especially those produced from petrochemical sources, can have a significant environmental impact due to the extensive resource extraction and energy-intensive manufacturing processes involved. These materials often rely on non-renewable resources, resulting in a high carbon footprint and contributing to overall environmental degradation [33,36]. In contrast, using waste-derived materials for insulation supports a circular economy by repurposing by-products from industrial processes, such as ash and slag, which would otherwise be disposed of in landfills [1,37]. By replacing virgin resources with reclaimed materials, the construction industry can significantly reduce both the volume of landfill waste and the demand for newly extracted raw materials [26,38]. This shift not only minimizes the carbon footprint associated with building insulation but also promotes more sustainable and responsible resource management practices across the sector.

The effectiveness of thermal insulation materials is largely determined by their thermal conductivity and R-value. Studies have shown that materials such as fly ash and aluminosilicate can exhibit low thermal conductivity, making them effective insulators [26,39]. These materials can also improve the energy efficiency of buildings by retaining heat during colder months and helping to maintain lower indoor temperatures during warmer periods. This leads to reduced energy consumption for heating and cooling, ultimately lowering operational costs. As the demand for eco-friendly construction practices increases, the development of thermal insulation materials based on waste from the fuel and energy sector is expected to continue [40]. These materials show great promise due to their cost-effectiveness, availability, and contribution to environmental sustainability. Further research and development are expected to improve the thermal properties and application methods of waste-derived insulation materials, enabling their broader adoption in both residential and commercial construction projects worldwide.

The use of waste from the fuel and energy sector, such as slag, fly ash, and aluminosilicate, as thermal insulation materials for technical buildings has attracted significant attention in recent years due to its environmental and economic benefits. With the growing global demand for sustainable construction solutions, repurposing industrial by-products offers a practical way to reduce landfill disposal and minimize the carbon footprint of the construction industry. These materials, often abundant and low-cost, provide favorable insulation properties by reducing thermal conductivity, which contributes to improved energy efficiency in buildings. By replacing traditional insulation materials that rely heavily on non-renewable resources, waste-derived insulations support a circular economy and contribute to sustainability goals across various sectors. This is a real scientific and practical challenge that the author addresses while also attempting to assess the economic effectiveness of the materials discussed. Table 2 presents a comparison of the key properties of various binder types used in the production of thermal insulation materials.

Table 2. Comparison of the properties of various types of binders used in thermal insulation materials.

Binder Type	Setting Speed	Water Resistance	Cost	Environmental Friendliness	Compatibility with Ashes
Cement	Medium	High	Medium	Medium	High
Gypsum	Fast	Low	Low	High	High
Lime	Slow	Medium	Low	High	High
Magnesium-Based	Fast	Medium	Medium	Medium	Medium
Alkaline (Geopolymer)	Medium/Fast	High	Higher	High	High
Aluminosilicate	Fast	Medium	High	Medium	High

Among the best-performing insulation materials currently used in the construction industry are polyisocyanurate (PIR) panels, polyurethane (PUR) foams, extruded polystyrene (XPS) boards, and vacuum insulation panels (VIPs). Their thermal conductivity ranges from approximately 0.020 to 0.035 W/m·K, making them highly effective in reducing heat loss.

At the same time, these materials have numerous limitations: they are produced from petrochemical raw materials, have low resistance to high temperatures, are difficult to recycle, and generate a high carbon footprint. In comparison, the materials examined in this study, obtained from fuel and energy sector waste, show thermal conductivity values in the range of 0.24–0.30 W/m·K which, although lower in absolute performance, may be sufficient for many technical applications.

Due to their good structural stability, mechanical strength, and resistance to thermal degradation, these materials may serve as a viable alternative to commercial insulators, particularly in the context of sustainable construction. In this regard, the present study complements the existing body of knowledge with a group of solutions that combine technical efficiency with environmental and economic benefits.

3. Materials and Methods

This chapter presents a detailed characterization of the materials used in the study, as well as a description of the applied analytical methods. The purpose of the analyses was to assess the suitability of selected inorganic substances for use in composites with thermal insulation properties. In this context, the chemical composition, physical properties, and phase structure of four different materials were compared: Portland cement CEM I

42.5R, fly ash, an ash–slag mixture, and synthetic aluminosilicate. The selection of these materials was based on their wide availability, diversity in physicochemical properties, and application potential in lightweight, low-emission insulating composites.

Four groups of samples with different compositions were prepared for testing:

- Group 1: contains cement (50 g) and fly ash (200 g);
- Group 2: contains cement (50 g) and ash–slag mixture (200 g);
- Group 3: contains cement (50 g), ash–slag mixture (100 g), and fly ash (100 g);
- Group 4: contains cement (50 g) and synthetic aluminosilicate (200 g).

For each sample, the following parameters were determined: sample weight, thickness, side length, volume, density, and thermal conductivity coefficient λ. The measurement results are presented in Table 3.

Table 3. Chemical composition and physical properties of cement and additives.

Oxides	CEM I 42.5R	Fly Ash	Ash–Slag	Aluminosilicate
		Chemical composition		
SiO_2	19.8	52	39.4	52.1
Al_2O_3	5.1	23	7.6	41.1
Fe_2O_3	2.6	13	1.3	4.3
CaO	64.1	6	42.1	0.07
MgO	1.6	2.5	6.1	0.2
SO_3	3.0	1.0	1.4	-
$Na_2O + K_2O$	0.8	2.9	1.1	0.9
		Physical properties		
Density, g/cm^3	3.11	2.35	2.90	2.60
Grain size, μm	10–50	1–100	<50	1–163
Specific surface area, m^2/kg	385	250–600	350–500	12,000–15,000
Shape	irregular	spherical	irregular	irregular

In this study, all samples of fly ash, ash–slag mixture, and synthetic aluminosilicate were sourced from a single controlled industrial facility, which allowed for uniform analysis conditions. However, it should be emphasized that the chemical and phase composition of industrial waste may vary significantly depending on the source, combustion technology, and the type of primary raw material used. Therefore, in the subsequent stages of the research, the authors plan to include samples from different locations in order to assess the impact of this variability on the thermal insulation and mechanical properties of the composite materials.

The applied research methods enabled a comprehensive evaluation not only of the elemental and phase composition of the investigated materials but also of their physical properties, including density, specific surface area, and particle morphology. Particular emphasis was placed on the presence of amorphous phases and microspheres in the ash structure, as these features are critical to the thermal conductivity and porosity of the resulting composites [41].

A comparative analysis of the chemical composition, physical characteristics, and phase structure of the four studied materials—Portland cement CEM I 42.5R, fly ash, ash–slag mixture, and synthetic aluminosilicate—provides valuable insights into their potential for use in thermal insulation composites. The chemical composition reveals notable differences in the content of major oxides, which directly influence the materials' performance under service conditions.

Fly ash and synthetic aluminosilicate exhibit the highest SiO_2 content, measured at 52.0% and 52.1%, respectively. This high silica content favors the development of a distinctive microstructure and contributes to the formation of porous zones with low thermal conductivity. The elevated Al_2O_3 content observed in aluminosilicate (41.1%) and fly ash (23.0%) further enhances their pozzolanic reactivity.

In contrast, Portland cement CEM I 42.5R is characterized by a high CaO content (64.1%), which imparts strong hydraulic properties but limits its effectiveness as a standalone component in thermal insulation systems. The ash–slag mixture, with a chemical composition of 42.1% CaO and 39.4% SiO_2, represents an intermediate formulation that may serve effectively as a supplementary component.

From a physical perspective, synthetic aluminosilicate is particularly notable for its exceptionally high specific surface area (12,000–15,000 m^2/kg), which supports the formation of porous structures with superior insulating performance. Fly ash also demonstrates a relatively large surface area (250–600 m^2/kg), and its low density (2.35 g/cm^3), coupled with the spherical morphology of its particles, has a favorable impact on the rheological behavior of the mixtures. In contrast, Portland cement, with its higher density (3.11 g/cm^3) and lower specific surface area (385 m^2/kg), primarily serves as a structural binder, while simultaneously reducing the thermal insulation capacity of the composite material.

An important aspect of the analysis concerns the phase composition, particularly in the case of silica-rich fly ash of conventional origin. This material consists of three primary phases: crystalline, glassy, and amorphous. The crystalline phase includes a broad spectrum of minerals such as quartz (in both α and β forms), mullite ($3Al_2O_3{\cdot}2SiO_2$), hematite, magnetite, calcite, dolomite, anhydrite, free CaO, normal spinel ($MgAl_2O_4$), iron spinel ($FeAl_2O_4$), dicalcium ferrites, corundum, rutile or anatase (TiO_2), ettringite, opal, and others. These minerals originate from unburned coal residues, crystallization products of the molten phase, and high-temperature physicochemical transformations. Their presence can influence the chemical stability, rheological behavior, and thermal conductivity of the material.

The most relevant phase in terms of thermal insulation performance is the glassy phase, which can constitute between 50 percent and up to 90 percent of the total mass in conventional fly ash. This phase is amorphous in nature, with a chemical composition similar to basalt, primarily consisting of SiO_4 and AlO_4 tetrahedra. It also contains modifying oxides such as CaO, MgO, Na_2O, and K_2O. Two principal types of glass are distinguished: Type I, characterized by low density, a high SiO_2/Al_2O_3 ratio, and a low content of modifying oxides; and Type II, characterized by higher density, a low SiO_2/Al_2O_3 ratio, and a high content of modifying oxides. Type I glass generally forms larger spherical particles, whereas Type II is usually found as smaller, hollow microspheres.

The presence of microspheres in fly ash is one of the key features that enhances its value as a component in thermal insulation materials. Owing to their hollow structure and spherical geometry, microspheres significantly reduce the bulk density of the material and improve its workability. Furthermore, the incorporation of microspheres increases porosity and lowers thermal conductivity, which is highly desirable in the development of lightweight composites with insulating properties.

The amorphous phase, present only in trace amounts in the analyzed conventional silica-based fly ash, is mainly composed of dehydrated residues of clay minerals and small quantities of amorphous carbon (soot). Although its content is minimal, it may influence the binding mechanisms and the evolution of the microstructure in mixtures containing fly ash with elevated surface reactivity.

Synthetic aluminosilicate, due to its exceptionally high specific surface area, elevated content of reactive phases, and low density, represents the most effective component for thermal insulation applications. Fly ash, with its considerable glassy phase content and presence of microspheres, serves well as a supporting component—both structurally and functionally. Ash–slag mixtures may be used as complementary ingredients in composite systems designed to balance strength and insulation capacity. In contrast, Portland cement CEM I 42.5R, due to its high CaO content and substantial density, should be limited in insulation formulations and utilized primarily as a structural and binding agent rather than an active insulating material [42].

This study adopts a practical experimental approach combined with an analytical methodology to evaluate the thermal conductivity of materials using the thermal conductivity coefficient λ. Through direct measurements, the thermal properties of each material were determined, allowing for a precise assessment of their applicability in insulation systems. The use of the λ coefficient facilitates a detailed understanding of each material's contribution to energy efficiency in technical buildings through heat transfer reduction. The thermal conductivity coefficient λ is calculated as follows [43]:

$$\lambda = \frac{H \cdot q}{T_{\mathrm{H}} - T_{\mathrm{x}}}$$

where

λ—coefficient of the effective thermal conductivity, W/m·K;
H—thickness of the measured sample, mm;
q—density of the steady-state heat flux passing through the measured sample, W/m^2;
T_{H}—temperature of the hot surface of the measured sample, K;
T_{x}—temperature of the cold surface of the measured sample, K.

To evaluate the effectiveness of utilizing waste from the fuel and energy sector as thermal insulation materials for technical buildings, five samples of varying compositions were prepared for each type of material under investigation. The test specimens were fabricated in standardized dimensions of 100 mm by 100 mm, with varying heights. These samples were produced by combining the selected materials with Portland cement as a binder, following the principles of kinetic and static similarity to ensure uniformity in composition and structural integrity [43].

The samples were compacted and processed in accordance with standardized procedures governing the initial and final setting times of cement-based binders. This careful preparation ensured consistency across all specimens and provided a reliable foundation for subsequent testing. The curing period lasted 28 days, during which the cement matrix was allowed to harden and stabilize—an essential condition for obtaining accurate and reproducible measurements [41,43]. Throughout the curing phase, the samples were closely monitored to confirm the proper setting and uniform internal structure. This approach ensured that the thermal insulation properties of the materials could be assessed under consistent and standardized conditions.

Thermal conductivity measurements were performed using the IPT-MG4 "100" apparatus, adhering to precise methodological protocols to guarantee the reliability and repeatability of the results. The instrument was calibrated according to industry standards, which enabled accurate readings for all samples. This methodology facilitated a thorough evaluation of each material's ability to minimize heat transfer—one of the key indicators of its effectiveness as a thermal insulator. By accurately measuring the thermal conductivity coefficient, the potential application of the tested materials for insulation purposes in technical buildings was effectively assessed. The results offer valuable insight into their performance and highlight their relevance for enhancing energy efficiency in a variety of construction contexts. Photographic documentation of the experimental process, including sample preparation and testing apparatus, is presented in Figure 1.

(a) (b) (c)

Figure 1. Photographic documentation of the research process: (**a**) ash–slag and fly ash used as the raw material; (**b**) mold for casting samples and prepared samples; (**c**) IPT-MG4 "100" device for measuring thermal conductivity, featuring an electronic control unit and a heating element.

To determine the statistical significance of the observed differences in thermal conductivity (λ) values among the tested mixtures, a one-way analysis of variance (ANOVA) was performed at a significance level of $\alpha = 0.05$. Subsequently, a Tukey post-hoc test was conducted to assess differences between group pairs. The results of the analysis indicated that the variations in the composition of the mixtures had a statistically significant effect on the thermal insulation properties of the tested materials.

The device operates based on the principle of establishing a steady-state heat flow through a flat sample of defined thickness, directed perpendicularly to its face surfaces. This process involves a series of precise measurements, including the sample's thickness, heat flux density, and the temperatures of the opposite surfaces. By achieving a steady-state thermal condition, the device can accurately evaluate the heat transfer through the material. This data is critical for determining the material's thermal conductivity, which is essential for assessing its insulation performance. Overall, this method provides valuable insights into the material's behavior under conditions representative of real-world applications.

The measurement system is configured in accordance with an asymmetrical scheme, as specified by the Ukrainian standard DSTU 7076 (DSTU B V.2.7-105-2000) [44]. The stationary system of the device consists of a heating control unit and a cooling unit based on Peltier elements, a heat flux meter, platinum resistance temperature sensors, a signal conversion unit for sensor outputs, and a power supply. The Peltier elements are cooled using an integrated fan system.

The side panels of the setup include a power switch, grounding terminal, fuse, and sockets for connecting both the electronic control unit and the main power supply cable, as well as an eccentric locking mechanism. The electronic unit is powered via a dedicated connection from the setup. At the top of the assembly, a pressure screw equipped with a measurement device is used to determine sample thickness, along with a dynamometric mechanism with a ratchet for applying uniform clamping force to the sample.

The key variables examined in this study included the density, thickness, and composition of the fabricated samples. These parameters were systematically adjusted to analyze their impact on the material's overall thermal behavior. The final focus of the study was the thermal conductivity coefficient (λ), which is fundamental for evaluating the material's effectiveness as thermal insulation. This coefficient enables an accurate assessment of the material's insulating capability across various application surfaces.

4. Results and Discussions

The research involved the preparation of samples (Figure 1b, right) using a disassemblable steel mold featuring two rectangular parallelepiped cavities, each with internal dimensions of 100 mm (Figure 1b, left). In the initial phase, the samples were allowed to reach an early setting state that permitted their removal from the mold without structural damage. This phase typically lasted between 2 and 3 days. The second phase consisted of a 28-day curing period, during which the samples developed stable mechanical strength. Once the required strength was achieved, the samples were subjected to a drying process, which continued until mass stabilization was confirmed—defined as a change of less than 0.1% after two consecutive drying intervals of 0.5 h each.

To ensure accurate thermal conductivity measurements, the upper surface of each sample was polished to meet stringent specifications: both contact faces, intended to interface with the device's measurement plates, were required to be flat and parallel, with deviations in parallelism not exceeding ±0.5 mm. This meticulous surface preparation was essential for ensuring consistent and complete contact with the measurement apparatus, thereby minimizing potential sources of error.

Sample thickness was measured at four distinct points: each located approximately 50 ± 5 mm from a corner vertex and at the center of each edge. These measurements were performed using a high-precision caliper with an accuracy of ±0.1 mm. Such detailed dimensional control was critical to ensuring that the measured thermal conductivity values accurately reflected the true properties of the material. By maintaining consistent sample geometry, the study ensured reliable and reproducible results across all measurements. The precision of these preparatory steps had a direct impact on the reliability of the calculated thermal conductivity values, confirming that careful sample preparation was a fundamental component of the experimental methodology.

Following this preparation process, four groups of five samples each were produced with the following compositions:

Cement (50 g) + Ash–slag (200 g);
Cement (50 g) + Fly ash (200 g);
Cement (50 g) + Ash–slag (100 g) + Fly ash (100 g);
Cement (50 g) + Aluminosilicate (200 g).

Before commencing the main tests, the functionality of the measurement device was verified using a control sample made of organic glass, which was provided as part of the instrument's standard calibration kit (Figure 1c). This step ensured the accuracy and reliability of the device, confirming that it was operating within the specified parameters.

For the thermal conductivity measurements, the prepared samples were carefully placed on the cooling plate, and the heating plate was then positioned on top using a micrometer screw integrated with a dynamometric unit. This setup enabled precise control of the applied force, maintaining a uniform pressure of 2.5 kPa across each sample to ensure consistent contact between the sample and the measurement surfaces. Additionally, the thickness of each sample was measured at multiple points using the micrometer screw gauge, which provided highly accurate readings with a precision of up to 0.05 mm. These precise thickness measurements were critical for the subsequent calculation of thermal conductivity values.

Prior to initiating the direct measurements, essential data were entered into the electronic control unit, including the actual sample thickness, the target temperature for the cooling plate (Tx), and the set temperature for the heating plate (T_H). These input values were selected based on the sample's physical characteristics and the operational guidelines provided in the device's user manual. The settings for Tx and T_H were determined according to recommended values that vary depending on the sample's thickness and the estimated range of its thermal conductivity coefficient. This ensured that the device would operate under optimal conditions, thereby enhancing the accuracy and reliability of the measurements.

Once the measurement process began, the device's software precisely controlled both the heating and cooling units to achieve and maintain the specified surface temperatures—

T_H (hot side) and Tx (cold side)—with a high degree of accuracy (±0.2 °C). Maintaining these temperature conditions was essential for establishing a thermal steady state conducive to precise measurement. The system continuously monitored the heat flow through the sample and automatically adjusted the operation to maintain equilibrium. Upon reaching a stable heat flux, the device automatically calculated the thermal conductivity coefficient (λ) and thermal resistance (R). These values were recorded in the device's data archive, with the entire measurement process governed by integrated software to ensure consistency and data integrity. An audible signal indicated the successful completion of each measurement, confirming that the sample's thermal properties had been accurately determined.

The results of the study are presented in Table 4, providing a comprehensive overview of the measured values. The table includes thermal conductivity coefficients along with other relevant parameters for each tested sample, enabling a detailed comparison of the materials and highlighting their respective thermal performance characteristics.

Table 4. The results of the study of thermal conductivity of samples of insulating materials depending on the change in geometric dimensions and physical properties. x_i is the current value obtained from each individual sample; $\bar{x}$ is the average value for similar samples of a certain identical composition.

#	Sample Composition				Measured Sample Parameters								
	Cement (Grade 500), g	Ash–Slag Mixture, g	Fly Ash, g	Alumo-Silicate, g	Weight, g		Thickness, mm		Side Length, mm	Volume, m³	Density, kg/m³	Thermal Conductivity Coefficient λ, W/m K	
					x_i	$\bar{x}$	x_i	$\bar{x}$		$\bar{x}$	$\bar{x}$	x_i	$\bar{x}$
					201		13.9					0.25	
					198		13.6					0.238	
2	50	200	0	0	210	204	14.4	14	100	0.00014	1457	0.257	0.249
					204		14					0.257	
					207		14.1					0.243	
					187		15.9					0.228	
					181		15.5					0.235	
1	50	0	200	0	184	184	15.7	15.7	100	0.000157	1172	0.24	0.24
					178		15.2					0.247	
					190		16.2					0.25	
					207		14.2					0.254	
					195		14.4					0.26	
3	50	100	100		198	201	13.8	14.1	100	0.000141	1426	0.268	0.268
					204		14.5					0.276	
					201		13.6					0.282	
					223		15.9					0.283	
					220		15.8					0.29	
4	50	0	0	200	231	227	16.4	16.2	100	0.000162	1401	0.299	0.299
					234		16.5					0.308	
					227		16.4					0.315	

The visualization of the research findings on the variation in the thermal conductivity coefficient as a function of insulation mixture composition is presented in Figure 2. The graphical data illustrate how different proportions of components within the insulation mixtures influence thermal conductivity values. Analysis of the graphs reveals discernible trends in the increase or decrease in thermal conductivity depending on the content of cement, ash–slag, fly ash, and aluminosilicates in each sample. Notably, samples with a higher content of aluminosilicate exhibited lower thermal conductivity coefficients, confirming its effectiveness as a component in insulation materials. Furthermore, the inclusion of fly ash was observed to reduce thermal conductivity, thereby enhancing the insulating properties, particularly when used in combination with cement. The results presented in Figure 2 demonstrate that optimizing the composition of the insulation mixture can lead to significant improvements in thermal performance, which is essential for the development of efficient insulation materials for technical applications.

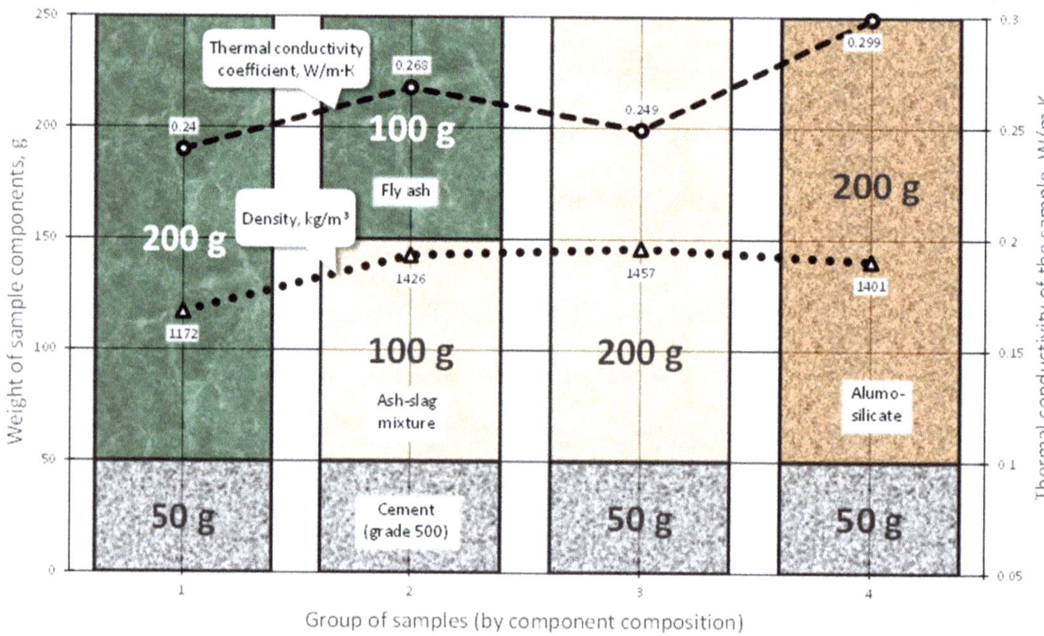

Figure 2. Dependence of thermal conductivity and sample density on the weight composition of components. All samples contain a constant amount of Portland cement (50 g) and a variable share of one of the following additives: fly ash, fly ash–slag mixture, or aluminosilicate. Thermal conductivity is represented by a dashed line, while density is indicated by a dotted line.

Visualizing the data through a chart provides a clear representation of the interdependence between sample composition, density, and thermal conductivity. The chart illustrates that the samples composed of fly ash and cement (Group 1) exhibit the lowest thermal conductivity, recorded at 0.24 W/m·K. Slightly higher thermal conductivity is observed in the samples made from ash–slag and cement (Group 3), with a value of 0.249 W/m·K. Interestingly, and somewhat counterintuitively, samples composed of equal parts fly ash, ash–slag, and cement exhibit an 8% higher thermal conductivity compared to the lowest values recorded in Groups 1 and 3. The highest thermal conductivity is observed in the samples made of cement and aluminosilicate (Group 4), reaching 0.299 W/m·K (Figure 3).

When analyzing density in relation to composition, the samples with a mixed filler (Group 2) demonstrate density values that fall between those of the single-component samples in Groups 1 and 3. This trend is consistent with the behavior of composite materials, whose properties typically reflect a balance between the characteristics of their individual constituents. Notably, a decrease in density observed in Group 4 samples, compared to the densities of Groups 2 and 3, is accompanied by a counterintuitive increase in thermal conductivity.

This phenomenon may be attributed to the influence of microstructural factors that extend beyond the material's bulk density. Thermal conductivity is affected by various factors, including the distribution and connectivity of pores, the presence of microspheres, the degree of cement hydration, and the quality of contact between components within the binding matrix. In some cases, materials with lower density may develop a more open and irregular porous structure, which can result in increased thermal conductivity. Thus, the overall insulating efficiency of a composite depends not only on its density but, more importantly, on the quality and architecture of its internal structure.

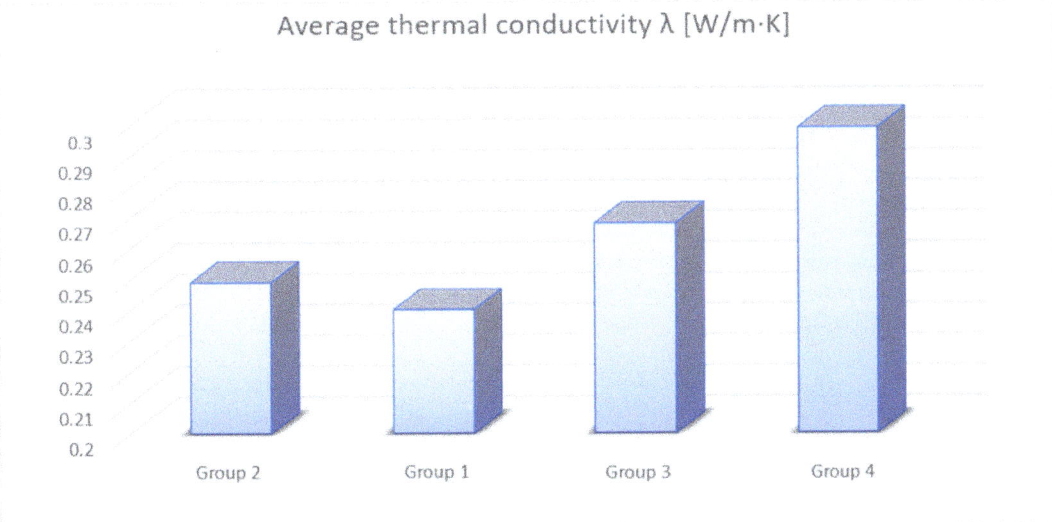

Figure 3. Comparison of the average thermal conductivity coefficient (λ) for different mixture compositions.

This unexpected outcome suggests that factors beyond density alone—such as microstructural configuration and specific interactions between components—may significantly influence the thermal properties of the tested samples. The improvement in the thermal insulation properties of the investigated materials results from the synergistic action of several physicochemical mechanisms. A key role is played by the presence of microspheres in fly ash, which, due to their hollow and spherical structure, reduce the bulk density of the composite and limit heat transfer. Additionally, glassy phases rich in silica and alumina contribute to the development of a porous and irregular microstructure, acting as a barrier to heat conduction. Moreover, the high specific surface area of aluminosilicates promotes an intense pozzolanic reaction with calcium ions released from cement hydration, leading to the formation of additional hydration products and the filling of microvoids. This process reduces the development of heat transport channels and improves the compactness of the structure. Collectively, these mechanisms define the overall thermal insulation efficiency of the composite.

Ash–slag, although the second-best performing material in terms of thermal conductivity, presents certain limitations due to its significantly higher density—over 22% greater than that of fly ash. This higher density may contribute to additional structural load, which can be a limiting factor in applications where lightweight materials are essential. Nonetheless, its comparable thermal performance makes ash–slag a viable alternative when weight constraints are not critical.

The results obtained in the present study are consistent with those reported by [43], who observed thermal conductivity values ranging from 0.26 to 0.31 W/m·K in panels composed of fly ash and lime. Similar values (0.22–0.28 W/m·K) were reported by Zhao et al. [43,45] for fly ash–perlite composites. In comparison, the best-performing samples in our study (fly ash + cement) achieved a thermal conductivity of 0.24 W/m·K, confirming the effectiveness of the proposed formulation.

Similarly, the findings of Kioupis et al. [46] for geopolymer mixtures with perlite indicate comparable thermal insulation properties, while emphasizing the critical influence of both micro- and macropore structures on thermal conductivity. The comparison of results underscores that industrial waste materials—particularly fly ash—can serve as a practical and effective alternative to commercial insulation products, offering both high performance and significant environmental advantages [46,47].

In future studies, the authors plan to expand their research by investigating the thermal conductivity of samples incorporating various additional materials, such as clay, sand, lime, lignin, and others. This exploration will provide a broader understanding of how these components influence the insulating properties of the resulting mixtures. Furthermore, experiments will be conducted using different mass fractions of these additives, allowing for a detailed assessment of how varying component ratios affect both thermal conductivity and overall material performance. The objective of these investigations is to identify optimal formulations for improved insulation efficiency, tailoring material compositions to meet specific application requirements.

In the context of long-term material durability, it is essential to consider potential risks associated with the use of waste-derived components such as fly ash, ash–slag mixtures, and aluminosilicates. The literature suggests that some of these additives may, over time, contribute to structural degradation through mechanisms such as ion leaching, microcracking, or uneven hydration processes. Although the present study did not include an analysis of aging effects, this issue will be addressed in subsequent research phases, particularly with regard to the performance of the material under technical operating conditions.

5. Conclusions

Based on the conducted research, it has been confirmed that the composition of the mixtures plays a crucial role in determining their thermal insulation properties. The lowest thermal conductivity (0.240 W/m·K) was recorded for samples containing fly ash and cement (Group 1), while the highest value (0.299 W/m·K) was observed in mixtures composed of cement and aluminosilicate (Group 4). This represents a nearly 20% difference, highlighting the significant influence of material composition on thermal performance.

In comparison, conventional concrete has a typical thermal conductivity of approximately 1.7 W/m·K, meaning that the Group 1 mixture provides a reduction in heat transfer of over 85%. This directly translates into significant energy savings and lower operational costs for buildings, particularly in terms of heating and cooling in technical or industrial applications.

The lowest recorded thermal conductivity coefficient was 0.24 W/m·K (fly ash with cement), while the highest was 0.30 W/m·K (aluminosilicate with cement). This demonstrates the significant influence of the type of industrial waste used and its proportion in the composite on the thermal insulation properties.

Structurally, fly ash exhibits low density and a porous structure, which enhances its thermal insulation properties. Although ash–slag-based mixtures demonstrated similar thermal conductivity to fly ash, their density was over 22% higher, which may be a limitation in applications where weight reduction is essential. Interestingly, some mixed compositions (Group 3) showed unexpectedly higher thermal conductivity values, suggesting that not all component combinations lead to beneficial synergistic effects.

Moreover, the results showed no clear correlation between density and thermal conductivity—in some cases, samples with lower density displayed higher thermal conductivity. This indicates that the interactions between components, rather than their individual properties, may have a dominant influence on overall thermal performance.

From an economic perspective, incorporating industrial by-products such as fly ash and ash–slag significantly reduces the cost of raw materials and contributes to a lower embodied energy in the material. The use of lightweight, insulating mixtures can reduce energy demand during building operations, leading to long-term financial savings. Additionally, such solutions support circular economy goals and reduce CO_2 emissions, making them not only cost-effective but also environmentally sustainable.

Given the above, fly ash emerges as the most promising component for the production of lightweight and thermally efficient insulating materials in construction. Future studies will focus on optimizing material formulations by introducing additional components such as clay, lime, sand, and lignin. These efforts aim to further enhance thermal performance and economic viability for a wide range of technical building applications.

In future research, the incorporation of lightweight additives of both natural and industrial origin is also planned, including expanded perlite, cellulose fibers, lignin-based biofillers, and amorphous silica. These materials exhibit very low thermal conductivity and have a beneficial effect on the porous microstructure of composites, without compromising their structural integrity. The use of perlite in combination with fly ash has been shown to achieve thermal conductivity values below 0.20 W/m·K. Such solutions could enable further optimization of the thermal insulation properties of the studied mixtures without the need for petrochemical-based additives.

Author Contributions: Conceptualization, A.P., S.D., R.D., D.S., R.D., O.A. and M.P.; methodology, A.P. and R.D.; software, S.D., O.A. and M.P.; validation, A.P., S.D., R.D., D.S., R.D., O.A. and M.P.; formal analysis, A.P., R.D., D.S., R.D., O.A. and M.P.; investigation, S.D., R.D., D.S., O.A. and M.P.; resources, A.P., S.D., D.S., R.D. and O.A.; data curation, A.P., R.D., O.A. and M.P.; writing—original draft preparation, A.P., S.D., R.D., D.S., R.D., O.A. and M.P.; writing—review and editing, A.P., S.D., R.D., D.S., R.D., O.A. and M.P.; visualization, S.D., R.D., O.A. and M.P.; supervision, A.P., R.D., D.S., O.A. and M.P.; project administration, A.P., S.D., R.D., D.S., R.D., O.A. and M.P.; funding acquisition, D.S. and M.P. All authors have read and agreed to the published version of the manuscript.

Funding: This work was supported by the subsidy of the Ministry of Education and Science for the AGH University of Krakow (Project No. 16.16.160.557).

Data Availability Statement: The original contributions presented in this study are included in the article. Further inquiries can be directed to the corresponding author.

Conflicts of Interest: The authors declare no conflict of interest.

References

1. Naumenkova, S.; Mishchenko, V.; Mishchenko, S. Key Energy Indicators for Sustainable Development Goals in Ukraine. *Probl. Perspect. Manag.* **2022**, *20*, 379–395. [CrossRef]
2. Khan, Z.; Khan, F.A.; Khan, A.U.; Hussain, I.; Khan, A.; Shah, L.A.; Khan, J.; Badrashi, Y.I.; Kamiński, P.; Dyczko, A.; et al. Climate-Streamflow Relationship and Consequences of Its Instability in Large Rivers of Pakistan: An Elasticity Perspective. *Water* **2022**, *14*, 2033. [CrossRef]
3. Beshta, O.; Fedoreyko, V.; Palchyk, A.; Burega, N. Independent Power Supply of Menage Objects Based on Biosolid Oxide Fuel Systems. *Power Eng. Control Inf. Technol. Geotech. Syst.* **2015**, *33*, 39. [CrossRef]
4. Cry, C.M. The Potential Use of Waste (Ash) Materials for EIFS Insulation and Related Components. *Fuel Energy Abstr.* **1996**, *37*, 187. Available online: https://www.researchgate.net/publication/294850101 (accessed on 2 May 2025). [CrossRef]
5. Kicki, J.; Jarosz, J.; Dyczko, A.; Puszcza, H. The Economic and Technical Aspects of Mine Closure in Poland. In Proceedings of the 14th International Symposium on Mine Planning and Equipment Selection (MPES 2005), Banff, AB, Canada, 1–3 November 2005.
6. Jankovic, L. Renewable Energy. In *Designing Zero Carbon Buildings*, 2nd ed.; Routledge: Abingdon, UK, 2024; pp. 195–208. [CrossRef]
7. Adekanye, O.G.; Davis, A.; Azevedo, I.L. Federal Policy, Local Policy, and Green Building Certifications in the U.S. *Energy Build.* **2020**, *209*, 109700. [CrossRef]

8. Yuan, J.; Farnham, C.; Emura, K. Optimal Combination of Thermal Resistance of Insulation Materials and Primary Fuel Sources for Six Climate Zones of Japan. *Energy Build.* **2017**, *153*, 403–411. [CrossRef]
9. Restuccia, L.; Reggio, A.; Ferro, G.A.; Tulliani, J.-M. New Self-Healing Techniques for Cement-Based Materials. *Procedia Struct. Integr.* **2017**, *3*, 253–260. [CrossRef]
10. Latif, E.; Bevan, R.; Woolley, T. *Thermal Insulation Materials for Building Applications*; ICE Publishing: London, UK, 2019. [CrossRef]
11. Loftus, J.J. *NBS Report No. 9988*; Noncombustibility of Mineral Wool and Glass Fiber Insulation Materials. National Bureau of Standards: Washington, DC, USA, 1969.
12. Tkaczewska, E.; Malata, G. Properties of the Cement, Slag and Fly Ash Mixture Composition Corresponding to CEM II/C-M and CEM VI. *Mater. Proc.* **2023**, *13*, 11. [CrossRef]
13. Chmura, D.; Jagodziński, A.M.; Hutniczak, A.; Dyczko, A.; Woźniak, G. Novel Ecosystems in the Urban-Industrial Landscape–Interesting Aspects of Environmental Knowledge Requiring Broadening: A Review. *Sustainability* **2022**, *14*, 10829. [CrossRef]
14. Abdrakhimov, V.Z. Environmental Management, Economic and Practical Aspects of the Use of Waste from the Fuel and Energy Complex in the Production of Thermal Insulation Materials. *Econ. Gov. Law Basis* **2021**, *1*, 11–16. [CrossRef]
15. Bieda, B.; Skalna, I.; Gaweł, B.; Grzesik, K.; Henclik, A.; Sala, D. Life Cycle Inventory Processes of the Integrated Steel Plant (ISP) in Krakow, Poland—Continuous Casting of Steel (CCS): A Case Study. *Int. J. Life Cycle Assess.* **2018**, *23*, 1274–1285. [CrossRef]
16. Rabbat, C.; Awad, S.; Villot, A. Towards the Production of High Added-Value Products from the Pyrolysis and Steam Pyro-Gasification of Five Biomass-Based Building Insulation Materials at End-of-Life. *Waste Biomass Valorization* **2023**, *14*, 2061–2083. [CrossRef]
17. Fedoreiko, V.S.; Rutylo, M.I.; Iskerskyi, I.S.; Zahorodnii, R.I. Optimization of Heat Production Processes in the Biofuel Vortex Combustion Systems. *Nauk. Visn. Nats. Hirnych. Univ.* **2020**, *6*, 83–88. [CrossRef]
18. Dychkovskyi, R.O. Forming the Bilayer Artificially Created Shell of Georeactor in Underground Coal Well Gasification. *Nauk. Visn. Nats. Hirnych. Univ.* **2015**, *5*, 37–42.
19. Fedoreiko, V. Distributed Energy Generation Based on Jet-Vortex Bioheat Generators. *E3S Web Conf.* **2024**, *567*, 01001. [CrossRef]
20. Norouzi, M.; Haddad, A.N.; Jiménez, L.; Hoseinzadeh, S.; Boer, D. Carbon Footprint of Low-Energy Buildings in the United Kingdom: Effects of Mitigating Technological Pathways and Decarbonization Strategies. *Sci. Total Environ.* **2023**, *882*, 163490. [CrossRef]
21. Latif, E.; Bevan, R.; Woolley, T. Retrofit, Renovation and Fuel Poverty Initiatives: Insulation Materials. In *Thermal Insulation Materials for Building Applications*; ICE Publishing: London, UK, 2019; pp. 155–180. [CrossRef]
22. Lewicka, D.; Zarębska, J.; Batko, R.; Tarczydło, B.; Woźniak, M.; Cichoń, D.; Pec, M. Circular Economy in the European Union. In *Circular Economy in the European Union: Organisational Practice and Future Directions in Germany, Poland and Spain*; Routledge: London, UK, 2023; pp. 21–267. [CrossRef]
23. Borowicz, M.; Paciorek-Sadowska, J.; Lubczak, J.; Czupryński, B. Biodegradable, Flame-Retardant, and Bio-Based Rigid Polyurethane/Polyisocyanurate Foams for Thermal Insulation Application. *Polymers* **2019**, *11*, 1816. [CrossRef]
24. Lewicka, B.; Lewicka, D. Environmental Risk Management in the Context of Environmental Management Systems for Agriculture Based on the ISO 14001:2015 Standard. *Acta Innov.* **2019**, *33*, 63–72. [CrossRef]
25. Smith, R.E.; Alcott, J.M.; Mazor, M.H. Design Considerations for Sustainable Extruded Polystyrene (XPS) Thermal Insulation. In *Next-Generation Thermal Insulation Challenges and Opportunities*; ASTM International: West Conshohocken, PA, USA, 2014; pp. 1–12. [CrossRef]
26. Aleksakhin, A.; Sala, D.; Golovin, K.; Kovalev, R. Reducing Energy Costs for Pipeline Transportation. *Transp. Res. Procedia* **2021**, *57*, 24–32. [CrossRef]
27. Laustsen, J. *Energy Efficiency Requirements in Building Codes, Energy Efficiency Policies for New Buildings*; International Energy Agency: Paris, France, 2008. Available online: https://iea.blob.core.windows.net/assets/3783f5e8-b14c-4c18-b04c-aab7c59d6e92/Building_Codes.pdf (accessed on 29 April 2025).
28. Beshta, O.S.; Fedoreiko, V.S.; Balakhontsev, O.V.; Khudolii, S.S. Dependence of Electric Drive's Thermal State on Its Operation Mode. *Nauk. Visn. Nats. Hirnych. Univ.* **2014**, *6*, 67–72.
29. Worbs, H. Utilization of Energy Code Compliance Procedures for the Prediction of Commercial Building Annual Fuel Consumption. In *Thermal Insulation: Materials and Systems*; ASTM International: West Conshohocken, PA, USA, 1987; pp. 8–20. [CrossRef]
30. Dyczko, A. Thin Coal Seams, Their Role in the Reserve Base of Poland. In *Technical, Technological and Economical Aspects of Thin-Seams Coal Mining. International Mining Forum 2007*; CRC Press: London, UK, 2007; pp. 81–87. [CrossRef]
31. Dudek, M. The Model for the Calculation of the Dispersed Iron Ore Resource Purchase Cost in the World Class Manufacturing (WCM) Logistics Pillar Context. *Metalurgija* **2014**, *53*, 567–570.
32. Fedoreiko, V.S.; Luchko, M.R.; Iskerskyi, I.S.; Zahorodnii, R.I. Enhancing the Efficiency of Energy Generation Systems Based on Solid Biofuels: Technical and Economic Aspects. *Nauk. Visn. Nats. Hirnych. Univ.* **2019**, *2*, 94–100. [CrossRef]

33. Osman, A.I.; Farghali, M.; Dong, Y.; Kong, J.; Yousry, M.; Rashwan, A.K.; Chen, Z.; Al-Fatesh, A.; Rooney, D.W.; Yap, P.-S. Reducing the Carbon Footprint of Buildings Using Biochar-Based Bricks and Insulating Materials: A Review. *Environ. Chem. Lett.* **2024**, *22*, 71–104. [CrossRef]
34. Dudek, M.; Pawlewski, P. Implementation of Network Oriented Manufacturing Structures. In *Lecture Notes in Computer Science*; Springer: Berlin/Heidelberg, Germany, 2010; pp. 282–291. [CrossRef]
35. Kosenko, A.; Khomenko, O.; Kononenko, M.; Myronova, I.; Pazynich, Y. Raises Advance Using Borehole Hydraulic Technology. *E3S Web Conf.* **2024**, *567*, 01008. [CrossRef]
36. Winkless, L. Biobased Polyurethane Foams Outperform Those Made from Petroleum. *Mater. Today* **2023**, *67*, 11–12. [CrossRef]
37. Dychkovskyi, R.; Dyczko, A.; Borojević Šoštarić, S. Foreword: Physical and Chemical Geotechnologies—Innovations in Mining and Energy. *E3S Web Conf.* **2024**, *567*, 00001. [CrossRef]
38. Dudek, M. Utilisation of Simulation Modelling to Coordinate of Distributed Logistic Resources. In *Congress Proceedings—CLC 2012: Carpathian Logistics Congress*; TANGER Ltd.: Ostrava, Czech Republic, 2012; pp. 151–159.
39. Srivastava, N.; Gupta, B.; Gupta, S.; Danquah, M.K.; Sarethy, I.P. Analyzing Functional Microbial Diversity. In *Microbial Diversity in the Genomic Era*; Elsevier: Amsterdam, The Netherlands, 2019; pp. 79–102. [CrossRef]
40. Polyanska, A.; Pazynich, Y.; Mykhailyshyn, K.; Babets, D.; Toś, P. Aspects of Energy Efficiency Management for Rational Energy Resource Utilization. *Rud. Geol. Naft. Zb.* **2024**, *39*, 13–26. [CrossRef]
41. Pyzalski, M.; Białoskórski, J.; Walasek, E. Reaction between Carbon Fibres and Molten Silicon: Heat Determination Using DTA. *J. Therm. Anal.* **1986**, *31*, 1193–1196. [CrossRef]
42. Durczak, K.; Pyzalski, M.; Pilarski, K.; Brylewski, T.; Sujak, A. The Effect of Liquid Slurry-Enhanced Corrosion on the Phase Composition of Selected Portland Cement Pastes. *Materials* **2021**, *14*, 1707. [CrossRef]
43. Zhao, P.; Ji, Y.; Ren, Q.; Li, X.; Vandeginste, V. On Thermal Insulation Properties of Various Foaming Materials Modified Fly Ash Based Geopolymers. *Polymers* **2023**, *15*, 3254. [CrossRef]
44. *DSTU B V.2.7-105:2000*; Building Materials. Materials and Products. Method for Determining Thermal Conductivity and Thermal Resistance Under Stationary Thermal Regime. State Committee for Construction, Architecture and Housing Policy of Ukraine: Kyiv, Ukraine, 2000. (In Ukrainian)
45. Mizuno, M.; Kokai, T.; Koho, J. Method and Apparatus for Thermal Decomposition of Waste Plastics for Fuel Manufacturing. *Fuel Energy Abstr.* **2003**, *44*, 121.
46. Kioupis, D.; Skaropoulou, A.; Tsivilis, S.; Kakali, G. Properties and Durability Performance of Lightweight Fly Ash Based Geopolymer Composites Incorporating Expanded Polystyrene and Expanded Perlite. *Ceramics* **2022**, *5*, 821–836. [CrossRef]
47. Saik, P.; Cherniaiev, O.; Anisimov, O.; Dychkovskyi, R.; Adamchuk, A. Mining of Non-Metallic Mineral Deposits in the Context of Ukraine's Reconstruction in the War and Post-War Periods. *Min. Miner. Depos.* **2023**, *17*, 91–102. [CrossRef]

energies

MDPI

Article

Theoretical and Simulation-Based Approach to BIPV Systems Integrated with Modular Building

Julia Brenk [1], Barbara Ksit [1] and Bożena Orlik-Kożdoń [2,*]

[1] Faculty of Civil and Transport Engineering, Poznan University of Technology, 61-139 Poznan, Poland; julia.brenk@student.put.poznan.pl (J.B.); barbara.ksit@put.poznan.pl (B.K.)

[2] Faculty of Civil Engineering, Silesian University of Technology, 44-100 Gliwice, Poland

* Correspondence: bozena.orlik-kozdon@polsl.pl

Abstract

This study presents a simulation-based analysis of a steel modular building that integrates technologies that support the energy transition in the built environment. The focus is placed on the implementation of building-integrated photovoltaics (BIPVs), with photovoltaic modules incorporated into the façade and balcony railings. Several modern photovoltaic façade systems were examined. In addition, the study considers the application of photovoltaic glazing enhanced with active quantum coatings. Seven distinct BIPV modules were analysed, each characterised by unique features, with particular emphasis on the influence of colour in tinted variants. A performance degradation analysis was conducted for railing-mounted modules with varying glass tints. The simulation results were correlated with the building's electricity demand.

Keywords: modular building; BIPV; nearly zero-energy building (nZEB); simulation; photovoltaic analysis

1. Introduction

In the face of the progress of climate change and the increase in energy demand, there is a growing need to implement solutions that improve the energy efficiency of buildings and contribute to climate neutrality [1–3]. Numerous researchers have employed numerical simulations to identify challenges and verify theoretical assumptions in practical contexts [4–8]. As part of the "Fit for 55" package, the European Union has committed to reducing CO_2 emissions by 55% by 2030 [9]. This policy accelerates the transformation of the construction sector towards energy-efficient and low-emission technologies, such as building-integrated photovoltaic (BIPV) and prefabricated systems. The application of photovoltaic technologies in residential construction is becoming a key component of sustainable development strategies. Proper integration of PV systems can significantly enhance the energy performance of buildings [10].

At the same time, modular construction technologies are developing dynamically, offering shorter investment timelines, improved quality control, and a reduced environmental impact of the construction process [11–13]. The use of prefabricated modules—particularly in steel structures—is justified by their repeatability, efficiency, and compatibility with modern building systems [14–16]. Moreover, prefabricated construction systems can significantly improve the efficiency and sustainability of building projects due to precise manufacturing and assembly processes [17]. Modular construction has evolved over decades in response to the demand for faster project delivery and improved quality assurance [18].

In Poland, the initial implementations of such solutions date back to the second half of the 20th century, primarily in the context of lightweight steel construction. Currently, there is a noticeable increase in interest in prefabrication, not only in residential buildings but also in temporary and commercial facilities [19]. Prefabrication encompasses various levels of organisation of the construction process, ranging from the production of individual components to preliminary assemblies and fully integrated building modules [20,21]. This approach reduces on-site work, increases the repeatability and quality of work, and shortens construction timelines [21,22]. It is being adopted in many countries and sectors, confirming its versatility and development potential [22].

The integration of modular construction with BIPV systems represents an innovative direction in the design and construction of energy-efficient buildings [23], especially in the context of multifamily and collective housing developments [24,25]. However, such solutions require detailed analysis in terms of technical, functional, and energy-related aspects. Key considerations include the appropriate selection of photovoltaic systems, their placement and orientation with respect to cardinal directions, shading conditions, cell efficiency, integration with the building envelope, and the overall impact on the energy balance [18,26–30]. This article focuses on the analysis of modern BIPV solutions and their integration with modular building structures, highlighting approaches that may represent an innovative direction in the design and construction of energy-efficient buildings in the future.

2. Integrated Photovoltaic Systems

One of the key directions in the development of contemporary construction is the integration of renewable energy sources directly into the building structure, as exemplified by building-integrated photovoltaics (BIPVs) [31].

BIPV systems enable the replacement of conventional building materials with components that generate electricity, while simultaneously fulfilling both functional and aesthetic functions [6,16,31–33]. As a result, they are gaining increasing popularity in modern residential and commercial building projects, particularly in the context of nearly-zero-energy building (nZEB) standards [33]. Despite numerous advantages, BIPV systems are associated with significant technical and economic limitations. The lack of effective ventilation leads to overheating of the modules and a reduction in efficiency. The complexity of wiring, resulting from the customised nature of each installation, increases costs and complicates maintenance. Additionally, the high initial cost [29], related to non-standard components and demanding installation procedures, poses a barrier to implementation. Over the long term, aesthetic degradation of the modules may also become an issue, negatively affecting the building's façade appearance and overall architectural perception. As noted in the literature, architects and designers rarely consider access to wiring or mounting elements during the conceptual design stage, which hampers future maintenance and replacement of damaged modules [31]. Moreover, BIPV systems, as external building envelopes, require regular cleaning—the frequency of which depends on climatic conditions and location—since maintaining a clean surface improves performance and slows down the rate of degradation [31].

2.1. Photovoltaic Ventilated Façade

A ventilated photovoltaic façade can be implemented in both newly constructed buildings and the thermal retrofitting of existing structures. The technical parameters of such systems are presented in Table 1. These modules can be integrated with various types of curtain wall systems. Ventilated façades consist of an insulation layer and an external cladding, separated by a ventilation gap. One of the main advantages of this system is the improvement of the thermal performance, which consequently reduces the risk of condensation of water vapour. Thus, a single component serves the dual function of providing weather protection and generating electricity for the building [24].

Table 1. Technical parameters of the ventilated photovoltaic façade [34].

Property	Value
Unit power	Up to 200 Wp/m^2
PV cells efficiency	Up to 25%
Maximum module size	3600 × 2020 mm
Structure colour	RAL palette
Module thickness	3 to 22 mm

2.2. Photovoltaic Curtain Wall Façade

The façade of the photovoltaic curtain wall is a skeletal structure composed of horizontal and vertical profiles. It enables the creation of an aesthetically pleasing façade for various types and allows for the complete or partial fulfilment of the building's energy demand of the building. These systems provide excellent thermal insulation, effectively protecting the interior from heat loss during winter and from overheating in summer. The technical parameters of the system are presented in Table 2. Due to the curtain wall façade, energy losses are reduced, resulting in lower heating and cooling demands and measurable energy savings. The system is characterised by low weight, good thermal insulation, and transparency. It is particularly recommended for large-area buildings but can also be applied in smaller structures, such as single-family houses [35].

Table 2. Technical parameters of the photovoltaic post-and-beam façade [34].

Property	Value
Unit power	Up to 200 Wp/m^2
PV cells efficiency	Up to 25%
Substructure material	acc. to system supplier
Maximum module size	acc. to system supplier

2.3. Photovoltaic Façade Louvers

Photovoltaic façade louvers are systems designed not only to protect the interior from excessive solar radiation and overheating but also to reduce carbon dioxide emissions, often associated with the operation of systems such as air conditioning.

Two main variants of photovoltaic louver systems can be distinguished: fixed and movable. They typically take the form of slats integrated with PV modules. The technical parameters of these systems are presented in Table 3. Photovoltaic cells are embedded in the glass surfaces of louvers, which are mounted on aluminium structures and can be installed either on curtain wall façades or directly on the building envelope.

Movable louver systems allow one to adjust the tilt angle of the PV modules in response to the direction of solar radiation. A key advantage of this system is its ability to maximise solar energy utilisation. The tilt angle can be adjusted manually or automatically through control systems. Sensors provide data on wind, precipitation, and solar irradiance. In particular, louvers can respond autonomously to prevailing weather conditions—for example, by automatically removing snow [35–37].

Table 3. Technical parameters of the photovoltaic sunshades [34].

Property	Value
Unit power	Up to 170 Wp/m^2
PV cells efficiency	Up to 25%
Support post width	50 mm
Max. Support post spacing	3000 mm
Lamella height	380 mm
Lamella thickness	1.5 to 20 mm
Lamella tilt adjustment	10 deg. Pitch

2.4. Photovoltaic Balustrades

Photovoltaic balustrades represent an innovative architectural solution that combines user safety with energy generation functionality. These transparent barriers, designed to prevent falls from height, comply with the requirements of TRAV (Technische Regeln für absturzsichernde Verglasungen) due to the use of symmetrically constructed tempered and laminated safety glass (ESG + VSG).

To ensure safety, glass–glass technology is employed, which involves the lamination of two glass panes. Balustrade elements can be mounted using various systems, including wedge, pressure, or point fixings, allowing flexible adaptation to specific design and technological requirements.

The electricity generated by the balustrades can be used locally to power lighting, equipment, or charging systems, or it can be directed to energy storage systems, thus increasing the building's energy autonomy of the building [35].

The technical parameters of this solution are presented in Table 4.

Table 4. Technical parameters of the photovoltaic balustrades [34].

Property	Value
Unit power	do 200 Wp/m^2
PV cells efficiency	do 22.5%
Maximum module size	acc. to system supplier
Structure colour	RAL palette

2.5. Glazing with an Active Quantum Dot Coating

An additional contemporary solution involves the use of glazing with an active quantum dot coating. Although it appears to be conventional, this type of glass performs all standard functions—such as providing thermal, wind, and precipitation protection, noise reduction, and natural daylighting—while also generating electricity from solar radiation.

A key advantage of this technology is that, visually, it is indistinguishable from traditional window glazing thanks to the application of the quantum dot layer. This coating functions as a spectral filter: it transmits visible light while converting harmful infrared and ultraviolet radiation into electrical energy.

The glass has a high thermal insulation coefficient and excellent light transmittance. The technical parameters of this solution are presented in Table 5.

An additional benefit includes the reduction in interior overheating and the mitigation of the urban heat island effect. This solution serves as a supplementary source of renewable solar energy [35,36].

Table 5. Technical parameters of the glass with an active quantum coating [34].

Property	Value
Maximum module size	1.2 × 2.2 m
Efficiency	From 30 W/m^2
Thermal insulation Ug value	From 0.4 W/m^{2}K

3. Materials and Methods

3.1. Subject of Research

The analysis covered a modular steel building by Climatic, located in Kobyłka, Poland (Figure 1). The external walls of the modules are constructed using cement-bonded particleboards with a minimum thickness of 16 mm or 24 mm. The internal partitions consist of Fermacell boards and are insulated with mineral wool. The thermal transmittance for partitions is as follows:

- $U \leq 0.17$ W/(m^2·K) walls.
- $U \leq 0.12$ W/(m^2·K) roof.
- $U \leq 0.23$ W/(m^2·K) floor.

Figure 1. Modular building structure in the analysed case; the arrow indicates the location of the wall-mounted panels [38].

The materials used in the modular construction exhibit high resistance to moisture, as confirmed by the following parameters:

The swelling after 24 h, according to EN 317 [39], does not exceed 1.5%.

- Durability under cyclic moisture tests, i.e., reduction in tensile strength after testing according to EN 319 and EN 321 [40,41], does not exceed 0.3 N/mm^2.
- The swelling after cyclic testing, according to EN 317 and EN 321 [39,41], does not exceed 1.5%.

In modular buildings, high-performance heating, cooling, and ventilation systems have been installed. In addition, alternative primary energy sources are utilised, complemented by energy-efficient LED lighting.

An analysis of the building was conducted using a ventilated photovoltaic façade installation (Figure 2), which proves to be highly suitable for implementation on an existing building.

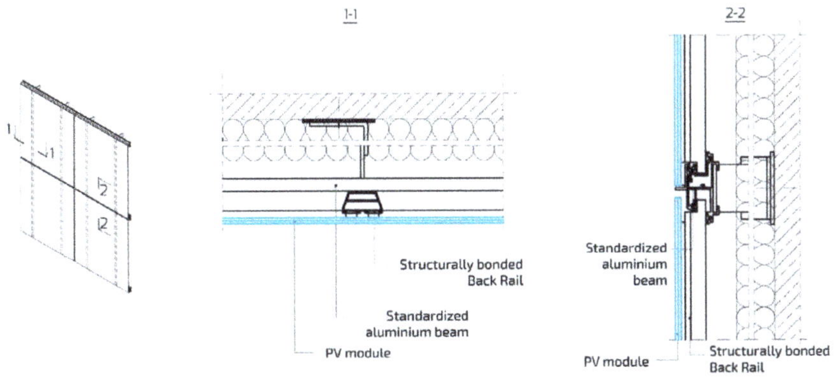

Figure 2. Diagram of ventilated photovoltaic façade installation [34].

An analysis was performed for the building, which was based on a comparison of annual energy yield, type and power of modules, surface area, and location of the system. Exposure to the cardinal directions, the impact of solutions on aesthetics, and the possibility of balancing energy in the building were also taken into account.

3.2. Assumptions Used in the Analysis

To evaluate the energy performance of the building-integrated photovoltaic (BIPV) system—taking into account the geometry, orientation, and local environmental conditions of the building—a three-stage computational approach was adopted, consisting of the following:

(1) Solar exposure analysis in an Autodesk RevitA 3D model of the building was utilised to perform solar exposure analysis, taking into account geometry and surrounding shading. This allowed identification of the most favourable surfaces for BIPV integration.
(2) Analytical calculations: The energy yield was estimated using a formula incorporating solar irradiation, PV module rated power, system losses, and Standard Test Conditions (STCs). This enabled a comparative assessment of different panel arrangement scenarios and their energy potential.

$$E_{actual} = \frac{insolation * correction\ factor * power * \text{efficiency ratio}}{radiant\ intensity(STC)} \quad (1)$$

(3) PVGIS System: The Photovoltaic Geographical Information System (PVGIS) was used to simulate the annual energy output based on real climatic data and system-specific parameters. The output served to validate the analytical calculations.

The combined model integrates geometric, climatic, and spatial analysis [31], allowing for a reliable comparison of the energy performance associated with various BIPV implementation strategies. This methodology is applicable both in the design phase [25] and in the evaluation of retrofit projects that involve existing buildings [42].

In the analysis, the boundary conditions were assumed to be uniform for each variant:

- System losses were assumed at the level of 15%;
- Mounting position was considered as building-integrated photovoltaics (BIPVs);
- Tilt angle of 90°;
- Solar irradiance under which the modules were tested: 1000 W/m^2.

Other boundary conditions depended on the orientation of the PV modules in relation to the cardinal directions. These included the following.

- Solar irradiation on a horizontal surface—obtained from solar irradiation maps: Photovoltaic Geographical Information System (PVGIS). Its value depends on the orientation of PV modules relative to the cardinal directions. The values adopted for the calculations were as follows:
    - Modules facing E 85°—658 kWh/m^2.
    - Modules facing NE 65°—533 kWh/m^2.
    - Modules facing SE 155°—944 kWh/m^2.
- Correction factor—obtained from the correction factor table; its value depends on the tilt and azimuth of the PV modules.
- Azimuth of the PV modules.

System losses were accounted for at a level of 15%. These losses primarily result from cable losses, inverter losses, temperature-related losses, losses due to operation under low solar irradiance, shading and soiling losses, current mismatch between modules, and losses on bypass diodes.

In the analysis, a surplus of photovoltaic system production in the range of 10–20% relative to annual energy consumption was assumed. This approach is commonly applied in design practice, as it allows for flexibility in case of future increases in energy demand, e.g., through the installation of a heat pump, an electric vehicle, or other high-consumption devices, without the need to expand the PV system itself. Moreover, it provides a buffer in the event of unforeseen system losses.

The 10–20% range represents a compromise between investment cost-effectiveness and minimising the risk of energy surpluses. Oversizing above 20% without energy storage may lead to significant feed-in of excess energy to the grid, extending the payback period and reducing the economic efficiency of the system.

In situations where PV system production exceeds the assumed threshold, the implementation of energy storage or the increase in self-consumption through the integration of additional loads may enable the economic utilisation of surplus energy without reducing the system's capacity.

An energy audit was conducted to determine the electricity consumption of a four-story steel-frame modular building designed for collective residential occupancy, with a capacity of up to 150 people. The total calculated annual energy consumption per unit area was 3.93 kWh/m^2·year. Based on the total floor area, the overall annual electricity demand was calculated as follows:

$$3.93\frac{\text{kWh}}{\text{m}^2 * \text{year}} * 1405.52\ \text{m}^2 = 5523.69\ \frac{\text{kWh}}{\text{year}}$$

- Energy Yield Calculation for the Considered Variants
    - **Variant 1**—PV modules with a rated power of 400 W.
    - **Variant 1A**—PV modules with a rated power of 400 W + PV railings with a rated power of 200 W.
    - **Variant 2**—Red PV modules with a rated power of 240 W.
    - **Variant 2A**—Red PV modules with a rated power of 240 W + PV railings with a rated power of 200 W.
    - **Variant 2B**—Red PV modules with a rated power of 240 W + PV railings with a rated power of 160 W.
    - **Variant 3**—Grey PV modules with a rated power of 290 W.
    - **Variant 3A**—Grey PV modules with a rated power of 290 W + PV railings with a rated power of 200 W.

- **Variant 3B**—Grey PV modules with a rated power of 290 W + PV railings with a rated power of 160 W.
- **Variant 3C**—Grey PV modules with a rated power of 290 W + PV railings with a rated power of 140 W.
- **Variant 4**—Green PV modules with a rated power of 290 W.
- **Variant 4A**—Green PV modules with a rated power of 290 W + PV railings with a rated power of 200 W.
- **Variant 4B**—Green PV modules with a rated power of 290 W + PV railings with a rated power of 260 W.
- **Variant 5**—PV modules with a rated power of 470 W.
- **Variant 6**—PV modules with a rated power of 480 W.

Table 6 presents the mechanical parameters of each type of PV module type, along with the calculated energy yields for the south-facing orientation, identified as having the highest solar potential. On the basis of the simulation results, the minimum total required module power was determined, enabling the calculation of the maximum feasible number of photovoltaic modules. It can be concluded that installing modules on the south-facing façade ensures compliance with the minimum energy yield requirement despite variations in module quantity. Within the set analysed, modules rated at 480 W are the least numerous, whereas green modules represent the majority. The variety of module types allows flexible configurations for integrating PV elements on both walls and balustrades, allowing for diverse installation arrangements.

Table 6. Table of mechanical parameters and energy yields for different PV modules. Own elaboration.

Variant	Name	Power [W]	Power [kW]	Area [m^2]	Module Colour	Module Efficiency [%]	Max Number of Modules [pcs]	Total Power [kW]	Energy Yield [kWh]	Minimum Energy Yield Condition
a)	Module 400 W	400	0.4	2.012	Black	19.7	22	8.80	5714.7	fulfilled
b)	Module 410 W	410	0.41	2.007	Black	20.4	21	8.61	5591.3	fulfilled
c)	Module 470 W	470	0.47	2.245	Black	20.93	19	8.93	5799.1	fulfilled
d)	Module 480 W	480	0.48	2.245	Black	20.93	18	8.64	5610.8	fulfilled
e)	Red module	240	0.24	2.011	Red	11.9	36	8.64	5610.8	fulfilled
f)	Gray module	290	0.29	2.011	Gray	14.4	30	8.70	5649.8	fulfilled
g)	Green module	200	0.2	1.711	Green	11.7	43	8.60	5584.8	fulfilled
h)	Glass–glass module	297	0.3	1.821	Black	16.3	29	8.61	5593.3	fulfilled

Table 7 summarises the key technical parameters of photovoltaic blinds, enabling an assessment of their energy potential and installation possibilities. These parameters serve as an important reference point when selecting the appropriate configuration of a BIPV system integrated with the building façade.

Analysing the energy yield revealed that, owing to the flexibility in the dimensions of individual louvers and adjustable power ratings, a wide range of configurations is possible, enabling versatile adaptation of module power. Consequently, it can be assumed that the energy yield requirements are satisfied.

The subsequent solution examined was photovoltaic glass featuring an active quantum dot coating, with its technical specifications detailed in Table 8.

Table 7. Table of mechanical parameters of the photovoltaic sunshades [34].

Property	Value
Unit power	Up to 170 Wp/m^2
PV cells efficiency	Up to 25%
Support post width	50 mm
Max. Support post spacing	3000 mm

Table 8. Technical parameters of the glass with an active quantum coating [34].

Property	Value
Maximum module size	1.2 × 2.2 m
Efficiency	From 30 W/m^2
Thermal insulation Ug value	From 0.4 W/m^{2}K

In the case of glass with an active quantum dot coating, because of the relatively low power output of these solutions, it was determined that they are better suited as an auxiliary source of solar energy rather than as a primary one. Taking into account the minimum required power for the PV modules, the necessary glazing area was calculated to allow the generation of free energy exclusively through this solution. The required area is 283.7 m^2, which is quite large but feasible. Such solutions could prove effective on glazed façades of high-rise buildings, such as office towers. In residential buildings, however, the total glazed area within the external envelope is usually too limited to allow for the installation of the required surface area of glass with an active quantum dot coating. However, for this type of façade, the previously discussed installation of photovoltaic modules represents a more efficient solution.

4. Research Results

4.1. Insolation

To enable analysis aimed at maximising solar energy capture, numerical simulations of solar exposure were conducted using Autodesk Revit software, version 2023. According to the site development plan, the front façades face southeast and northeast, while the courtyard-facing walls are orientated toward the southwest and northwest. The window bays protruding from the façade are oriented eastward. The highest solar energy gains can be achieved from the south, whereas the lowest are from the north [36,43,44]. The analysis assumptions included, among others, the building's location, its orientation relative to the cardinal directions, geographic coordinates, date, and time, as well as the sun's angle and azimuth for the specified day and hours. Figures 3 and 4 illustrate the solar exposure of the building at noon on both the longest and shortest days of the year.

According to the analysis, the most favourable location for installing photovoltaic modules is the front wall facing southeast. However, due to the arrangement of the window openings on this facade, the installation of photovoltaic cladding is not feasible. Therefore, other façades must be considered.

The next best option in terms of potential solar energy yield is the wall facing the parking area; nevertheless, this wall also does not permit the installation of photovoltaic cladding.

Another viable option is the front wall facing northeast, which—despite having numerous window openings—provides a portion of surface area suitable for such an installation.

Photovoltaic modules could be mounted on the left section of this wall, covering an approximate area of $12 \times 6 = 72\ m^2$.

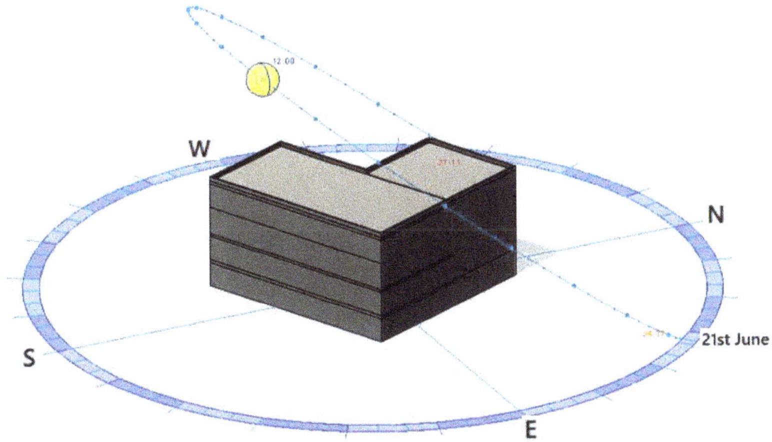

Figure 3. Building solar exposure on the longest day of the year. Solar analysis of the building on the summer solstice (21 June), performed using Autodesk Revit. The model presents the trajectory of the sun over the building throughout the day—from sunrise (04:17) to sunset (21:11). The position of the sun at 12:00 PM is marked. This analysis enables the assessment of daylight availability, the potential for implementing BIPV systems, and the impact of shading on individual building façades. Geographical directions (N, S, E, W) are visible, allowing the model to be oriented in accordance with the actual site location. Analysis performed using Autodesk Revit software.

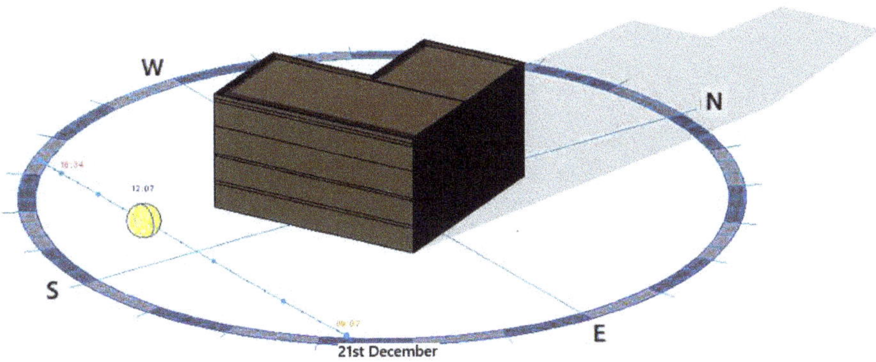

Figure 4. Building solar exposure on the shortest day of the year. Solar analysis of the building on the winter solstice (21 December), performed using Autodesk Revit. The model presents the trajectory of the sun over the building throughout the day—from sunrise (09:07) to sunset (16:34), with the sun's position marked at 12:00 PM. On this day, the sun follows the lowest path in the sky throughout the year, resulting in a shorter duration of sunlight and longer shadows cast by the building. The analysis enables the assessment of the impact of limited daylight access on individual façades, as well as the effectiveness of passive design strategies and BIPV systems under winter conditions. Geographical directions (N, S, E, W) are visible, allowing the analysis to be aligned with the actual site location. Analysis performed using Autodesk Revit software.

The area of the wall suitable for installing the photovoltaic cladding is approximately 72 m^2. A ventilated photovoltaic façade system can be applied to the analysed wall. Modules with various power ratings are used for this type of façade, and an energy yield analysis was performed for each variant of the module. The number of PV modules selected for each variant depends on the available wall area, assuming maximum utilisation of the surface for aesthetic purposes and to ensure the highest possible energy yield. In the case of photovoltaic balustrades, their quantity depends on the number of windows where the balustrades can be installed.

4.2. Energy Yield

4.2.1. Variant 1—Modules with a Power Rating of 400 W

A maximum of 36 modules can be installed, producing energy with a nominal power of 14.4 kW. The actual energy yield is $E_{actual} = 5219.14$kWh; however, the minimum required energy yield was not achieved. Figure 5 presents the graph of energy yield as a function of the month of the year. The highest energy yield occurs in June, reaching approximately 900 kWh.

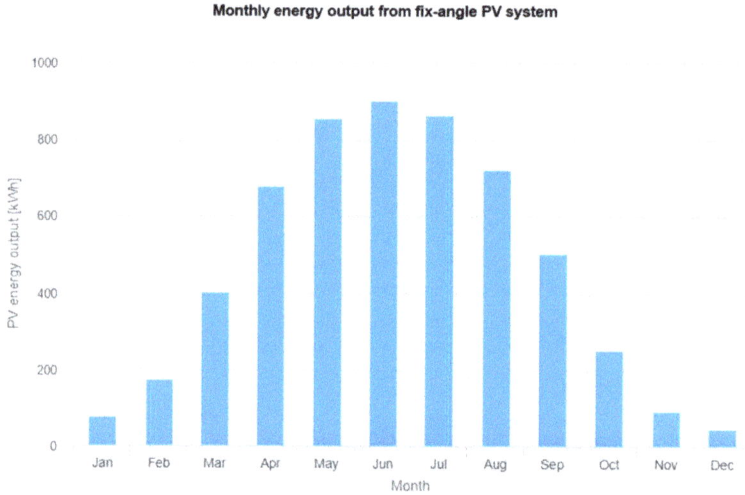

Figure 5. Performance results of grid-connected PV. The vertical axis represents the PV energy yield per month throughout the year (kWh), while the horizontal axis represents the successive months of the year. The results, presented in the form of a bar chart, indicate the energy yield value for each month of the year in kWh/month. Source: Photovoltaic Geographical Information System (PVGIS).

To meet the required minimum energy yield, photovoltaic balustrades can be employed. So-called "French balconies" equipped with these photovoltaic balustrades can be installed on the building.

4.2.2. Variant 1A—400 W Modules and Balustrades with a Power Density of 2.0 kW/m^2

French balconies (also known as portfenetres) can be installed on tall windows located on the front façades facing both the northeast and southeast directions. Due to the possibility of manufacturing balustrades in custom dimensions up to a maximum of 3.5 m, balustrades measuring 1.80 × 1.10 m have been selected for installation on four windows on the southeast side, while balustrades measuring 1.20 × 1.10 m will be installed on nine windows on the southeast side and eleven windows on the northeast side. The balustrades are designed with a maximum power density of 2.0 kW/m^2.

The maximum power output of the modules integrated into the balustrades is 6.86 kW. The total actual energy yield from the photovoltaic modules on the balustrades facing southeast (SE 155°) amounts to $E_{actual} = 2542.00$ kWh, whereas the yield from those facing northeast (E 85°) is $E_{actual} = 1297.58$ kWh, resulting in a combined total of 3839.58 kWh. The total energy produced by the modules on the northeast façade and the photovoltaic balustrades is $E_{actual} = 9058.72$ kWh, thereby meeting the minimum energy production requirements.

The total generated energy exceeds the minimum required by 63%. Consequently, several optimisation solutions can be considered.

Discussion on the selection of the optimal solution for the significant energy surplus: The energy surplus exceeds 20%, making the application of photovoltaic modules in this configuration economically unviable. The energy surplus generated by photovoltaic (PV) modules should generally fall within the range of 10% to 20%, providing a technical buffer that compensates for potential losses caused by factors such as partial shading, soiling, or long-term module degradation. Maintaining this margin helps to ensure consistent system performance and prevents underproduction during less favourable conditions. When the surplus reaches approximately 20%, the excess energy can be either exported to the electrical grid or stored in an energy storage system (ESS), without causing operational inefficiencies or overgeneration issues.

To preserve the aesthetic quality regarding the placement and number of PV modules, it is recommended to reduce the power rating of the photovoltaic modules. In this case, tinted glass can be used, which will particularly enhance the visual appearance. Any glass colour can be applied; however, three variants with parameters specified in the manufacturer's datasheet will be analysed.

4.2.3. Variant 2—Red Modules, 240 W

A maximum of 36 modules can be installed, producing energy with a rated power of 8.64 kW. The actual energy yield amounts to $E_{actual} = 3131.48$ kWh; however, the minimum energy requirement is not met. Figure 6 presents a graph illustrating the monthly energy yield throughout the year. The highest energy yield is achieved in June, exceeding 500 kWh.

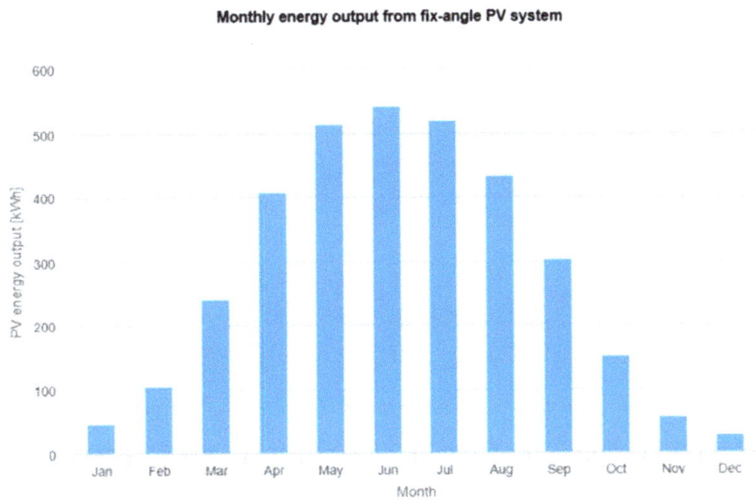

Figure 6. Performance results of grid-connected PV. The vertical axis represents the PV energy yield per month throughout the year (kWh), while the horizontal axis represents the successive months of the year. The results, presented in the form of a bar chart, indicate the energy yield value for each month of the year in kWh/month. Source: Photovoltaic Geographical Information System (PVGIS).

4.2.4. Variant 2A—Red Modules 240 W Combined with Balustrades Rated at 2.0 kW/m^2

Taking into account the photovoltaic balustrades the total actual energy produced by the modules on the northeast façade and the photovoltaic balustrades amounts to $E_{actual} = 6971.06$ kWh, thereby meeting the minimum energy yield requirement.

The total generated energy exceeds the minimum required by 21%.

This surplus should be reduced to improve the economic feasibility of the photovoltaic system.

The optimal solution is to decrease the power rating of the modules installed on the balustrades.

4.2.5. Variant 2B—Red Modules 240 W and Balustrades with a Power Density of 1.60 kW/m^2

The maximum power of the modules is 5.49 kW. The total actual energy yield of photovoltaic modules on the balustrades facing southeast (SE 155°) is E_{actual} = 2009.45 kWh, while for those facing northeast (E 85°) it is E_{actual} = 1038.06 kWh, resulting in a combined total of 3047.51 kWh. The total energy produced by the modules on the northeast façade and the photovoltaic balustrades with a power density of 1.60 kW/m^2 amounts to E_{actual} = 6178.99 kWh, thereby meeting the minimum energy yield requirement.

The total generated energy exceeds the minimum required by 11%, which is considered an optimal surplus.

4.2.6. Variant 3—Gray Modules, 290 W

A maximum of 36 modules can be used, producing energy with a rated power of 10.44 kW. The actual energy yield is E_{actual} = 3783.87 kWh, which does not meet the minimum energy requirement. Figure 7 presents a graph showing the monthly energy yield throughout the year. The highest energy yield is achieved in June, with a value of approximately 650 kWh.

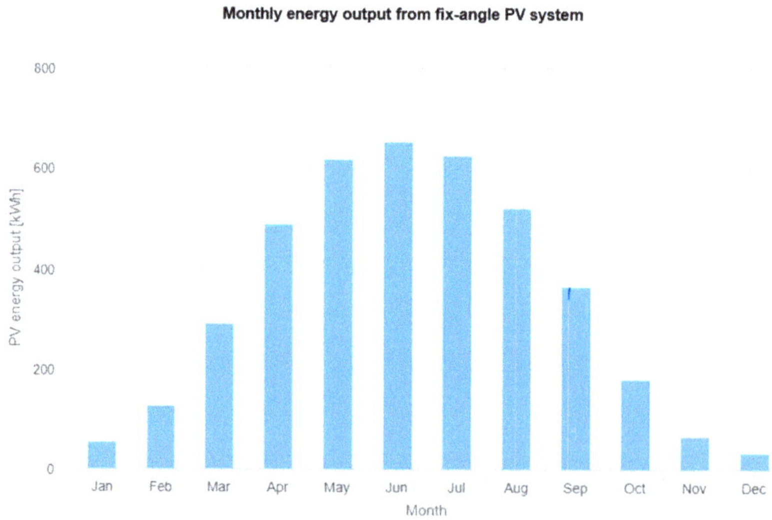

Figure 7. Performance results of grid-connected PV. The vertical axis represents the PV energy yield per month throughout the year (kWh), while the horizontal axis represents the successive months of the year. The results, presented in the form of a bar chart, indicate the energy yield value for each month of the year in kWh/month. Source: Photovoltaic Geographical Information System (PVGIS).

4.2.7. Variant 3A—Grey Modules 290 W and Balustrades with a Power Density of 2.0 kW/m^2

Taking into account photovoltaic balustrades, the total actual energy produced by the modules in the northeast façade and the photovoltaic balustrades amounts to E_{actual} = 7623.45 kWh, fulfilling the minimum energy yield requirement. The total generated energy exceeds the minimum required by 38%. This surplus should be reduced to improve the economic feasibility of the photovoltaic system. The optimal solution is to reduce the power rating of the modules installed on the balustrades.

4.2.8. Variant 3B—Grey Modules 290 W and Balustrades with a Power Density of 1.60 kW/m^2

The total actual energy produced by the modules in the northeast façade and the photovoltaic balustrades is E_{actual} = 6831.38 kWh, meeting the minimum energy yield requirement.

The total generated energy exceeds the minimum required by 24%. This surplus should be reduced to improve the economic efficiency.

4.2.9. Variant 3C—Grey Modules 290 W and Balustrades with a Power Density of 1.40 kW/m^2

Assuming that the maximum power photovoltaic modules for balustrades can produce is 1.40 kW/m^2, the maximum module power is 4.80 kW. The total actual energy yield of photovoltaic modules on the balustrades facing southeast (SE 155°) is E_{actual} = 1755.89 kWh, while on those facing northeast (E 85°) it is E_{actual} = 908.30 kWh, which together totals 2664.19 kWh. The total energy produced by the modules on the northeast façade and the photovoltaic balustrades with a power density of 1.40 kW/m^2 amounts to E_{actual} = 6448.06 kWh, thus meeting the minimum energy yield requirement. The total generated energy exceeds the minimum required by 14%. This level of surplus is considered optimal.

4.2.10. Variant 4—Green Modules 200 W

A maximum of 42 modules can be installed, yielding a rated power of 8.40 kW. The actual energy yield amounts to E_{actual} = 3044.50 kWh, which fails to meet the minimum energy requirement. Figure 8 presents a graph that illustrates the monthly energy yield throughout the year, with the highest yield recorded in June, exceeding 500 kWh.

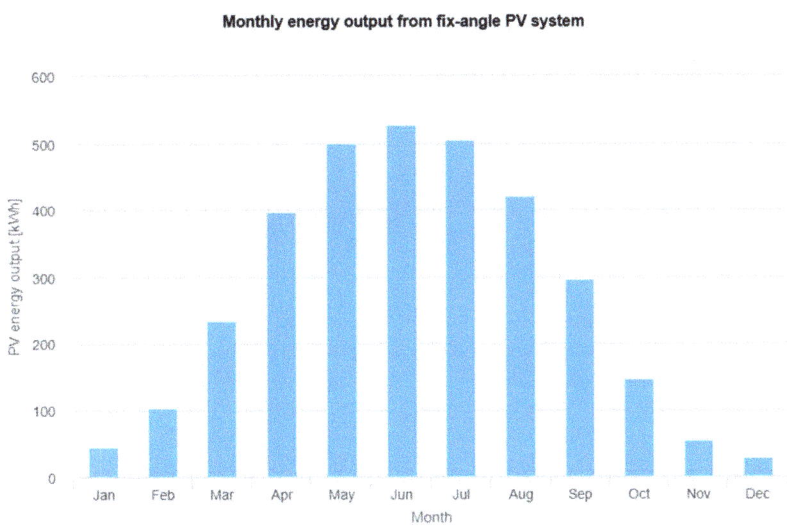

Figure 8. Performance results of grid-connected PV. The vertical axis represents the PV energy yield per month throughout the year (kWh), while the horizontal axis represents the successive months of the year. The results, presented in the form of a bar chart, indicate the energy yield value for each month of the year in kWh/month. Source: Photovoltaic Geographical Information System (PVGIS).

4.2.11. Variant 4A—Green Modules 200 W and Balustrades with a Power Density of 2.0 kW/m^2

Taking into account photovoltaic balustrades, the total actual energy produced by the modules on the northeast facing wall and the photovoltaic balustrades amounts to $E_{actual} = 6884.08$ kWh, fulfilling the minimum energy yield requirement.

The total energy produced is 25% higher than the minimum required. The surplus should be reduced to improve the cost-effectiveness of the photovoltaic system. The best solution is to reduce the power of the modules on the balustrades.

4.2.12. Variant 4B—Green Modules 200 W and Balustrades with a Power Density of 1.60 kW/m^2

The total energy produced by the modules on the wall and the photovoltaic balustrades is $E_{actual} = 6092.01$ kWh, meeting the minimum energy yield requirement.

The total produced energy is 10% higher than the minimum required. This level of surplus is optimal.

4.2.13. Variant 5—Modules with a Power of 470 W

A maximum of 32 modules can be installed, providing a total nominal power of 15.04 kW. The actual annual energy yield is $E_{actual} = 5451.10$ kWh, which satisfies the minimum energy yield requirement. Figure 9 presents a graph that illustrates the variation in monthly energy yield throughout the year. The highest monthly energy production is observed in June, reaching approximately 900 kWh.

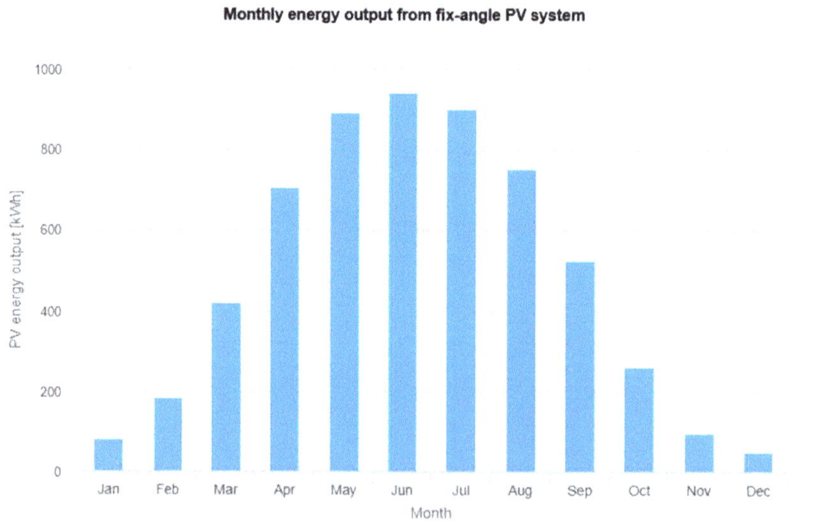

Figure 9. Performance results of grid-connected PV. The vertical axis represents the PV energy yield per month throughout the year (kWh), while the horizontal axis represents the successive months of the year. The results, presented in the form of a bar chart, indicate the energy yield value for each month of the year in kWh/month. Source: Photovoltaic Geographical Information System (PVGIS).

Due to the failure to meet the minimum energy requirement based on manual calculations, it is necessary to increase the rated power of the photovoltaic modules. However, the application of photovoltaic balustrades in this configuration would result in a significant energy surplus, rendering the solution economically unviable. Therefore, an alternative scenario was analysed using modules with a rated power of 480 W.

4.2.14. Variant 6—480 W Module

A maximum of 32 modules can be installed, resulting in a total nominal power of 15.36 kW. The actual annual energy yield amounts to $E_{actual} = 5567.08$ kWh, thereby fulfilling the minimum energy requirement. Figure 10 presents a graph that illustrates the monthly distribution of energy production throughout the year. The highest energy yield is observed in June, approaching 1000 kWh.

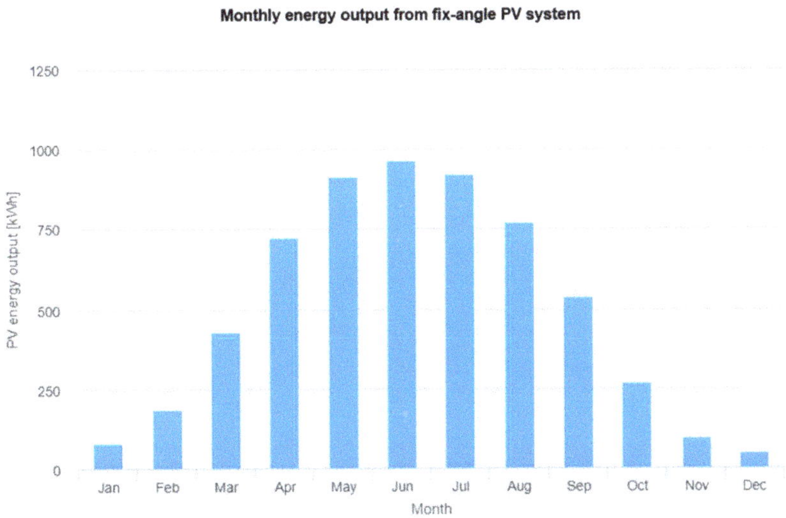

Figure 10. Performance results of grid-connected PV. The vertical axis represents the PV energy yield per month throughout the year (kWh), while the horizontal axis represents the successive months of the year. The results, presented in the form of a bar chart, indicate the energy yield value for each month of the year in kWh/month. Source: Photovoltaic Geographical Information System (PVGIS).

5. Analysis of the Results

As part of this study, a comparative analysis of various layout and configuration options for photovoltaic modules within a BIPV system was conducted. The analysis included both vertically mounted façade-integrated modules and additional modules incorporated into balcony railings. A total of 14 variants were examined, involving different types (standard and coloured), rated power outputs (ranging from 200 W to 480 W), and numbers of installed units. Table 9 presents a summary of the BIPV variants, reflecting the influence of the module type on the performance of the system. For each configuration, the total installed capacity and the estimated annual electricity yield were calculated, taking into account contributions from modules mounted on the façade and integrated with the rails. This comparative approach enables an evaluation of the impact of aesthetic choices and technical configurations on the overall energy performance of the BIPV installation.

Figure 11 presents a graph illustrating the significant variation in total installed system capacity across the analysed BIPV variants. Base variants (1–6), which include only façade-mounted modules, provide installed capacities ranging from 8.4 kW to 15.36 kW, depending on the type and quantity. On the contrary, the extended variants—incorporating additional balcony railing modules (variants 1A, 2A, etc.)—increase the total capacity to more than 21 kW, as observed in variant 1A.

Table 9. Comparison of BIPV installation variants. Own elaboration.

Variant	Wall Module Power [W]	Balustrade Module Power [W/m²]	Number of Wall Modules [pcs]	Number of Balustrade Modules [pcs]	Wall Module Power [kW]	Balustrade Module Power [kW]	Total Power [kW]	Wall Energy Yield [kWh]	Balustrade Energy Yield [kWh]	Total Energy Yield [kWh]
1	400	0	36	0	14.4	0	14.4	5219.136	0	5219.136
1A	400	200	36	24	14.4	6.86	21.26	5219.136	3839.58	9058.716
2	240	0	36	0	8.64	0	8.64	3131.482	0	3131.482
2A	240	200	36	24	8.64	6.86	15.5	3131.48	3839.58	6971.06
2B	240	160	36	24	8.64	5.49	14.13	3131.482	3047.51	6178.992
3	290	0	36	0	10.44	0	10.44	3783.874	0	3783.874
3A	290	200	36	24	10.44	6.86	17.3	3783.874	3839.58	7623.454
3B	290	160	36	24	10.44	5.49	15.93	3783.874	3047.51	6831.384
3C	290	140	36	24	10.44	4.8	15.24	3783.874	2664.19	6448.064
4	200	0	42	0	8.4	0	8.4	3044.5	0	3044.5
4A	200	200	42	24	8.4	6.86	15.26	3044.5	2829.58	5874.08
4B	200	160	42	24	8.4	5.49	13.89	3044.5	3047.51	6092.01
5	470	0	32	0	15.04	0	15.04	5451.098	0	5451.098
6	480	0	32	0	15.36	0	15.36	5567.078	0	5567.078

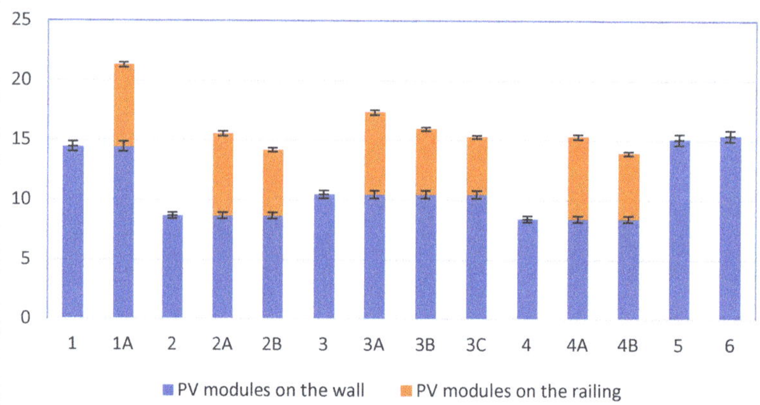

Figure 11. Installed capacity of the BIPV system. Total installation power for individual variants. The vertical axis represents the total power (kW) of the analysed variants, while the horizontal axis represents the individual variants.

The most substantial increase in installed capacity was recorded in variants with integrated railing modules (variants labelled A, B, and C), where the balcony components contributed more than 30% of total system capacity. A standard deviation of 3% error was assumed.

Figure 12 presents a graph illustrating the projected annual energy yield for each variant analysed. The highest yield was achieved by variant 1A, reaching 9058.7 kWh per year, resulting from the use of high-efficiency 400 W modules on the façade combined with additional 200 W modules integrated into the balustrades. On the contrary, variants 2 and 4 (featuring red and green modules) without balustrades produced total yields not exceeding 3100–3200 kWh annually.

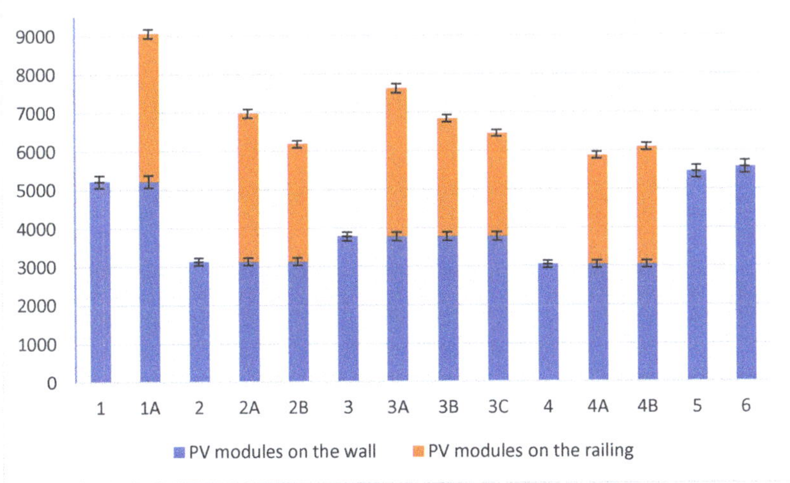

Figure 12. Annual energy yield. Annual energy yield for individual variants. The vertical axis represents the total energy yield (kWh) of the analysed variants, while the horizontal axis represents the individual variants.

The inclusion of balustrades increased energy production by an average of 40–60% compared to base variants. For variants 2A–4B (coloured modules), the additional balustrade modules partially compensated for efficiency losses due to the lower optical performance of these modules. A standard deviation of 3% error was assumed.

To conduct a more detailed investigation, an uncertainty analysis was performed to assess the potential shading of the PV modules. Shading of the module surface can significantly affect system efficiency. It is assumed that shading at the level of a few percent can reduce efficiency by up to 50%. Accordingly, calculations were carried out for efficiency reductions of 25%, 40%, and 50%. The results are presented in Table 10.

Table 10. Comparison of BIPV installation variants considering different efficiency losses due to shading. Own elaboration.

Variant	Total Power [kW]	Total Energy Yield [kWh]	25% Efficiency Loss [kWh]	Minimum Energy Yield Condition	Energy Efficiency	40% Efficiency Loss [kWh]	Minimum Energy Yield Condition	Energy Efficiency	50% Efficiency Loss [kWh]	Minimum Energy Yield Condition	Energy Efficiency
1	14.4	5219.136	4605.12	unfulfilled	NO	3684.096	unfulfilled	NO	3070.08	unfulfilled	NO
1A	21.26	9058.716	8444.7	fulfilled	NO	7523.676	fulfilled	NO	6909.66	fulfilled	NO
2	8.64	3131.482	2763.072	unfulfilled	NO	2210.458	unfulfilled	NO	1842.048	unfulfilled	NO
2A	15.5	6971.06	6602.652	fulfilled	YES	6050.038	fulfilled	NO	5681.628	fulfilled	NO
2B	14.13	6178.992	5810.582	fulfilled	NO	5257.968	unfulfilled	NO	4889.558	unfulfilled	NO
3	10.44	3783.874	3338.712	unfulfilled	NO	2670.97	unfulfilled	NO	2225.808	unfulfilled	NO
3A	17.3	7623.454	7178.292	fulfilled	NO	6510.55	fulfilled	YES	6065.388	fulfilled	NO
3B	15.93	6831.384	6386.222	fulfilled	YES	5718.48	fulfilled	NO	5273.318	unfulfilled	NO
3C	15.24	6448.064	6002.902	fulfilled	NO	5335.16	unfulfilled	NO	4889.998	unfulfilled	NO
4	8.4	3044.5	2686.32	unfulfilled	NO	2149.056	unfulfilled	NO	1790.88	unfulfilled	NO
4A	15.26	5874.08	5515.9	unfulfilled	NO	4978.636	unfulfilled	NO	4620.46	unfulfilled	NO
4B	13.89	6092.01	5733.83	fulfilled	NO	5196.566	unfulfilled	NO	4838.39	unfulfilled	NO
5	15.04	5451.098	4809.792	unfulfilled	NO	3847.834	unfulfilled	NO	3206.528	unfulfilled	NO
6	15.36	5567.078	4912.128	unfulfilled	NO	3929.702	unfulfilled	NO	3274.752	unfulfilled	NO

The analysis assessed the impact of shading at 25%, 40%, and 50% on meeting the minimum required energy yield and optimal production (10–20% surplus). The results indicate that only variants 1A, 2A, and 3A meet the minimum yield in all loss scenarios, with the optimal level of surplus achieved in variant 2A at 25% losses, in variant 3A at 40% losses, and in variant 3B at 25%. Although variant 1A meets the requirements, it generates a surplus above the optimal range in each configuration, suggesting system oversizing. Most of the remaining variants fail to achieve the minimum yield in any loss scenario, indicating insufficient system power under shading conditions.

Based on the presented table and chart, the results of the cost analysis and annual energy performance of various photovoltaic (PV) module variants can be summarised as follows:

Table 11 provides a comparison of the approximate purchase costs of PV modules for the analysed variants. These variants differ in the power rating of the wall-mounted and balustrade modules, as well as in their quantities. Costs have been converted to Polish zloty (PLN) and US dollars (USD). Additionally, the energy efficiency of each variant is indicated, where "YES" denotes an energy-efficient variant, and "NO" indicates a non-efficient one.

Table 11. Comparison of approximate purchase costs of PV modules for the analysed variants. Own elaboration.

Variant	Wall Module Power [W]	Balustrade Module Power [W/m^2]	Number of Wall Modules [pcs]	Number of Balustrade Modules [pcs]	Total Power [kW]	Total Energy Yield [kWh]	Approximate Costs, PLN	Approximate Costs, USD	Energy Efficiency
1	400	0	36	0	14.4	5219.136	10,080.00 zł	$2739.13	NO
1A	400	200	36	24	21.26	9058.716	70,080.00 zł	$19,043.48	NO
2	240	0	36	0	8.64	3131.482	135,000.00 zł	$36,684.78	NO
2A	240	200	36	24	15.5	6971.06	195,000.00 zł	$52,989.13	NO
2B	240	160	36	24	14.13	6178.992	185,400.00 zł	$50,380.43	YES
3	290	0	36	0	10.44	3783.874	135,000.00 zł	$36,684.78	NO
3A	290	200	36	24	17.3	7623.454	195,000.00 zł	$52,989.13	NO
3B	290	160	36	24	15.93	6831.384	185,400.00 zł	$50,380.43	NO
3C	290	140	36	24	15.24	6448.064	181,800.00 zł	$49,402.17	YES
4	200	0	42	0	8.40	3044.5	157,500.00 zł	$42,798.91	NO
4A	200	200	42	24	15.26	5874.08	217,500.00 zł	$59,103.26	NO
4B	200	160	42	24	13.89	6092.01	207,900.00 zł	$56,494.57	YES
5	470	0	32	0	15.04	5451.098	10,560.00 zł	$2869.57	NO
6	480	0	32	0	15.36	5567.078	11,040.00 zł	$3000.00	NO

Out of the 14 analysed variants, only three (2B, 3C, 4B) were marked as energy efficient (green shading denotes variants that satisfy the energy optimality criterion, whereas red shading denotes variants that do not satisfy this criterion), indicating that their calculated energy yield relative to the generated surplus is the most favourable. Variants with balustrade modules rated at 160 W (e.g., 2B, 3C, 4B) exhibit higher efficiency despite higher purchase costs compared to variants without balustrade modules or with lower-power modules.

The bar chart (Figure 13) shows the annual energy performance (in USD on the vertical axis) for the individual variants (on the horizontal axis). Variants 2B, 3C, and 4B achieve high energy performance values at moderate costs. The cheapest variants (1, 1A, 6) do not necessarily translate into high energy efficiency, highlighting the need for thorough profitability analysis when investing in PV systems.

In summary, variants with balustrade modules of 160 W (2B, 4B) and 140 W (3C) offer the best compromise between cost and annual energy yield, making them the most energy-efficient solutions among the analysed options. Among them, variant 3C (marked in green colour) meets the condition of energy profitability while having the lowest approximate cost. The remaining variants, despite their lower purchase costs, exhibit lower energy efficiency and may be less profitable during long-term operation.

The analysis was conducted based on the study [45], which demonstrated that, assuming the PV system operates over a 30-year period, the maintenance cost of the photovoltaic installation amounts to USD 3.55/kWh. The results were compared with the average annual electricity purchase cost, which for the analysed building totals USD 1689.06.

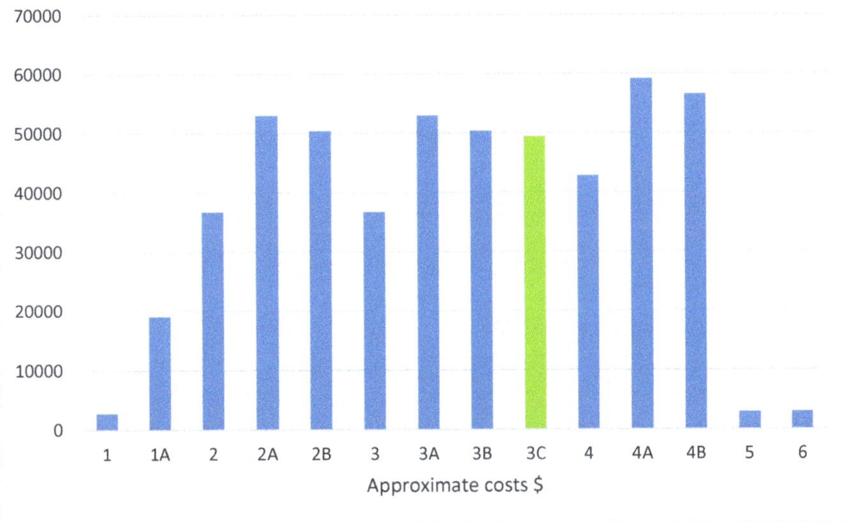

Figure 13. Annual energy yield for individual variants. The vertical axis represents the approximate costs USD of the analysed variants, while the horizontal axis represents the individual variants.

The lifecycle cost (LCC) analysis of the photovoltaic installations, presented in Table 12, indicated that all installation variants provide savings compared to conventional electricity purchase (USD 1689.06/year). The highest savings (78.67%) were achieved with variant 4, which includes exclusively 200 W wall-mounted modules, whereas variants with additional balustrade modules exhibited lower savings, e.g., variant 1A—36.54%. Considering the most energy-efficient variants, namely 2A, 3C, and 4B, the savings amount to approximately 54%.

From the analysis carried out, the following conclusions can be drawn:

- Using only 400 W modules on the northeast facing façade does not meet the minimum energy yield requirement.
- The addition of photovoltaic balustrades on tall windows significantly exceeds the minimum energy production, resulting in an excessive surplus.
- Coloured photovoltaic modules provide more aesthetically favourable options but have lower energy yields; however, balustrades with a power density of 2.0 kW/m^2 surpass the necessary energy production.
- Reducing the balustrade power density to 1.6 kW/m^2 for the red and green modules and to 1.4 kW/m^2 for the grey modules yields an optimal surplus, improving economic viability.
- Using 470 W modules alone on the northeast façade almost meets the requirement (only 1.3% shortfall); adding balustrades would create a significant surplus.
- Using 480 W modules alone on the northeast façade meets the minimum requirement, but with a margin below 1%, making further balustrade modules economically unjustified.
- Variants with balustrade modules rated at 160 W and 140 W (2B, 3C, 4B) provide the best compromise between purchase cost and annual energy performance.
- Variant 3C stands out as the most cost-effective option while maintaining high energy efficiency.

- Cheaper variants exhibit lower efficiency, which may lead to reduced savings over longer operational periods.
- Analysing the selection of the optimal variant with respect to potential shading of the installation, it can be observed that variant 3A performs best, as its energy production under shaded conditions remains optimal even with losses of up to 40%.
- Variant 3A can be considered optimal for losses of 25%, even in cases of a 23% surplus in generated energy, due to its favourable performance under 40% loss conditions.
- Considering the estimated costs of PV modules, the investment for variant 3A is approximately USD 3500 higher compared to variant 3C, which generates energy under standard conditions (i.e., without additional shading).

Table 12. Comparison of lifecycle costs of photovoltaic installations with energy purchased from suppliers. Own elaboration.

Variant	Wall Module Power [W]	Balustrade Module Power [W/m^2]	Total Power [kW]	Total Energy Yield [kWh]	Lifecycle Cost USD	Cost Savings [%]
1	400	0	14.4	5219.136	$617.60	63.44
1A	400	200	21.26	9058.716	$1071.95	36.54
2	240	0	8.64	3131.482	$370.56	78.06
2A	240	200	15.5	6971.06	$824.91	51.16
2B	240	160	14.13	6178.992	$731.18	56.71
3	290	0	10.44	3783.874	$447.76	73.49
3A	290	200	17.3	7623.454	$902.11	46.59
3B	290	160	15.93	6831.384	$808.38	52.14
3C	290	140	15.24	6448.064	$763.02	54.83
4	200	0	8.4	3044.5	$360.27	78.67
4A	200	200	15.26	5874.08	$695.10	58.85
4B	200	160	13.89	6092.01	$720.89	57.32
5	470	0	15.04	5451.098	$645.05	61.81
6	480	0	15.36	5567.078	$658.77	61.00

6. Discussion

The results of the conducted analysis confirm a significant dependence between location, module type, their power, and the total energy yield of BIPV systems. The highest efficiency was observed in variants with additional modules mounted on balustrades (e.g., variant 1A), which aligns with findings from other authors that highlight the importance of maximising PV exposure area in residential projects while maintaining a coherent architectural integration strategy.

Studies [30,33] indicate that the use of coloured modules—despite their efficiency limitations—can be justified from an aesthetic perspective, especially in projects with high architectural demands. In the variants 2 to 4 analysed (red, grey, and green modules), a significant decrease in energy yield was observed compared to variants with classic black modules (variants 1, 5, and 6). However, additional balustrade installations (e.g., variants 2A–4B) partially compensated for these differences, increasing energy production by 40% to 60%. In the study [1], the authors emphasise that BIPV should continue to develop as a key technology to accelerate the energy transition and achieve the NZEB targets.

The conclusions drawn from the conducted analysis are consistent with the observations made by the authors in [25], who highlighted a particular synergy between BIPV technology and modular construction in their review of the literature. They emphasise that the prefabrication of PV-integrated elements not only reduces assembly time and improves workmanship quality, but also facilitates the standardisation and scalability of solutions [36]. In the context of the modular building variants, incorporating balustrade elements with prefabricated PV modules aligns with the development trends identified in the literature, serving as a solution that promotes both energy efficiency and technological optimisation. This supports the view of prefabricated BIPV integration not only as an aesthetic addition but as an integral component of the building's energy system.

Effective integration of BIPV systems requires not only considering potential energy yields but, more importantly, balancing them with user demand and the daily energy consumption profile of the building [46]. Therefore, in the variants analysed, the optimisation of the system should account for both solar exposure conditions and functional layout alongside the characteristics of the energy loads [47–49]. The results of the analysis may serve as practical support for engineers and architects designing modular buildings with integrated BIPV systems. The use of photovoltaic modules as façade elements requires consideration of both technical and aesthetic aspects, particularly in the context of prefabrication and the repeatability of building modules. The differences in performance identified in the study, depending on the type and arrangement of the modules, can be used to optimise design solutions, allowing for better adaptation of the BIPV system to the functional and visual requirements of a given building.

7. Conclusions

The drive toward climate neutrality imposes new requirements on zero-energy buildings. The development of integrated modular construction, recognised for being both cost-effective and environmentally friendly, should be harmonised with BIPV systems. The analysis confirms that the effectiveness of photovoltaic systems depends on several factors:

- The highest energy yields are achieved with configurations utilising classic (black), high-power modules combined with additional balustrade surfaces.
- Coloured modules, despite their lower efficiency, can be applied effectively in projects where aesthetic considerations are prioritised, provided the installation area remains reasonable.
- An integrated design approach that considers technical, architectural, and functional parameters is essential for the effective integration of BIPV in residential construction; this is also supported by recent international studies.

Additionally, simulations of the variants indicated that the optimal solution is a combination of photovoltaic balustrades on the windows and coloured photovoltaic modules on the walls. To further this research, it would be advisable to analyse modules in other colours, given the impact of the colour of the PV module on power output.

The conducted analyses indicate that the variants incorporating balustrade modules (2B, 3C, 4B) offer the optimal trade-off between initial investment cost and annual energy performance. Among these, variant 3C emerges as the most cost-efficient solution while still delivering high energy efficiency. Lower-cost alternatives exhibit reduced performance, which may lead to diminished savings over the long term. Moreover, when considering the estimated costs of PV modules, variant 3A is approximately USD 3500 more expensive than 3C, despite generating a comparable amount of energy under standard conditions (i.e., without additional shading).

Moreover, future research directions on BIPV systems should consider not only visual aspects but also technical and operational issues. In particular, it is worth exploring the integration with energy storage systems, the performance analysis of modules under varying climatic conditions, and the application of artificial intelligence-based optimisation algorithms for module layout planning. Such approaches may significantly enhance the energy efficiency and reliability of BIPV systems in engineering practice.

Note that the expected power output is not always attainable, as demonstrated by the analysis of various photovoltaic module solutions and configurations. The study revealed that each module type can be used in varying quantities, which consequently affects overall costs. The application of photovoltaic balustrades could have a particularly broad use in urban construction, given the prevalence of multifamily buildings and balconies.

These efforts align with the broader goals of transforming the energy sector within construction and advancing low-emission systems through modular building technologies. The results obtained may serve as a basis for further research into the integration and utilisation of building-integrated photovoltaic (BIPV) technology in next-generation modular systems.

Author Contributions: Conceptualisation, J.B. and B.K.; investigation, J.B.; data curation J.B. and B.K.; formal analysis, B.O.-K.; methodology, J.B. and B.K.; resources, J.B. and B.K.; software, J.B.; supervision, B.O.-K.; validation, J.B.; visualisation, J.B., B.K. and B.O.-K.; writing—original draft preparation J.B. and B.K.; writing—review and editing, J.B., B.K. and B.O.-K. All authors have read and agreed to the published version of the manuscript.

Funding: This research received no external funding.

Data Availability Statement: The original contributions presented in the study are included in the article, further inquiries can be directed to the corresponding author.

Conflicts of Interest: The authors declare no conflicts of interest.

References

1. Walczak, Z.; Ksit, B.; Szymczak-Graczyk, A. Decision-support tools for diagnosing and selecting the optimal method of repairing buildings. *Bull. Pol. Acad. Sciences. Tech. Sci.* **2024**, *72*, e151382-1. [CrossRef]
2. Siewczyńska, M.; Ksit, B. Effectiveness of Screens Shading Opaque Facades in Terms of Building Thermal Modernisation. *Civ. Environ. Eng. Rep.* **2024**, *34*, 126–137. [CrossRef]
3. Nowogońska, B. Diagnoses in the Aging Process of Residential Buildings Constructed Using Traditional Technology. *Buildings* **2019**, *9*, 126. [CrossRef]
4. Stefaniak, K.; Wierzbicki, J.; Ksit, B.; Szymczak-Graczyk, A. Biocementation as a Pro-Ecological Method of Stabilizing Construction Subsoil. *Energies* **2023**, *16*, 2849. [CrossRef]
5. Smoczyk, M.; Ksit, B.; Szymczak-Graczyk, A. Numerical analysis of heat transfer and surface condensation risk in the plinth zone of basement-included building. *Energy Build.* **2025**, *344*, 116001. [CrossRef]
6. Smoczyk, M.; Ksit, B.; Szymczak-Graczyk, A. Numerical analysis of the ground temperature function depending on edge thermal insulation parameters for shallow slab foundations. *Energy* **2025**, *314*, 134221. [CrossRef]
7. Pokorska-Silva, I.; Kadela, M.; Orlik-Kożdoń, B.; Fedorowicz, L. Calculation of Building Heat Losses through Slab-on-Ground Structures Based on Soil Temperature Measured In Situ. *Energies* **2022**, *15*, 114. [CrossRef]
8. Orlik-Kożdoń, B.; Radziszewska-Zielina, E.; Fedorczak-Cisak, M.; Steidl, T.; Białkiewicz, A.; Żychowska, M.; Muzychak, A. Historic Building Thermal Diagnostics Algorithm Presented for the Example of a Townhouse in Lviv. *Energies* **2020**, *13*, 5374. [CrossRef]
9. Fedorczak-Cisak, M.; Radziszewska-Zielina, E.; Orlik-Kożdoń, B.; Steidl, T.; Tatara, T. Analysis of the Thermal Retrofitting Potential of the External Walls of Podhale's Historical Timber Buildings in the Aspect of the Non-Deterioration of Their Technical Condition. *Energies* **2020**, *13*, 4610. [CrossRef]
10. EEA. Communication from the Commission to the European Parliament, The Council, The European Economic and Social Committee and the Committee of The Regions. In *'Fit For 55': Delivering the Eu's 2030 Climate Target on The Way to Climate Neutrality*; European Commission: Brussels, Belgium, 2021; pp. 2–15.
11. Mahmud, D.M.; Sule, E.E.; Wang, T.; Thomas, V. Towards clean energy: Design and integration strategy of photovoltaics for residential electrification. *J. Eng. Res.* **2025**, *13*, 1329–1343. [CrossRef]
12. Rocha, P.F.; Ferreira, N.O.; Pimenta, F.; Pereira, N.B. Impacts of Prefabrication in the Building Construction Industry. *Encyclopedia* **2023**, *3*, 28–45. [CrossRef]

13. Liu, S.; Li, Z.; Teng, Y.; Dai, L. A dynamic simulation study on the sustainability of prefabricated buildings. *Sustain. Cities Soc.* **2022**, *77*, 103551. [CrossRef]
14. Zohourian, M.; Zohourian, M.; Kermanshachi, S.; Almaskati, D. Modular Construction: A Comprehensive Review. *Buildings* **2025**, *15*, 2020. [CrossRef]
15. Farajian, M.; Sharafi, P.; Bigdeli, A.; Eslamnia, H.; Rahnamayiezekavat, P. Experimental Study on the Natural Dynamic Characteristics of Steel-Framed Modular Structures. *Buildings* **2022**, *12*, 587. [CrossRef]
16. Guaygua, B.; Sánchez-Garrido, A.J.; Yepes, V. Life cycle assePREssment of seismic resistant prefabricated modular buildings. *Heliyon* **2024**, *10*, e39458. [CrossRef]
17. Habib, O.; Al Houri, A.; Alibrahim, B.; Habib, M. Behavior and sustainability benefits of modular steel buildings. *Int. J. Sustain. Eng.* **2025**, *18*, 2459715. [CrossRef]
18. Chippagiri, R.; Gavali, H.R.; Bras, A.; Ralegaonkar, R.V. Performance Evaluation of a Sustainable Prefabricated System Using Small-Scale Experimental Model Technique. *Buildings* **2022**, *12*, 2000. [CrossRef]
19. Tomanek, M. *Budownictwo Modułowe Czasowe*; Wydawnictwo Naukowe Śląsk: Katowice, Poland, 2006; ISBN 978-83-7164-499-3.
20. Goodier, C.I.; Gibb, A.; Mancini, M.; Turck, C.; Gjepali, O.; Daniels, E. Off-site Fabrication: Prefabrication, Pre-assembly and Modularisation. *ICE Proc. Civ. Eng.* **2019**, *172*, 1–45.
21. Kazeem, K.O.; Olawumi, T.O.; Adam, J.J.; Lam, E.W.-M. Integration of Building Services in Modular Construction: A PRISMA Approach. *Appl. Sci.* **2024**, *14*, 4151. [CrossRef]
22. Lawson, R.M.; Ogden, R.G.; Bergin, R. Application of modular construction in high rise buildings. *J. Archit. Eng.* **2012**, *18*, 148–154. [CrossRef]
23. Runkiewicz, L.; Szulc, J.; Sieczkowski, J. The evolution of prefabricated construction in Poland. *Build. Rev.* **2020**, *91*, 12–15.
24. Kong, J.; Dong, Y.; Poshnath, A.; Rismanchi, B.; Yap, P.-S. Application of Building Integrated Photovoltaic (BIPV) in Net-Zero Energy Buildings (NZEBs). *Energies* **2023**, *16*, 6401. [CrossRef]
25. Bao, Y.; Xiang, C. Integration of BIPV technology with modular prefabricated building—A review. *J. Build. Eng.* **2025**, *102*, 111940. [CrossRef]
26. Batista, F.; Guimarães, A.S.; Palmero-Marrero, A.I. Building Integrated Photovoltaics: A multi-level design review for optimized implementation. *Renew. Sustain. Energy Rev.* **2025**, *220*, 115837. [CrossRef]
27. Carbone, R.; Borrello, C. A Building-Integrated Bifacial and Transparent PV Generator Operated by an "Under-Glass" Single Axis Solar Tracker. *Energies* **2023**, *16*, 6350. [CrossRef]
28. Vodapally, S.N.; Ali, M.H. A Comprehensive Review of Solar Photovoltaic (PV) Technologies and Integration Methods. *Energies* **2023**, *16*, 319.
29. Silva, L.A.C.; Ruiz, L.G.B.; Criado-Ramón, D.; Bessa, J.G.; Micheli, L.; Pegalajar Jiménez, M.C. Assessing the impact of soiling on photovoltaic efficiency using supervised learning techniques. *Expert Syst. Appl.* **2023**, *231*, 120816. [CrossRef]
30. Pelle, M.; Causone, F.; Maturi, L.; Moser, D. Opaque Coloured Building Integrated Photovoltaic (BIPV): A Review of Models and Simulation Frameworks for Performance Optimisation. *Energies* **2023**, *16*, 1991. [CrossRef]
31. Siraki, A.G.; Pillay, P. Study of optimum tilt angles for solar panels in different latitudes for urban applications. *Solar Energy* **2012**, *86*, 1920–1928. [CrossRef]
32. Gholami, H.; Røstvik, H.N.; Steemers, K. The Contribution of Building-Integrated Photovoltaics (BIPV) to the Concept of Nearly Zero-Energy Cities in Europe: Potential and Challenges Ahead. *Energies* **2021**, *14*, 6015. [CrossRef]
33. Liu, Y.; Wang, X.; Chen, J. Advances in colored building-integrated photovoltaics: Energy performance and aesthetic applications. *Energies* **2023**, *1–5*, 10–14.
34. Available online: https://mlsystem.pl/ (accessed on 1 April 2025).
35. Bonomo, P.; Frontini, F. Building Integrated Photovoltaics (BIPV): Analysis of the Technological Transfer Process and Innovation Dynamics in the Swiss Building Sector. *Buildings* **2024**, *14*, 151. [CrossRef]
36. N'tsoukpoe, K.E. Effect of orientation and tilt angles of solar collectors on their performance: Analysis of the relevance of general recommendations in the West and Central African context. *Sci. Afr.* **2022**, *15*, e01069. [CrossRef]
37. Yu, G.; Yang, H.; Luo, D.; Cheng, X.; Ansah, M.K. A review developments and researches of building integrated photovoltaic (BIPV) windows and shading blinds. *Renew. Sustain. Energy Rev.* **2021**, *149*, 111355. [CrossRef]
38. Available online: https://climatic.pl/pl/ (accessed on 1 April 2025).
39. *PN-EN 317:1999*; Particleboards and Fiberboards—Determination of Swelling in Thickness After Immersion in Water. European Committee for Standardization: Brussels, Belgium, 1999.
40. *PN-EN 319:1999*; Particleboards and Fiberboards—Determination of Tensile Strength Perpendicular to the Plane of the Board. Polish Committee for Standardizatio: Warsaw, Poland, 1999.
41. *PN-EN 321:2004*; Wood-Based Panels—Determination of Resistance to Cyclically Changing Humidity. iTeh Standards: San Francisco, CA, USA, 2004.

42. Pero, C.D.; Leonforte, F.; Aste, N. Building-Integrated Photovoltaics in Existing Buildings: A Novel PV Roofing System. *Buildings* **2024**, *14*, 2270. [CrossRef]
43. Ebhota, W.S.; Tabakov, P.Y. Impact of Photovoltaic Panel Orientation and Elevation Operating Temperature on Solar Photovoltaic System Performance. *Int. J. Renew. Energy Dev.* **2022**, *11*, 591–599. [CrossRef]
44. Mostafa Habibi1, Mohammadhossein Azizibabani Evaluation of the orientation of residential buildings in Mashhad based on direct solar radiation reception. *Civ. Eng. Archit.* **2025**, *24*, 45–56.
45. Huang, B.; Wang, Y.; Huang, Y.; Xu, X.; Chen, X.; Duan, L.; Yu, G.; Li, Z.; Liu, H.; Ku, H.W.; et al. Life cycle cost analysis of solar energy via environmental externality monetization. *Sci. Total Environ.* **2023**, *856*, 158910. [CrossRef]
46. Zhou, Z.; Yuan, Z.; Yi, Z.; Xue, O.; Li, N.; Huang, F. Progress of semitransparent emerging photovoltaics for building integrated applications. *Green Energy Environ.* **2024**, *9*, 992–1015. [CrossRef]
47. Habibi, S.; Valladares, O.P.; Peña, D.M. Sustainability performance by ten representative intelligent Façade technologies: A systematic review. *Sustain. Energy Technol. Assess.* **2022**, *52*, 102001. [CrossRef]
48. Karthick, A.; Murugavel, K.; Kalaivani, L.; Babu, U.S. Performance study of building integrated photovoltaic modules. *Adv. Build. Energy Res.* **2017**, *12*, 178–194. [CrossRef]
49. Barrutieta, X.; Kolbasnikova, A.; Irulegi, O.; Hernández, R. Decision-making framework for positive energy building design LOKAthrough key performance indicators relating geometry, localization, energy and PV system integration. *Energy Build.* **2023**, *297*, 113442. [CrossRef]

energies

MDPI

Article

Safe Reinforcement Learning for Buildings: Minimizing Energy Use While Maximizing Occupant Comfort

Mohammad Esmaeili [1], Sascha Hammes [2], Samuele Tosatto [1], David Geisler-Moroder [2,*] and Philipp Zech [1]

[1] Department of Computer Science, University of Innsbruck, 6020 Innsbruck, Austria; mohammad.esmaeili@uibk.ac.at (M.E.); samuele.tosatto@uibk.ac.at (S.T.); philipp.zech@uibk.ac.at (P.Z.)

[2] Unit of Energy Efficient Building, University of Innsbruck, 6020 Innsbruck, Austria; sascha.hammes@uibk.ac.at

* Correspondence: david.geisler-moroder@uibk.ac.at

Abstract

With buildings accounting for 40% of global energy consumption, heating, ventilation, and air conditioning (HVAC) systems represent the single largest opportunity for emissions reduction, consuming up to 60% of commercial building energy while maintaining occupant comfort. This critical balance between energy efficiency and human comfort has traditionally relied on rule-based and model predictive control strategies. Given the multi-objective nature and complexity of modern HVAC systems, these approaches fall short in satisfying both objectives. Recently, reinforcement learning (RL) has emerged as a method capable of learning optimal control policies directly from system interactions without requiring explicit models. However, standard RL approaches frequently violate comfort constraints during exploration, making them unsuitable for real-world deployment where occupant comfort cannot be compromised. This paper addresses two fundamental challenges in HVAC control: the difficulty of constrained optimization in RL and the challenge of defining appropriate comfort constraints across diverse conditions. We adopt a safe RL with a neural barrier certificate framework that (1) transforms the constrained HVAC problem into an unconstrained optimization and (2) constructs these certificates in a data-driven manner using neural networks, adapting to building-specific comfort patterns without manual threshold setting. This approach enables the agent to almost guarantee solutions that improve energy efficiency and ensure defined comfort limits. We validate our approach through seven experiments spanning residential and commercial buildings, from single-zone heat pump control to five-zone variable air volume (VAV) systems. Our safe RL framework achieves energy reduction compared to baseline operation while maintaining higher comfort compliance than unconstrained RL. The data-driven barrier construction discovers building-specific comfort patterns, enabling context-aware optimization impossible with fixed thresholds. While neural approximation prevents absolute safety guarantees, reducing catastrophic safety failures compared to unconstrained RL while maintaining adaptability positions this approach as a developmental bridge between RL theory and real-world building automation, though the considerable gap in both safety and energy performance relative to rule-based control indicates the method requires substantial improvement for practical deployment.

Keywords: HVAC system; reinforcement learning; safe reinforcement learning; neural barrier certificates

1. Introduction

Heating, ventilation, and air conditioning (HVAC) controls typically rely on standardized comfort criteria like Fanger's thermal comfort model, which forms the basis for key standards, including EN ISO 7730 [1] and ASHRAE 55 [2–4]. This model employs the Predicted Mean Vote (PMV) and Predicted Percentage of Dissatisfied (PPD) indices to quantify comfort boundaries. Human thermal comfort depends on physical activity, clothing insulation, air temperature, mean radiant temperature, and air velocity [5,6]. The PMV model, based on extensive field studies, quantifies how these variables impact thermal perception [3]. Significant variations exist based on age and gender [7–9], and psychological factors contribute to subjective differences in thermal perception [10,11]. The PPD extends the PMV by predicting the percentage of thermally dissatisfied occupants. Comfort limits vary seasonally, with the temperature range between 20 and 25 °C generally identified as acceptable across seasons [10], forming a subset of specifications outlined by Nikam et al. [12].

Maintaining thermal comfort corridors is essential for system acceptance, as user intervention in energy-optimized controls can undermine system performance. Comfort preferences fluctuate due to psychological factors and environmental conditions [10], necessitating adaptive solutions—particularly in shared spaces where consensus is needed. HVAC systems are crucial for ensuring thermal comfort but come with a significant energy demand. HVAC systems are particularly energy-intensive [13], accounting for 40% of energy use in non-residential buildings [14] and 75% in residential buildings [15]. Buildings account for 30–40% of global energy consumption and a third of CO_2 emissions [15–18]. Therefore, improving the energy efficiency of HVAC operation is essential to reducing overall energy use and emissions.

Climate change concerns have intensified research on balancing thermal comfort with energy efficiency [10], making the reduction in HVAC energy consumption while maintaining comfort a critical challenge for energy management systems. Learning systems offer the potential to continuously optimize both comfort and energy efficiency by incorporating evolving user preferences. Balancing comfort priorities, however, is critical: insufficient attention to comfort affects occupant health and productivity, while over-prioritization increases energy costs. This balance is particularly significant given current climate challenges and the fact that people spend approximately 90% of their time indoors [5,19]. An acceptable combination of environmental conditions according to PMV and PPD methodologies remains essential [13].

In recent years, reinforcement learning (RL) [20] has received considerable attention for data-driven learning of building control policies [21]. RL is a computational approach to learning from interaction, where an agent learns to make decisions by receiving rewards or penalties from its environment through trial and error [20]. Due to the complexity of modern building control policies in orchestrating competing objectives (cf. energy consumption, occupant comfort), RL offers a feasible approach to efficient data-driven learning of (near-)optimal control policies [22]. Moreover, due to the complexity of modern building environments, providing an accurate model of the building or the HVAC system is an intractable task that may result in inaccurate dynamics, which compromises controller efficiency and leads to sub-optimal control performance. RL can tackle this problem due to its model-free nature and provide a controller that can act (near-)optimally without having any knowledge of the system dynamics [22].

1.1. Research Problem and Contributions

The BOREALIS project [23] aims to develop RL solutions for creating greener, more energy-efficient buildings. While BOREALIS primarily focuses on optimizing artificial lighting control systems to reduce energy consumption while maintaining occupant comfort, the underlying challenge of balancing energy efficiency with user comfort extends across all building control systems. This paper addresses a parallel challenge in HVAC systems, where similar trade-offs between energy consumption and thermal comfort create complex optimization problems that can benefit from the RL methodologies being developed within the BOREALIS framework.

Many RL studies for HVAC controls incorporate both energy consumption and thermal comfort in their reward functions, creating complexity due to these conflicting objectives [21,24]. Balancing these goals with weighted reward functions raises questions about proper quantification and relative importance [25]. Our approach utilizes neural barrier certificates [26] to restrict the action space to options that are both optimal and respect thermal comfort constraints. This data-driven method eliminates the need for explicit thermal comfort modeling while transforming the constrained RL problem into an unconstrained formulation. By maintaining adherence to comfort boundaries, the agent becomes more efficient at optimizing the single-objective function of energy consumption. Moreover, this approach eliminates the requirement for explicit system dynamics in formulating the forward invariance property of barrier certificates [27], transforming it into a purely data-driven learning problem that attempts to achieve improved safety performance while indicating comparable energy efficiency without prior knowledge of system or constraint dynamics. This methodology, while applied here to HVAC systems, offers insights that can inform the broader BOREALIS objective of developing generalizable RL techniques for sustainable building control across multiple domains.

In addressing our research problem, we are guided by the following research questions:

- RQ1 What is the applicability and effectiveness of safe reinforcement learning with neural barrier certificates in the building control domain?
- RQ2 What is the impact of pre-training, random and non-random episode generation, and the length of training on a safe RL agent's performance?
- RQ3 To what extent does safe RL control with neural barrier certificates improve upon or match the performance of baseline controllers?

We thereby deliver the following contributions:

- C1 Demonstrates the applicability of safe RL with neural barrier certificates in the building control domain, validating its effectiveness for handling thermal comfort constraints while optimizing energy consumption.
- C2 Provides empirical evidence of the method's performance across multiple BOPTEST test cases, different control levels (air handling unit (AHU) duct pressure and supply temperature), and varying episode lengths and learning intervals.
- C3 Demonstrates that pre-trained barrier functions improve learning efficiency and performance compared to non-pretrained approaches in HVAC control applications.
- C4 Shows that safe RL with neural barrier certificates achieves energy and safety performance comparable to a Proximal Policy Optimization (PPO) agent and a Rule-based Controller (RBC), establishing its viability as an alternative control approach for buildings.

1.2. Research Methodology

This study follows an experimental research methodology, employing controlled experiments to establish cause-and-effect relationships and validate theoretical concepts with empirical evidence [28]. We implement a safe RL controller with neural barrier certificates for HVAC systems and systematically evaluate its performance across multiple BOPTEST [29] test cases with varying control configurations. The investigation encompasses different episode lengths, learning intervals, and the impact of pre-training barrier functions on controller performance. The results are benchmarked against established baselines, including a PPO agent and an RBC controller. Based on the identified research gap that we will discuss in Section 2, we formulate the following hypothesis:

Hypothesis (H1). *Safe RL with neural barrier certificates achieves energy efficiency comparable to or better than classical controllers while maintaining thermal comfort constraints through learned barrier certificates.*

This is investigated against the null hypothesis:

Hypothesis (H0). *Safe RL with neural barrier certificates does not achieve competitive energy efficiency compared to classical controllers and/or fails to maintain thermal comfort constraints effectively.*

Our experimental design allows us to test whether the neural barrier certificate approach can successfully balance energy optimization with constraint satisfaction in real building control scenarios, while also examining the practical benefits of pre-training for deployment efficiency.

Article Organization: Following the introduction, Section 2 reviews existing literature on RL-based building control and safe reinforcement learning approaches. Section 3 provides foundational concepts in reinforcement learning and its application to building control systems. Section 4 presents the neural barrier certificate framework and formulates its application to constrained HVAC control problems. Section 5 details the experimental design, including BOPTEST test cases, control configurations, and evaluation metrics, then presents results comparing the safe RL controller against baseline controllers while analyzing the impact of pre-training on performance. Finally, Section 6 summarizes the contributions and suggests future research directions.

2. Related Work

Model-based approaches have been extensively explored for HVAC control, particularly through proportional–integral–derivative (PID) controllers [30–33] to manage building energy and air quality. Optimal control formulations for HVAC systems have also been investigated [34–36], aiming to lower energy costs and enhance occupant comfort. In addition, fuzzy control strategies have been proposed [37,38] to handle uncertainties arising from external and internal variations. Model predictive control (MPC) further advances these efforts by leveraging predictive models to balance energy consumption and comfort [39–41], where adaptive frameworks and grey-box models are employed. Iterative learning control (ILC) has likewise been used to refine MPC output [42].

Model-based approaches for HVAC control face significant challenges due to the complexity of underlying building dynamics, varying occupant behavior, and external environmental factors, often necessitating simplified models that fail to capture nuanced interactions. They also struggle to accommodate changing building characteristics and occupancy patterns while remaining highly sensitive to uncertainties such as weather, occupancy estimates, and internal heat gains [43]. RL addresses these limitations by learning optimal strategies through direct interaction with the environment, requiring no explicit model and allowing continuous adaptation to evolving conditions, thereby overcoming the rigidity of traditional model-based approaches [43].

Various RL methods have been leveraged to optimize HVAC energy usage and maintain occupant comfort [24]. Deep deterministic policy gradients (DDPG) have been used to manage HVAC control [44,45], relying on deep neural networks to capture occupant comfort and energy interactions. Azuatalam [46] proposed a PPO controller that integrates thermal comfort as a penalty to balance energy efficiency and demand response. Asynchronous advantage actor–critic (A3C) was employed to reduce energy consumption while maintaining multi-zone thermal comfort [47], with comfort penalties embedded in the reward. Deep Q-networks (DQN) [48,49] have accounted for occupant-driven factors (e.g., behavior and indoor air quality) while minimizing energy use. Clustering-based RL was introduced [50] to segment data for more robust training. Q-learning under model uncertainty, aided by Bayesian-convolutional neural networks (BCNNs), was investigated by Lork et al. [51] to refine set-point decisions while upholding comfort requirements. The combination of MPC and RL has also been explored [52]. Nguyen et al. [53] used phasic policy gradient (PPG), which is an actor-critic method. The reward function was designed such that the agent is penalized for both high energy consumption and for violating the occupant comfort temperature range. They used a conditional reward function, which only considers comfort penalties during working hours, allowing the agent to prioritize energy savings at other times. One of the main challenges in deep reinforcement learning (DRL) applications is real-world deployment. To directly address this, Solinas et al. [54] utilized Imitation Learning to solve the cold-start problem. This enabled the agent to be deployed directly online and begin improving energy efficiency from day one, avoiding the high costs of white-box models and the poor initial performance of standard online learning. To enable proactive and cost-aware HVAC control, Liu et al. [55] proposed a multi-step predictive DRL (MSP-DRL) framework. Their primary contribution is the integration of a highly accurate weather prediction model, named GC-LSTM, directly into the decision-making process of a DDPG agent. The GC-LSTM model was specifically designed with a Generalized Correntropy loss function to robustly handle the non-Gaussian nature of real-world temperature data.

Kadamala et al. [56] investigated the effectiveness of transfer learning for HVAC control. Their primary contribution is a systematic evaluation of pre-training and fine-tuning methodology. They first trained a foundation model on a source building and then transferred its core network weights to new agents, which were fine-tuned for different target environments. To enhance the proactivity of DRL-based HVAC controllers, Shin et al. [57] proposed a hybrid framework named WDQN-temPER that integrates weather forecasting directly into the learning process. Their primary contribution is twofold: first, they use a gated recurrent unit (GRU) to predict future outdoor temperatures, which is then fed as a state variable to a DQN agent, enabling it to make forward-looking decisions. Second, they introduce temPER, a novel prioritized experience replay technique that gives higher learning priority to transitions where a large change in outdoor temperature occurred. Ding et al. [58] proposed CLUE, a model-based reinforcement learning framework. Their approach utilizes a Gaussian process (GP) to model building dynamics, which explicitly quantifies predictive uncertainty. To mitigate the GP's high data requirements, they employ a meta-kernel learning technique that pre-tunes the model on data from other buildings. This uncertainty estimate is then used by the controller to discard unsafe action trajectories and activate a rule-based fallback mechanism to ensure robust performance under high model uncertainty. To enable personalized and safely deployable HVAC control, Chen et al. [59] proposed a human-in-the-loop framework that integrates meta-learning with model-based offline reinforcement learning. Their method first employs meta-learning to train a thermal preference model on a diverse dataset of human feedback, allowing for rapid adaptation to a new occupant's subjective comfort with only a few interactions. Concurrently, a building dynamics model is learned from a static offline dataset using an ensemble of neural networks, where the disagreement among models serves as an uncertainty penalty in the reward function. The key contribution is a practical control strategy that avoids risky online exploration by constraining the agent to known, safe operating regions, while still delivering a highly personalized thermal environment. This dual approach effectively decouples the challenges of fast user adaptation and safe policy learning, presenting a viable pathway for deploying user-centric, intelligent HVAC systems.

Cui et al. [60] proposed a novel hybrid framework that synergistically combines reinforcement learning with active disturbance rejection control (ADRC). Their core contribution is the decomposition of the control problem into a main-auxiliary structure. A twin delayed deep deterministic policy gradient (TD3) agent is designated as the main controller, responsible for optimizing thermal comfort by managing zone-specific supply airflows. Simultaneously, an ADRC-based auxiliary controller manages indoor air quality by regulating the fresh air ratio, while actively estimating and compensating for system-wide disturbances. Gehbauer et al. [61] proposed a grid-interactive RL controller that jointly actuates electrochromic (EC) facades and HVAC using a standardized functional mock-up interface (FMI) based co-simulation toolchain. Their distributed distributional deterministic policy gradient agent (D4PG) consumes 4 h forecasts (weather, occupancy, and TOU prices) to minimize combined energy and demand costs while respecting comfort, and is pre-trained across multiple climate zones before virtual deployment. Benchmarking on simple and high-fidelity Modelica emulators against MPC and heuristic baselines showed RL achieving lower costs than heuristic control and competitive demand reduction, while the open-source functional mock-up interface—machine learning center (FMI-MLC) framework enables rapid interfacing with third-party simulators for practical off-site training. Xue et al. [62] proposed a hybrid multi-agent DRL controller, a genetic algorithm multi-agent DDPG (GA–MADDPG) for multi-zone HVAC. Each zone agent sets its temperature set-point; the reward combines an energy term with zone-specific PMV comfort terms. A key idea is allocating the shared cooling energy to zones via weights and using a genetic algorithm (GA) to optimize these weights to maximize the sum of agents' average rewards. Li et al. [63] developed a model-free control strategy using reinforcement learning to optimize temperature set-points. Their approach employs the PPO algorithm to learn an adaptive policy that dynamically adjusts thermostat settings based on time-of-use (TOU) electricity prices, occupant comfort, and environmental conditions. The primary contribution is the integration of a policy-based RL agent with an active thermal energy storage (ATES) system, which was validated on a high-fidelity TRNSYS-MATLAB co-simulation platform. Yu et al. [64] introduced online comfort-constrained control via feature transfer (OCFT), a deployment-time safety wrapper for policy transfer in HVAC. OCFT learns expressive nonlinear feature maps from source buildings and, on the target, adapts a lightweight parametric surrogate via online set-membership updates combined with nested convex body chasing. At each control step, a robust action filter minimally perturbs the transferred policy's proposal so the model-predicted next-step temperature remains within tightened comfort bounds, with slack variables to handle temporary infeasibility due to disturbances and actuator limits. The method further provides finite-time safety guarantees, including an upper bound on the number of comfort violations under bounded worst-case disturbances and, under mild stochastic assumptions, a finite stopping time beyond which all constraints are satisfied. While OCFT provides strong safety assurances, it assumes the availability of historical source-building data (or a surrogate model trained on such data); if these data are unavailable or of poor quality, applicability is reduced, and performance—particularly during initial deployment—may deteriorate. Although these methods successfully optimize energy efficiency and maintain occupant comfort, they require complex reward functions that encode and weigh multiple objectives. Crafting such functions is non-trivial, as the relative importance of energy versus comfort can be subjective and difficult to quantify, often leading to a tedious tuning process [42]. Controllers based on ADRC [60] or meta-learning [59] effectively manage disturbances and personalization but still frame the problem as a multi-objective optimization. Similarly, approaches using PPO for demand response [63] embed comfort as a weighted penalty. This reliance on weighting and penalties means that comfort is treated as a "soft" constraint, which can be violated if the agent determines that doing so maximizes its cumulative reward. While safety is a growing concern, many solutions rely on probabilistic models or offline datasets to indirectly ensure safe operation [58,59,64]. In contrast, our approach requires no historical data or pretraining—neither for initializing the RL agent nor for learning safety constraints—enabling fully online deployment from scratch, without building-specific logs. To streamline the learning process and provide stronger safety guarantees, we adopt a safe RL approach in which the agent is only tasked with maximizing energy efficiency, while occupant comfort constraints are enforced separately as "hard" constraints. Additionally, this data-driven approach enables neural networks to learn constraints as barrier certificates without requiring explicit mathematical models. We thus remove the need to encode or weight comfort in the reward function, ensuring optimal energy usage without compromising occupant comfort. Table 1 summarizes and contrasts various safe RL and reward weighting approaches, highlighting their distinct advantages and disadvantages and thereby motivating our adopted safe RL method.

Table 1. Comparison of safe RL and reward weighting approaches for HVAC control.

Category	Approach	Pros	Cons
Safe RL	CLUE [58]	- Explicit uncertainty quantification via GP - Rule-based fallback for safety - Meta-kernel learning for efficiency - Discards unsafe trajectories	- High data requirements for GP - Computational complexity - Requires data from other buildings
	Human-in-the-loop [59]	- Fast adaptation to user preferences - Avoids risky online exploration - Personalized thermal comfort - Constrains to safe operating regions	- Requires diverse offline dataset - Complex ensemble modeling - Limited to previously seen regions
	OCFT [64]	- Deployment-time safety wrapper - Finite-time safety guarantees - Minimal-adjustment action filter - Online set-membership adaptation	- Needs historical source-building data
	This paper (Hard Constraints)	- No reward tuning needed - Attempts to enforce hard constraints - Simplified learning process - Data-driven barrier certificates	- May sacrifice some energy efficiency - Relying on neural networks' prediction of safety
Reward Weighting	DDPG [44,45]	- Deep networks capture complex interactions - Continuous action space - Well-established algorithm	- Requires careful reward weighting - Comfort as soft constraint - Sample inefficiency
	PPO [46,63]	- Stable training - Integrates demand response - Works with ATES systems	- Comfort as weighted penalty - Subjective weight selection - Potential comfort violations
	A3C [47]	- Multi-zone thermal comfort - Parallel training efficiency - Asynchronous updates	- Complex reward engineering - Comfort penalties in reward - Difficult multi-objective balance
	DQN [48,49]	- Accounts for occupant behavior - Considers air quality - Discrete action space	- Multiple factors to weigh - Soft comfort constraints - Action discretization limitations
	Q-learning with BCNNs [51]	- Handles model uncertainty - Bayesian approach - Refines set-point decisions	- Comfort as soft requirement - Computational complexity - Requires reward weighting
	PPG [53]	- Conditional reward function - Context-aware (working hours) - Dynamic prioritization	- Still requires weight tuning - Comfort violations possible - Complex reward structure
	TD3-ADRC [60]	- Separation of concerns - Active disturbance rejection - Specialized controllers	- Multi-objective optimization - Complex architecture - Controller coordination overhead
	EC facade + HVAC (D4PG) co-simulation [61]	- Joint actuation of electrochromic facades and HVAC - Uses short-term forecasts (weather, occupancy, TOU) - FMI-MLC co-simulation enables off-site training - Competitive cost/demand reduction vs. baselines	- Comfort as soft, weighted penalty - Pretraining and hyperparameter tuning overhead
	GA–MADDPG (multi-agent) zone-energy allocation [62]	- Optimizes per-zone energy weights (α_i) via GA - Multi-agent setpoint control with inter-zone coupling - TRNSYS–MATLAB co-simulation; thermal comfort (PMV) reward - Lower energy than MPC and plain MADDPG (meets comfort)	- Comfort as soft, weighted penalty - GA outer loop

3. Background and Preliminaries

A reinforcement learning problem consists of an agent and an environment in which the agent aims to learn an optimal policy by interacting with the environment. The agent takes an action $a_t \in \mathcal{A}$, where $\mathcal{A}$ is the set of all possible actions, called the action space, based on the current state $s_t \in \mathcal{S}$, where $\mathcal{S}$ is the set of all states (observations), called the state space, and the current policy, which is the mapping from current state to an action $\pi(a_t|s_t)$. After taking the action a_t, the agent receives the reward r_t and the next state s_{t+1} according to the transition probability $P(s_{t+1}|s_t, a_t)$ from this interaction with the environment. The reward function is the feedback from the environment on how good or bad the action taken by the agent was.

For a given trajectory $\tau = (s_0, a_0, \ldots)$, the discounted return is defined as

$$R(\tau) = \sum_{t=0}^{\infty} \gamma^t r_t, \tag{1}$$

where $\gamma \in [0, 1)$ is the discount factor that determines the importance of future rewards [20]. Thus, the central RL problem can be expressed as follows:

$$\pi^* = \arg\max_{\pi} J(\pi) \tag{2}$$

where π^* is the optimal policy and

$$J(\pi) = \mathbb{E}_{\tau \sim \pi}[R(\tau)] \tag{3}$$

is the expected return. Value functions measure the expected cumulative discounted reward of taking an action in a given state and acting according to a specific policy after that. There are two types of value functions: (i) the on-policy value function determines the expected return starting from state s and acts according to a policy π, while (ii) the on-policy action-value function gives the expected return of starting from state s, takes an action a, and acts according to the policy π.

$$\begin{aligned} V^{\pi}(s) &= \mathbb{E}_{\tau \sim \pi}[R(\tau)|s_0 = s] \quad \text{On-policy value function} \\ Q^{\pi}(s,a) &= \mathbb{E}_{\tau \sim \pi}[R(\tau)|s_0 = s, a_0 = a] \quad \text{On-policy action-value function} \end{aligned} \tag{4}$$

The value function and action-value function follow the Bellman principle: the value of a state (or state–action pair) equals the immediate expected reward plus the discounted value of subsequent states [65].

The advantage function $A(s, a)$ measures how much better a specific action a is compared to an average action that would be taken in state s under the current policy π:

$$A^{\pi}(s,a) = Q^{\pi}(s,a) - V^{\pi}(s). \tag{5}$$

A positive advantage value indicates that an action a is better than the policy's expected performance from state s. In practice, the true advantage function $A^{\pi}(s,a)$ is unknown because it depends on the true value functions Q^{π} and V^{π}, which are typically intractable to compute exactly. Moreover, in policy gradient methods, advantage estimates are crucial for reducing variance while maintaining unbiased gradient estimates. Therefore, we must estimate advantages from sampled trajectories using function approximation and empirical returns. Consider the finite-horizon return from time t:

$$\hat{R}_t = \sum_{k=0}^{T-t-1} \gamma^k r_{t+k}, \tag{6}$$

where T is the length of the trajectory.

Monte-Carlo (MC) Estimation: The MC advantage estimator uses the empirical return:

$$\hat{A}_t^{MC} = \hat{R}_t - V(s_t), \tag{7}$$

This estimator is unbiased but suffers from high variance due to the stochasticity of the entire trajectory.

Temporal Difference TD(0) Estimation: The one-step TD advantage estimator uses bootstrapping:

$$\hat{A}_t^{TD(0)} = r_t + \gamma V(s_{t+1}) - V(s_t). \tag{8}$$

This estimator has lower variance but introduces bias through the value function approximation.

Generalized Advantage Estimation (GAE): GAE [66] provides a principled trade-off between bias and variance:

$$\begin{aligned} \hat{A} &= \sum_{l=0}^{\infty} (\gamma\lambda)^l \delta_{t+l} \quad \lambda \in [0,1] \\ \delta_t &= r_t + \gamma V(s_{t+1}) - V(s_t) \end{aligned} \tag{9}$$

where $\delta_t = r_t + \gamma V(s_{t+1}) - V(s_t)$ is the TD residual and $\lambda \in [0,1]$ controls the bias-variance trade-off. The geometric weighting $(\gamma\lambda)^l$ exponentially decays the contribution of future TD residuals. When $\lambda = 0$, GAE reduces to TD(0); when $\lambda = 1$, it becomes equivalent to the MC estimator. GAE's key advantage lies in its ability to smoothly interpolate between these extremes, allowing one to balance bias and variance.

Proximal Policy Optimization

Policy gradient methods work by directly optimizing the policy parameters θ in the direction that maximizes the expected cumulative reward, $J(\theta)$. The gradient is estimated as:

$$\nabla_\theta J(\pi_\theta) = \mathbb{E}\big[\nabla_\theta \log \pi_\theta(a_t|s_t) \hat{A}^{\pi_\theta}(s_t, a_t)\big] \tag{10}$$

where $\pi_\theta(a_t|s_t)$ is the policy and $\hat{A}_t$ is an estimate of the advantage function. A major issue with this approach is that a single update step can be excessively large. A large gradient update can drastically change the policy, leading it to a "bad" region of the policy space from which it may never recover. This makes the training process highly unstable. PPO introduces a first-order method to constrain the policy update. Instead of a hard constraint, PPO modifies the objective function to penalize large changes to the policy. The ratio (Equation (11)) indicates how much more likely the new action a_t is under the new policy $\pi_\theta(a_t|s_t)$ compared to the old one $\pi_{\theta_{old}}(a_t|s_t)$. PPO ensures that the ratio stays close to 1, preventing the new policy from deviating too far from the old one. PPO uses clipping to discourage the policy from changing aggressively.

$$\rho(\theta) = \frac{\pi_\theta(a_t|s_t)}{\pi_{\theta_{old}}(a_t|s_t)} \tag{11}$$

The objective function of the PPO [67] is as follows:

$$\begin{aligned} L(\theta) = \frac{1}{|D|} \sum_{s,a \in D} \min \quad &\Big(\rho(\theta) A^{\pi_{\theta_k}}(s,a), \\ &\text{clip}(\rho(\theta), 1-\epsilon, 1+\epsilon) A^{\pi_{\theta_k}}(s,a)\Big), \end{aligned} \tag{12}$$

where ϵ determines the size of the trust region. If the advantage is positive, indicating a better-than-average action, the objective is capped to prevent the agent from making an overly large and potentially destabilizing policy change. Conversely, if the advantage is negative for a worse-than-average action, the objective is also clipped to stop the agent from over-reacting and making that action extremely unlikely, which could harm future exploration. By always taking the minimum of the unclipped and clipped objectives, the algorithm guarantees that each policy update remains within a trusted "proximal" region around the previous policy, promoting a smooth and reliable learning process.

4. Methodology and Problem Formulation

In the following, we detail the underlying methodology of our proposal, viz., safe RL by capitalizing on learned neural barrier certificates to eventually learn control policies that ensure occupant comfort.

4.1. Safe Reinforcement Learning for HVAC Control

The HVAC control problem is modeled as a Markov decision process (MDP) $M = (\mathcal{S}, \mathcal{A}, P, R)$ solved through RL. In this framework, the environment provides states $s_t \in \mathcal{S}$ to the agent, consisting of continuous HVAC system variables including operative zone temperature and occupancy $s_t = [T_t, O_t]^\top$, where O is either 0 or 1 (no presence or presence of people, respectively) and T is a real number. The reward function serves as feedback from the environment, evaluating the quality of the agent's actions. The reward function is defined as $r_t(s_t, a_t) = -(C_t + E_t)$, where C_t, E_t are the HVAC system cost and energy, respectively. The agent selects an action (e.g., temperature set-points or heat pump modulation signals) $a_t \in \mathcal{A}$ according to a policy $\pi(a|s)$, which maps states to actions. The objective is to find the policy that maximizes the expected cumulative reward $J(\pi) = \mathbb{E}[\sum_{t=0}^{\infty} \gamma^t r_t(s_t, a_t)]$. Both the state and action spaces can be either discrete or continuous, with further details discussed in Section 5.

While the agent aims to minimize HVAC energy consumption, it must also maintain thermal comfort. Though comfort depends on multiple factors, including temperature, humidity, and clothing levels [68], this study focuses primarily on indoor temperature with respect to thermal comfort bound between 18 to 24 degrees. This constraint transforms the RL problem into a constrained RL problem. Figure 1 illustrates the process flow for the thermal management system control mechanism. A deep RL agent is trained to control the HVAC system; simultaneously, a barrier certificate is trained in a data-driven manner to prevent the agent from violating the safety constraints (thermal comfort). During early training, safety violations occur frequently since the barrier and RL agent learn simultaneously. As the barrier improves with more data, violations progressively decrease. The barrier function manipulates the RL agent's objective function by penalizing violations of occupant comfort to advance policy improvement in the direction of safe and optimal actions.

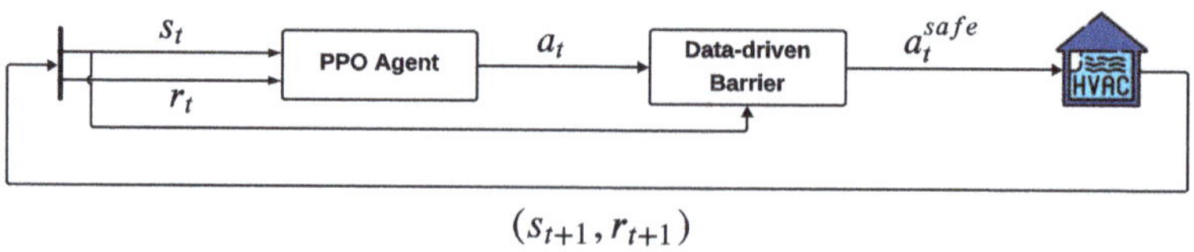

Figure 1. Block diagram of safe RL agent for HVAC control. The PPO agent receives state s_t and reward r_t to decide on actions a_t. The data-driven safety barrier (e.g., the learned neural barrier certificate) amends these actions a_t if necessary to produce safe actions a_t^{safe} for HVAC control.

This approach ensures that the agent's learning is driven towards actions that maintain occupant comfort while minimizing energy consumption, allowing us, in turn, to train an agent specifically to minimize energy consumption while maintaining occupant comfort.

4.2. Safe Reinforcement Learning Through Neural Barrier Certificates

Consider the dynamical system $s_{t+1} = f(s_t, a_t)$, where $s \in \mathcal{S}$ and $a \in \mathcal{A}$ are the state and control input, respectively. The constraint function is defined as $C(s_t) \leq 0$ and we define the set of safe states as $\mathcal{S}_{safe} = \{s | C(s_t) \leq 0\}$ and the set of unsafe states as $\mathcal{S}_{unsafe} = \{s | C(s_t) > 0\}$, respectively; the set of initial states is defined as $\mathcal{S}_{init} \subset \mathcal{S}$. We assume that $\mathcal{S}_{init} \subseteq \mathcal{S}_{safe}$.

Barrier certificates are mathematical constructs that define forward invariant sets in state space. The forward invariant property of the barrier certificates can be used to guarantee the safety of dynamical systems [27]. The function $h^\pi : \mathcal{S} \rightarrow \mathbb{R}$ is a barrier certificate [26] of a given policy $\pi(a_t|s_t)$ if it satisfies

$$\begin{aligned} &\forall \quad s \in \mathcal{S}_{init} : h^\pi(s) \leq 0, \\ &\forall \quad s \in \mathcal{S}_{unsafe} : h^\pi(s) > 0, \\ &\forall \quad s \in \mathcal{S} : h^\pi(s_{t+1}) - h^\pi(s_t) \leq -\alpha(h^\pi(s_t)), \end{aligned} \tag{13}$$

where α is a strictly increasing function with $\alpha(0) = 0$; typically $\alpha(h^\pi) = \lambda h^\pi$ for some $0 < \lambda < 1$. If a barrier certificate h^π exists for a system under policy π with an initial state $s_0 \in \mathcal{S}_{init}$, then the system's trajectory remains within the safe set due to forward invariance [27]. Since the system transition function $f(\cdot)$ is needed but unknown, we approximate the barrier certificate using a neural network parameterized by Φ. The dataset D collected from agent–environment interactions is partitioned into D_{init} (initial states) and D_{unsafe} (unsafe states). The loss function of the neural network is designed to penalize violations of the properties of the control barrier functions [26]. The safety loss corresponds to the first property of Equation (13),

$$J_{\text{Safe}}(\Phi) = \frac{1}{|D_{init}|} \sum_{s \in D_{init}} \max(0, h^\pi(s; \Phi)), \tag{14}$$

where D_{init} is the set of initial state samples. The unsafe loss function corresponds to the second property,

$$J_{\text{Unsafe}}(\Phi) = \frac{1}{|D_{unsafe}|} \sum_{s \in D_{unsafe}} \max(0, -h^\pi(s; \Phi)), \tag{15}$$

where D_{unsafe} is the set of unsafe state samples. The invariant loss corresponds to the forward invariant property (third property),

$$J_{\text{Invt}}(\Phi) = \frac{1}{|D|} \sum_{s,s' \in D} \max(0, h^\pi(s'; \Phi) - (1 - \lambda) h^\pi(s; \Phi)), \tag{16}$$

where D is the set of all state samples. The barrier loss is the sum of all three loss functions.

$$J_h(\Phi) = J_{\text{Safe}}(\Phi) + J_{\text{Unsafe}}(\Phi) + J_{\text{Invt}}(\Phi). \tag{17}$$

The parameter Φ is updated by minimizing Equation (17) using gradient descent.

4.3. Policy Learning

Barrier certificate constraints are applied to the policy optimization problem. In this paper, the PPO is chosen to solve the optimization problem. The objective function of the PPO [67] is Equation (12). The policy constraint is derived from the invariant property of the barrier certificate, which ensures that the agent remains within the safe region. The policy optimization problem then becomes

$$\max_{\theta} L(\theta) \quad \text{s.t.} \quad \rho(\theta) \cdot J_{\text{Invt}}(\Phi) \leq 0. \tag{18}$$

Importance sampling is applied since the invariant loss is computed using data from the old policy. The optimization problem in Equation (18) can be transformed into an unconstrained problem using the Lagrangian dual [26],

$$\min_{\nu \geq 0} \max_{\theta} F(\nu, \theta) = L(\theta) - \nu\rho(\theta) J_{\text{Invt}}(\Phi). \tag{19}$$

The value function is also learned by minimizing the loss function

$$J_V = \frac{1}{|D|} \sum_{s \in D} (V_\psi(s_t) - \hat{R}_t)^2, \tag{20}$$

where ψ is the critic network parameter and $\hat{R}_t$ is the estimated return.

4.4. Safety Mechanism Overview

This framework can be understood as creating an invisible boundary around comfortable temperature zones. The barrier certificate $h(s;\Phi)$, which is itself a neural network with parameters Φ, learns to approximate the constraint function directly from building operation data rather than requiring manual design of the barrier function. The parameterized barrier $h(s;\Phi)$ functions as a learned topographical map where negative values indicate safe (comfortable) temperatures and positive values signal unsafe (uncomfortable) conditions. Crucially, unlike traditional hard constraints that define fixed temperature thresholds, $h(s;\Phi)$ learns a smooth, continuous safety landscape that adapts to the specific thermal dynamics of each building. The learning process occurs through three complementary objectives. When the system starts at a comfortable temperature, Equation (14) ensures $h(s;\Phi) \leq 0$, confirming that the barrier recognizes this as safe. As the temperature approaches unsafe regions, Equation (15) trains $h(s;\Phi)$ to output positive values. Moreover, Equation (16) enforces the forward invariance property, ensuring that once comfortable, the system tends to remain comfortable. When the PPO agent proposes an action that would lead the system toward uncomfortable temperatures, the term $\nu\rho(\theta)J_{\text{Invt}}(\phi)$ in Equation (19) becomes large, effectively preventing unsafe actions during policy optimization.

5. Experiments and Discussion

In this section, we present a comprehensive training and evaluation of the safe reinforcement learning controller through systematic testing across diverse operational scenarios. Our experimental design aims to assess both the performance characteristics and practical viability of the safe RL-based approach and to compare it to conventional control strategies. The section includes: (1) experimental setup for single-zone and multi-zone building test cases, (2) training results showing learned policies and barrier certificates, (3) performance comparisons with PPO and RBC controllers for energy consumption and cost, (4) safety violation analysis during training and evaluation phases, and (5) discussion of key findings and practical implementation considerations.

5.1. Experimental Design

We evaluate our safe RL approach using two BOPTEST [29,69] test cases representing different control complexity levels. For Experiments 1–5, we use the BESTEST Hydronic Heat Pump scenario modeling a 192 m^2 residential home with five occupants (present

07:00–20:00 weekdays, 24 h weekends; only Experiment 5 considers occupancy state). The RL agent observes normalized zone temperature (where $T = -1$ represents the lower setpoint, $T = +1$ represents the upper setpoint, with valid range $T \in (-\infty, +\infty)$) and controls heat pump modulation through 12 discrete levels. For Experiments 6–7, we employ a five-zone Chicago office building with variable air volume (VAV) system, totaling 1655 m^2. The agent observes zone temperatures, internal heat gains, and outdoor temperature, with Experiment 6 controlling AHU parameters and Experiment 7 managing core zone setpoints. All variables are normalized to $[-1, 1]$ for learning efficiency, with rewards based on BOPTEST KPIs targeting energy and cost reduction [29]. Table 2 details the experimental configurations and their mapping to research questions from Section 1.

Table 2. Experimental configurations and research question mapping.

Exp.	Description	Building Type	Floor Area	Control Variable(s)	Comfort Zone	RQs
1	Without pre-trained barrier; random episodes	Single-zone (Brussels)	192 m^2	Heat pump (12 bins)	18–21 °C	RQ1, RQ2
2	Pre-trained barrier (20 episodes)	Single-zone (Brussels)	192 m^2	Heat pump (12 bins)	18–21 °C	RQ1, RQ2
3	Training with diverse random episodes	Single-zone (Brussels)	192 m^2	Heat pump (12 bins)	18–21 °C	RQ1, RQ2
4	Training with repeated identical episodes	Single-zone (Brussels)	192 m^2	Heat pump (12 bins)	18–21 °C	RQ1
5	Occupancy-aware barrier adaptation	Single-zone (Brussels)	192 m^2	Heat pump (12 bins)	18–21 °C (occ.) 15–30 °C (unocc.)	RQ1
6	Multi-zone AHU system control	5-zone office (Chicago)	Total: 1655 m^2 Core: 985 m^2 N/S: 208 m^2 ea. E/W: 131 m^2 ea.	Supply air temp. Duct pressure	18–24 °C	RQ1
7	Multi-zone core zone control	5-zone office (Chicago)	(Same as Exp. 6)	Heating setpoint Cooling setpoint	18–24 °C	RQ1, RQ3

Table 3 shows all hyperparameters for the PPO agent and barrier neural network.

Table 3. Hyperparameters for our experiments. The actor and critic networks share the same architecture parameters.

Number Of Hidden Layers	2	Size of Hidden Layers	256
Learning Rate	0.0003	GAE parameter	0.95
Policy Clip	0.2	PPO Activation Function	ReLU
Barrier Activation Function	*tanh*	Lagrangian Initial Value	1
Optimizer	Adam		

5.2. Experimental Results

For the first experiment (Experiment 1), the length of the episode is considered to be 2 days, and episodes are randomly generated. The agent interacts with the environment every 30 min and collects data for training. Figure 2a shows the learned policy after training for 188 days, which indicates that the agent learned the best possible policy with respect to the barrier certificate that is shown in Figure 2b. The trained barrier certificate clearly distinguishes between safe and unsafe temperature regions, showing negative values (below zero) only within the comfortable temperature range of 18–21 °C and positive values in the unsafe regions.

Next, we investigated the impact of a pre-trained barrier certificate on the previous experimental setup (Experiment 2). The barrier certificate was trained separately for 20 episodes. The agent successfully learned the safe policy (Figure 2c) within 150 days of training; the corresponding learned barrier certificate is displayed in Figure 2d.

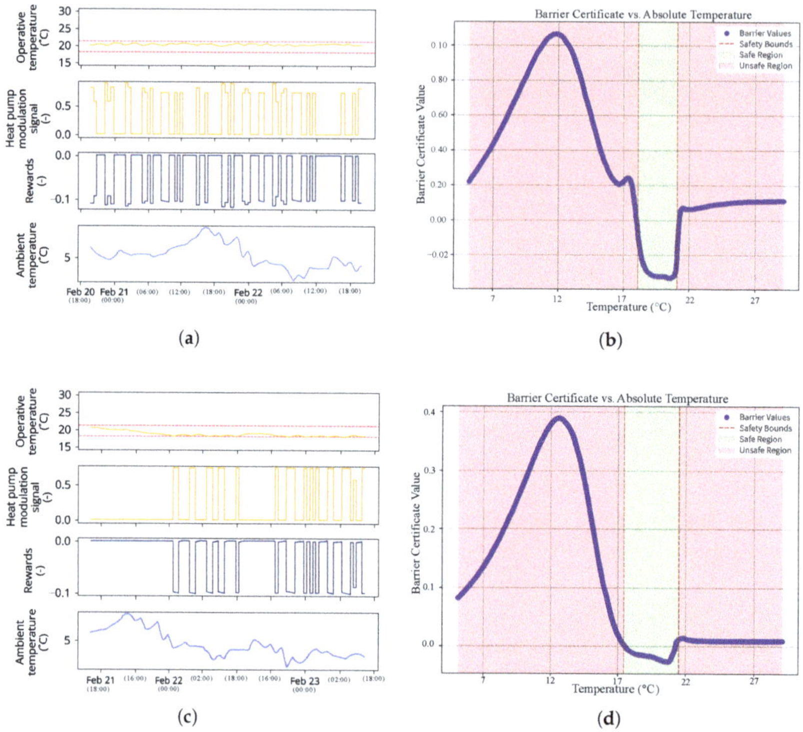

Figure 2. End-of-training results for random 2-day episodes. Figure 2a,c show the learned policy trajectories (operative temperature in °C and normalized heat pump modulation signal a_t). Figure 2b,d plot the learned barrier certificate value $h(s;\Phi)$, where $h \leq 0$ denotes safe states and $h > 0$ denotes unsafe states. Figure 2a,b are trained without a pre-trained barrier certificate; Figure 2c,d use a 20-episode pre-trained barrier certificate. The pre-trained barrier certificate yields a less conservative policy with small violations. (**a**) Learned policy (final episode) after 188 days. (**b**) Learned barrier certificate after 188 days. (**c**) Learned policy (final episode) after 150 days. (**d**) Learned barrier certificate after 150 days.

Figure 3 shows training results using 5-day episodes with different training methods: Figure 3a,b show the agent trained on repeated identical days (Experiment 3), while Figure 3c,d show performance with randomly generated daily conditions (Experiment 4). The difference in using either 2- or 5-day episodes shows that in the latter case, learned policies are less conservative in that they tend to violate the barrier more frequently to maximize the reward.

In Experiment 5, we trained an occupancy-aware barrier certificate (pre-trained for 20 episodes) that enforces different comfort bounds: 18–21 °C during occupied hours and 15–30 °C during unoccupied hours. Initially, the agent disregarded occupancy status (Figure 4a), but after further training, it learned to minimize actions during occupied hours (Figure 4b), improving energy efficiency while maintaining comfort. Figure 5 illustrates the barrier certificates for both occupied and unoccupied states. For the next set of experiments (6 and 7) on multi-zone test cases, the safety constraint for the agent is that the temperature for each zone must be within the range of 18–24 °C, and the reward function is for the agent to reduce the cost and energy. In Experiment 6, the episode duration is 15 days, and the agent takes an action every 30 min; the learning frequency is every 1.5 days. The agent is tasked to reduce the cost and energy consumption and keep the temperature within the safety constraints for all zones by controlling the pressure setpoint and supply temperature setpoint. Figure 6 shows the learned policy during the training for the agent and final episode of the training (Figure 6a) and one evaluation episode (Figure 6b) in this environment. The agent managed to keep the temperature within the safe zone despite having some violations during the final episodes of training. In the evaluation episode, the agent had some violations but managed to keep the temperature within the safe bounds during the episode.

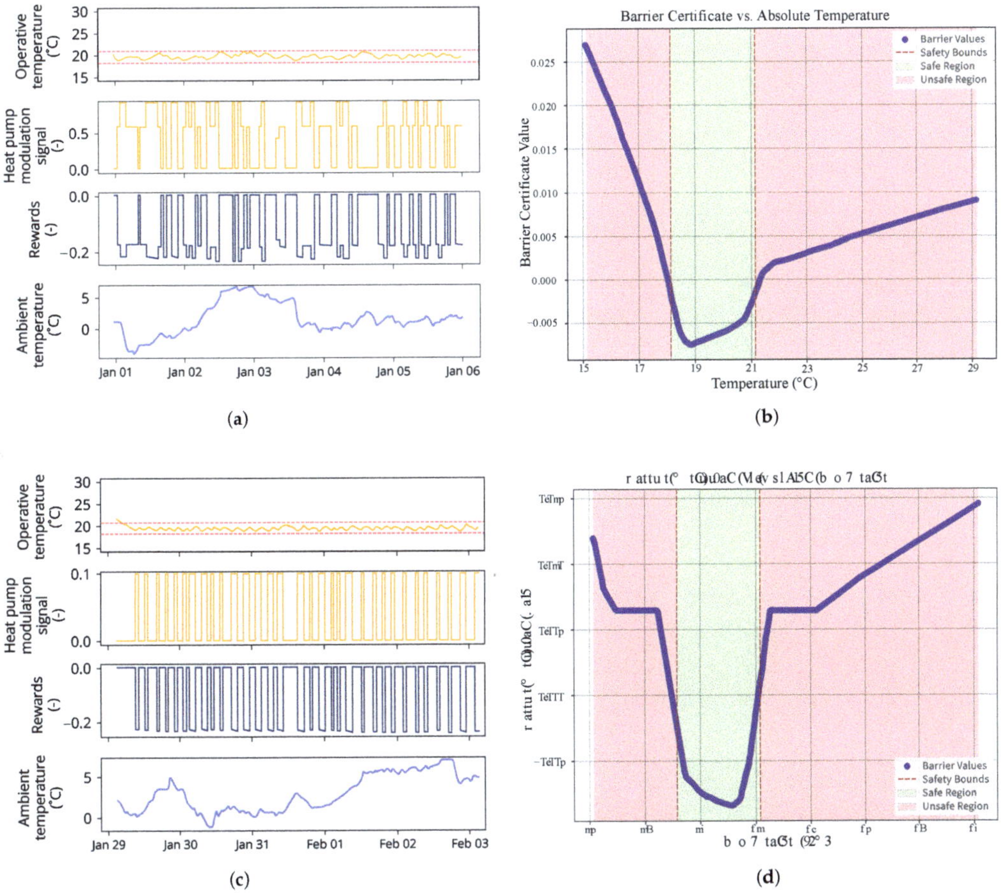

Figure 3. End-of-training results for 5-day episodes. Figure 3a: learned policy trajectories (operative temperature in °C and normalized heat pump modulation signal a_t) without random episode generation; Figure 3b: corresponding learned barrier certificate value $h(s;\Phi)$ (where $h \leq 0$ indicates safe states). Figure 3c: learned policy with random episode generation; Figure 3d: corresponding barrier certificate. Randomized episodes reduce overfitting relative to repeated-day training, at the cost of slower convergence. (**a**) Learned policy (final episode) after 50 days. (**b**) Learned barrier certificate after 50 days. (**c**) Learned policy (final episode) after 125 days. (**d**) Learned barrier certificate after 125 days.

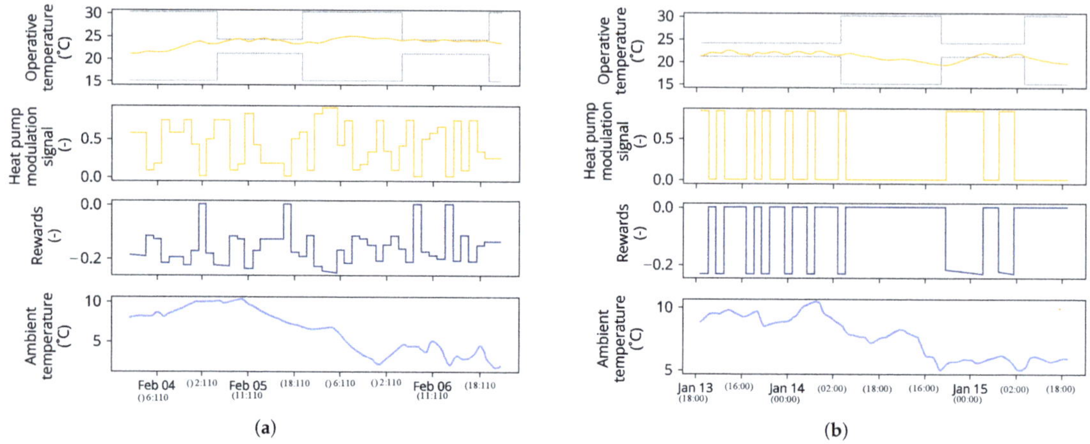

Figure 4. Training progression with an occupancy-aware barrier certificate. Figure 4a: Day 80 (partially learned policy); Figure 4b: Day 160 (converged policy). Grey bands indicate the comfort-temperature bounds: 18–21 °C (occupied) and 15–30 °C (unoccupied). Curves show operative temperature (°C) and normalized heat pump modulation signal a_t. (**a**) Learned policy after 80 days of training. (**b**) Learned policy after final (160 days) episode of training.

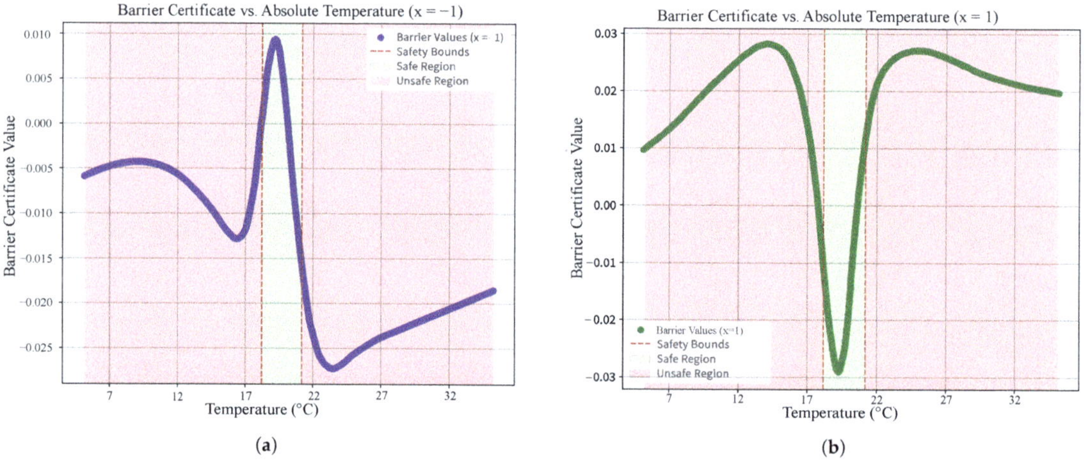

Figure 5. Occupancy -aware barrier after 160 training days. Figure 5a shows the learned barrier certificate during unoccupied periods; Figure 5b shows it during occupied periods. Plots report barrier value $h(s;\Phi)$ (where $h \leq 0$ denotes safe states) versus temperature (°C). (**a**) Barrier certificate for unoccupied hours ($x = -1$). (**b**) Barrier certificate for occupied hours ($x = 1$).

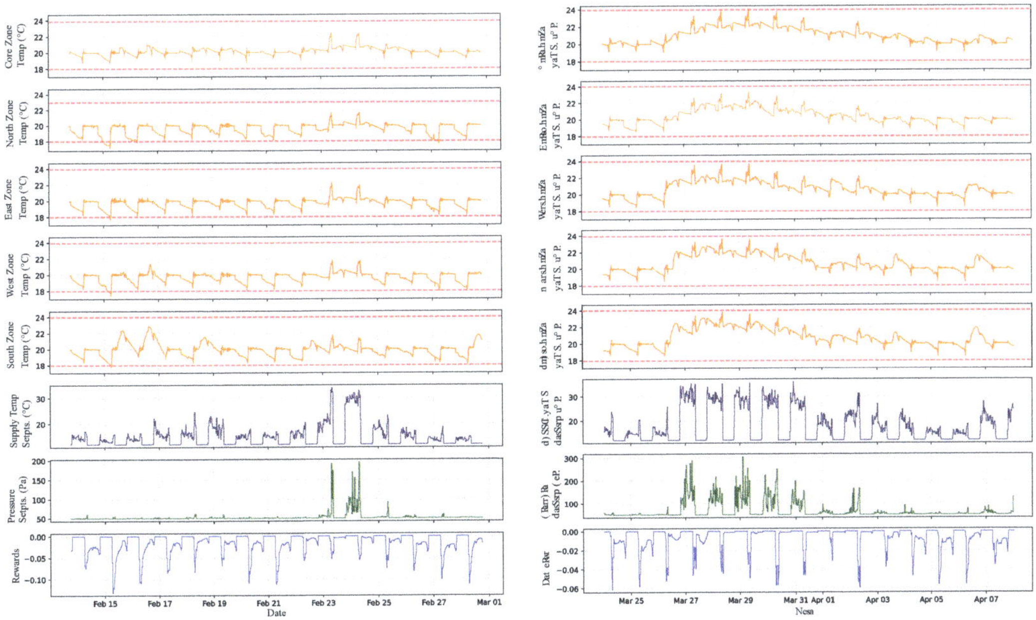

(**a**) (**b**)

Figure 6. Multi-zone office with 15-day episodes. Figure 6a: Final training episode showing learned policy controlling AHU supply-air temperature setpoint (°C) and duct static-pressure setpoint (Pa). Figure 6b: held-out evaluation episode. All zone temperatures (°C) remain within the safety band of 18–24 °C with only minor violations. (**a**) Learned policy for the agent; (**b**) Agent evaluated in the environment.

In the next experiment, the agent controls heating and cooling setpoints with an extended 30-day episode duration, maintaining the same 30 min sampling interval but with learning updates every 2 days. The objective focuses on maintaining the core zone temperature within 18–24 °C through setpoint manipulation. Figure 7 presents the final training episode (Figure 7a) and an evaluation episode (Figure 7b) from the multi-zone office environment. The agent successfully maintained the designated temperature bounds with only minor violations throughout the training episodes. Figure 8 shows the agent's performance under both summer and winter conditions. While the agent successfully maintained safety bounds during cold winter periods, it violated safety constraints under summer cooling conditions.

Our experimental evaluation progressed through a series of increasingly complex scenarios to assess the safe RL controller's capabilities. Initial experiments with 2-day episodes demonstrated that the agent could learn effective control policies while respecting temperature constraints (18–24 °C), with the barrier certificate clearly delineating safe and unsafe regions. Notably, pre-training the barrier certificate reduced learning time by approximately 20% (from 188 to 150 days), suggesting computational benefits for this approach. When we extended episode lengths to 5 days, an interesting pattern emerged: the agent developed less conservative policies, more frequently approaching barrier boundaries to optimize rewards. This behavior was particularly pronounced when training with randomly generated daily conditions versus repeated patterns.

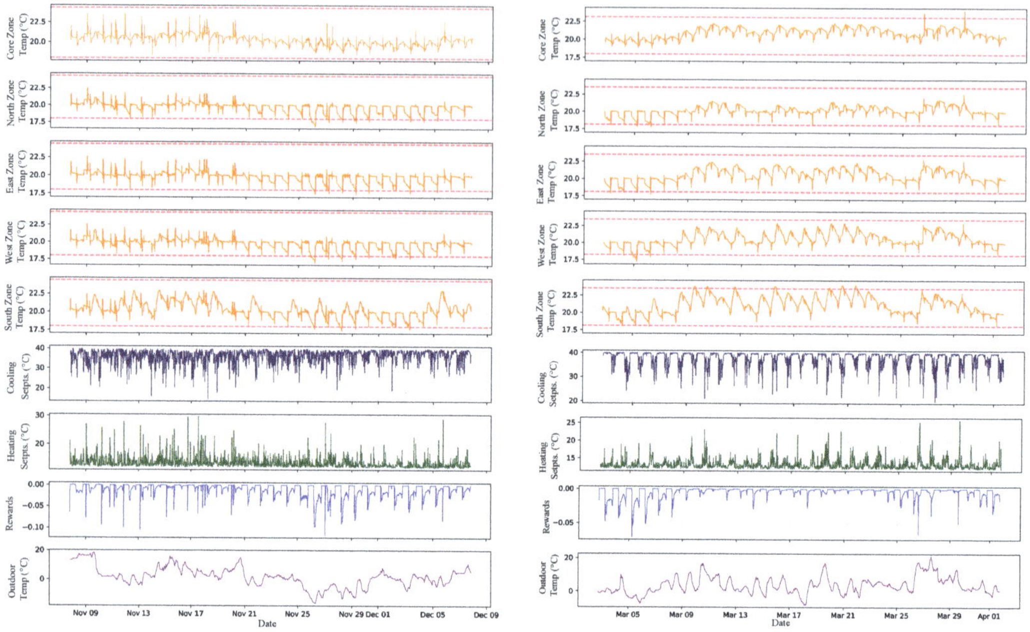

(a) (b)

Figure 7. Multi-zone office with 30-day episodes. Figure 7a: Final training episode; Figure 7b: Held-out evaluation episode. The agent manipulates the core zone heating and cooling setpoints to keep the core zone temperature within the 18–24 °C safety band. (**a**) Learned policy for the agent; (**b**) agent evaluated in the environment.

The introduction of occupancy-aware barriers marked a significant advancement in the controller's sophistication. By allowing relaxed temperature bounds during unoccupied hours (15–30 °C versus 18–21 °C when occupied), we observed the agent's ability to learn context-dependent strategies. Initially, the agent ignored occupancy status, but through continued training, it developed an energy-efficient policy that minimized control actions during occupied periods while exploiting the relaxed constraints during unoccupied times. This learning progression was successfully scaled to multi-zone buildings with 15 and 30-day episodes, where the agent managed multiple zones simultaneously while maintaining zone-specific temperature constraints (18–24 °C) and optimizing for cost reduction. Throughout all experiments, the 30 min control interval remained constant, providing consistent temporal resolution for the agent's decision-making process.

The final sets of experiments showcased the agent's ability to perform in more complex and real-world-like scenarios, where inter-zone thermal dynamics and multiple control variables significantly increased the challenge. In Experiment 6, the agent successfully managed a five-zone office building by controlling centralized AHU parameters, demonstrating its capability to handle system-level control decisions that affect all zones simultaneously. Experiment 7 explored an alternative control strategy where the agent manipulated zone-level heating and cooling setpoints, providing more direct but localized control. These experiments not only validated the scalability of the safe RL approach from single-zone residential to multi-zone commercial buildings but also demonstrated its flexibility in adapting to different control architectures.

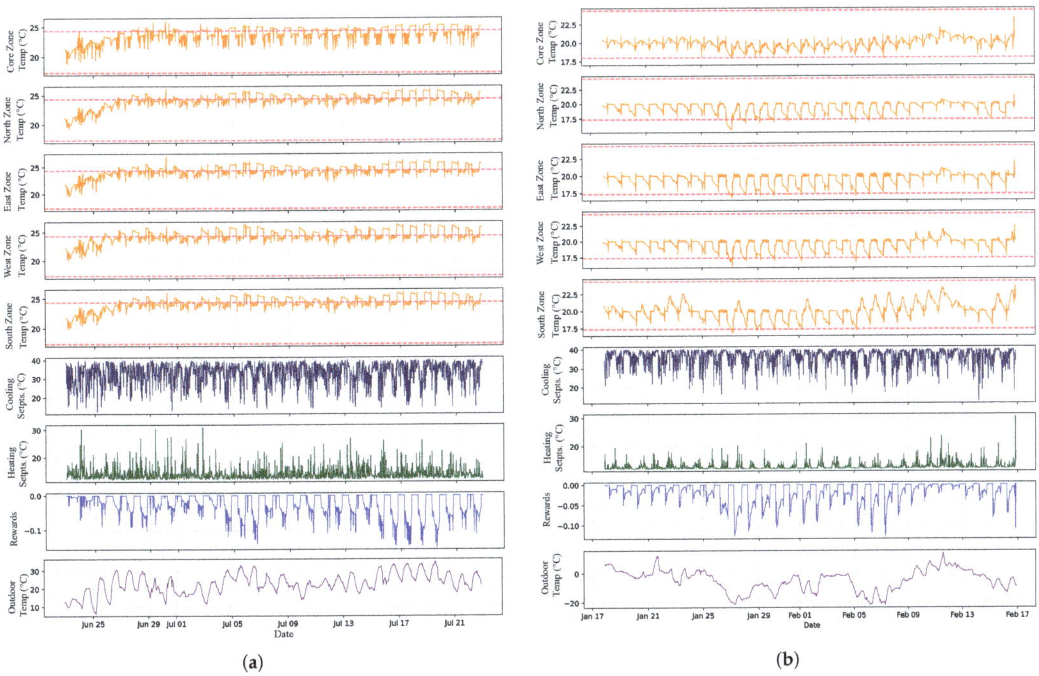

Figure 8. Seasonal operating conditions. Figure 8a: Cooling-dominated period with the highest rate of comfort-constraint violations. Figure 8b: Heating-dominated period with fewer violations. Plots show core zone temperature (°C) relative to the safety band and the agent's setpoint (cooling and heating) actions. (a) Agent evaluated in cooling-dominated scenario; (b) agent evaluated in heating-dominated scenario.

5.3. Comparative Study

To enhance the robustness of the evaluation of our safe RL agent, we conducted an additional set of comparative experiments using the configurations of experiments (7) as outlined in Table 2, but instead, each controller was evaluated in each season of the year to better investigate its energy and safety performance. This approach allows for more comprehensive data collection and demonstrates stronger evidence of performance. The safe PPO agent is compared against two baselines: (1) a PPO agent trained with reward weighting, (2) RBC with static setpoints, as described in [70]. The RBC controller acts as a supervisory controller for setting the set-points for local PID controllers in the test case. PPO uses a comparable reward function to the safe RL agent but incorporates an additional BOPTEST [29] KPI for thermal violation penalties, tuned according to [70]. This test case was chosen for the comparative study since it enables us to investigate the agent's performance in both cooling and heating scenarios. Our evaluation methodology employs a more rigorous generalization test than typical approaches in the literature. While previous work, such as [70], evaluates agents on full annual cycles with stochastic weather variations, we randomly sample episodes throughout the year. This approach ensures that our agent performs well across diverse seasonal conditions rather than optimizing for average annual performance. This evaluation strategy provides a more realistic assessment of deployment readiness, as real-world controllers must maintain efficiency at all points during the year.

According to Table 4, PPO with barrier certificate demonstrates consistent performance across seasons. The controller achieves a median winter consumption of 2.25 $kWhm^{-2}$ (IQR: 2.15–2.40) at USD 0.21/m^2 (IQR: 0.20–0.23), and spring consumption of 1.06 $kWhm^{-2}$

(IQR: 0.77–1.20) at USD 0.10/m^2 (IQR: 0.07–0.11). Summer shows the highest consumption among all controllers at 3.82 kWhm^{-2} (IQR: 3.41–3.97) costing USD 0.36/m^2 (IQR: 0.32–0.37), indicating a conservative cooling strategy prioritizing comfort. Autumn performance reaches 1.14 kWhm^{-2} (IQR: 0.83–1.23) at USD 0.11/m^2 (IQR: 0.08–0.12), showing competitive efficiency with relatively narrow IQRs suggesting stable control.

PPO without barrier shows more aggressive cost optimization, achieving 10% lower summer consumption at 3.45 kWhm^{-2} (IQR: 3.15–3.74) costing USD 0.33/m^2 (IQR: 0.29–0.35), saving USD 0.03/m^2 compared to the barrier version. Winter performance matches the barrier version at 2.25 kWhm^{-2} (IQR: 1.95–2.35) and USD 0.21/m^2 (IQR: 0.18–0.22), though with wider variability. Spring consumption of 1.04 kWhm^{-2} (IQR: 0.73–1.34) shows the widest IQR span (0.61 kWhm^{-2}), indicating higher operational uncertainty. Autumn performance deteriorates to 1.28 kWhm^{-2} (IQR: 0.94–1.40) at USD 0.12/m^2 (IQR: 0.09–0.13), representing 12% higher energy use than the barrier version. The RBC excels in mild conditions, achieving the best spring performance at 0.79 kWhm^{-2} (IQR: 0.62–0.95) costing USD 0.08/m^2 (IQR: 0.06–0.09), representing 24–25% energy savings and 20% cost reduction versus PPO variants. Summer performance of 3.28 kWhm^{-2} (IQR: 2.66–3.76) at USD 0.31/m^2 (IQR: 0.25–0.35) positions it between the two PPO approaches, though with the widest summer IQR (1.10 kWhm^{-2}). Winter consumption of 2.31 kWhm^{-2} (IQR: 2.15–2.39) matches all controllers' cost at USD 0.21/m^2 (IQR: 0.20–0.23). Autumn shows 1.29 kWhm^{-2} (IQR: 1.19–1.35) at USD 0.12/m^2 (IQR: 0.11–0.13), with the narrowest autumn energy IQR (0.16 kWhm^{-2}) indicating predictable transitional season performance. Figures 9 and 10 show the energy and cost performance of each controller in each season.

Table 4. Energy consumption and cost analysis by season.

Controller	Metric	Seasonal Performance			
		Winter	Spring	Summer	Autumn
PPO w/ Barrier	Energy (kWhm^{-2})	2.25 [2.15–2.40]	1.06 [0.77–1.20]	3.82 [3.41–3.97]	1.14 [0.83–1.23]
	Cost (USD/m^2)	0.21 [0.20–0.23]	0.10 [0.07–0.11]	0.36 [0.32–0.37]	0.11 [0.08–0.12]
PPO w/o Barrier	Energy (kWhm^{-2})	2.25 [1.95–2.35]	1.04 [0.73–1.34]	3.45 [3.15–3.74]	1.28 [0.94–1.40]
	Cost (USD/m^2)	0.21 [0.18–0.22]	0.10 [0.07–0.13]	0.33 [0.29–0.35]	0.12 [0.09–0.13]
Rule–Based	Energy (kWhm^{-2})	2.31 [2.15–2.39]	0.79 [0.62–0.95]	3.28 [2.66–3.76]	1.29 [1.19–1.35]
	Cost (USD/m^2)	0.21 [0.20–0.23]	0.08 [0.06–0.09]	0.31 [0.25–0.35]	0.12 [0.11–0.13]

As for the safety performance analysis (Table 5), RBC demonstrates the best safety compliance, maintaining 97.8% (spring) and 100% (autumn) safety rates with minimal variability, indicating highly predictable comfort delivery. Even during challenging summer conditions, it sustains 89.3% compliance with narrow (IQR: 89.0–89.5%), suggesting its conservative fixed rules prioritize occupant comfort over energy savings. This aligns with its slightly higher energy consumption but provides reliable comfort assurance.

The safe PPO shows moderate safety performance with higher variability. Spring safety at 99.9% (IQR: 90.5–99.9%) and autumn at 99.6% (IQR: 81.3–99.6%) indicate that the barrier function provides meaningful constraint enforcement, though the wide IQR ranges suggest inconsistent protection across different operating conditions. The dramatic summer degradation to 58.1% (IQR: 57.4–58.6%) reveals that even with barrier constraints, the controller struggles to maintain comfort during peak cooling demands, potentially explaining its higher summer energy consumption as it attempts to recover from constraint violations.

Table 5. Safety performance analysis by season.

Controller	Median Safety Rate (%) [IQR: Q1–Q3]			
	Winter	Spring	Summer	Autumn
Rule-Based	100.0 [100.0–100.0]	97.8 [97.4–100.0]	89.3 [89.0–89.5]	100.0 [98.7–100.0]
PPO w/ Barrier	100.0 [100.0–100.0]	99.9 [90.5–99.9]	58.1 [57.4–58.6]	99.6 [81.3–99.6]
PPO w/o Barrier	100.0 [100.0–100.0]	71.1 [60.6–85.2]	25.0 [22.2–28.1]	85.6 [59.3–100.0]

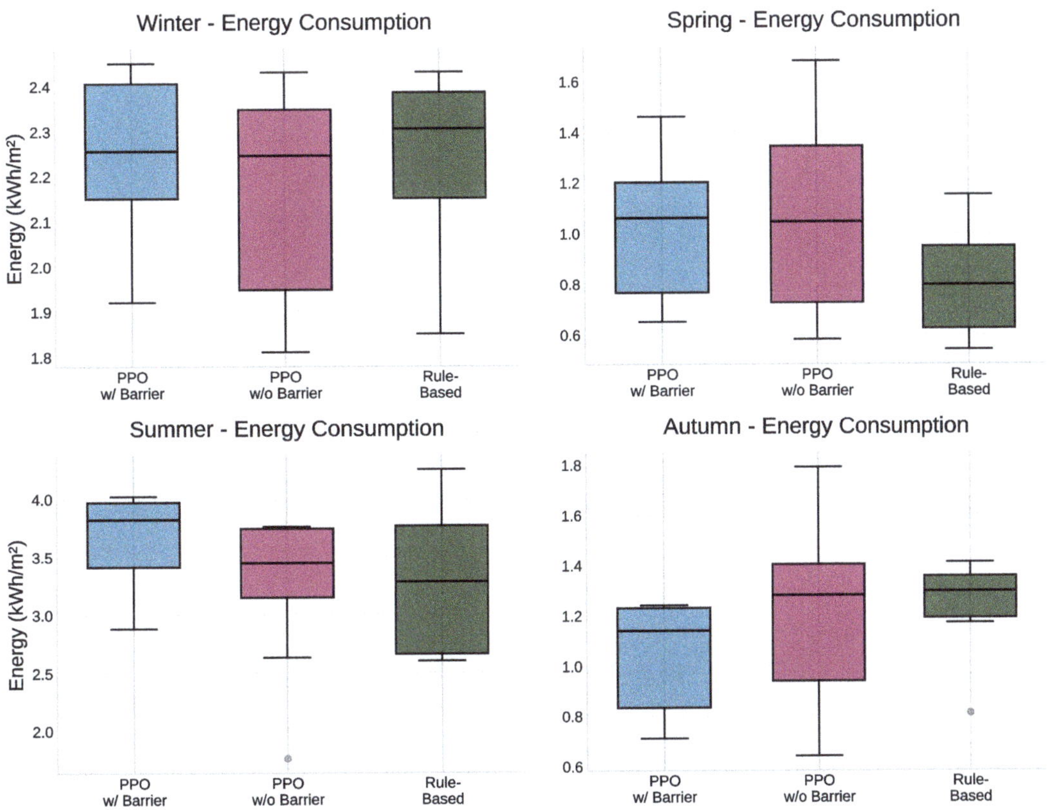

Figure 9. Direct energy consumption comparison of each controller for each season.

PPO without barrier exhibits the poorest safety performance, particularly the alarming 25.0% summer compliance (IQR: 22.2–28.1%), meaning that comfort constraints are violated 75% of the time during peak cooling season. Spring performance shows 71.1% compliance (IQR: 60.6–85.2%), with an IQR span of 24.6 percentage points, indicating high variability. Autumn demonstrates even greater unpredictability at 85.6% median compliance (IQR: 59.3–100.0%), where the 40.7 percentage point IQR span suggests performance can range from barely acceptable (59.3%) to perfect compliance. This aggressive optimization explains its energy savings but reveals the critical trade-off: the 10% summer energy reduction compared to barrier PPO comes at the cost of acceptable comfort conditions. During winter, all controllers achieve perfect 100% safety compliance (IQR: 100.0–100.0%), reflecting the wider comfort bounds typical of heating seasons. This relaxed constraint environment reveals interesting performance dynamics: the unconstrained PPO achieves competitive energy efficiency at 2.25 kWhm^{-2} (matching the barrier variant), while the RBC consumes slightly more at 2.31 kWhm^{-2}. The PPO-with-barrier's identical winter consumption to its unconstrained counterpart (2.25 kWhm^{-2} at USD 0.21 m^{-2}) suggests that winter comfort bounds are sufficiently relaxed that the barrier function rarely activates. However, the barrier variant maintains tighter IQR bounds (2.15–2.40 vs. 1.95–2.35 kWhm^{-2}), demonstrating more consistent control behavior. This consistency reflects the barrier function's design to prioritize safety assurance over energy optimization as per Equation (18), maintaining predictable operation even when natural conditions ensure safety. In summary, the seasonal analysis reveals critical trade-offs between energy efficiency and occupant comfort across different weather conditions. The RBC provides the most reliable comfort, with median safety rates ranging from 89.3% (summer) to 100% (winter/autumn), maintaining narrow IQRs, indicating predictable performance. PPO without barrier achieves lower energy costs but suffers from unacceptable safety performance, particularly in the summer (25.0% compliance, IQR: 22.2–28.1%). PPO with barrier offers a compromise with better safety than unconstrained PPO, yet still exhibits not perfect summer safety rates of 58.1% (IQR: 57.4–58.6%), while maintaining high compliance in other seasons (99.6–100%). Figure 11 shows the safety performance of each controller in each season.

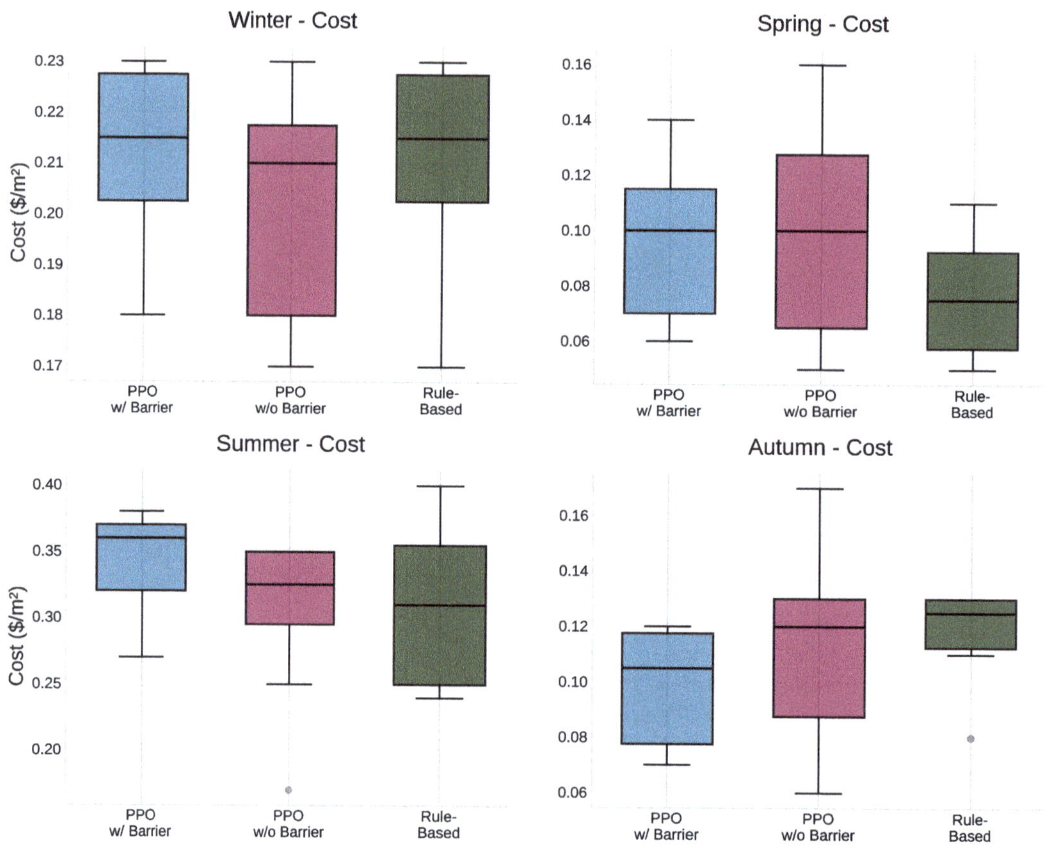

Figure 10. Direct cost consumption of each controller for each season.

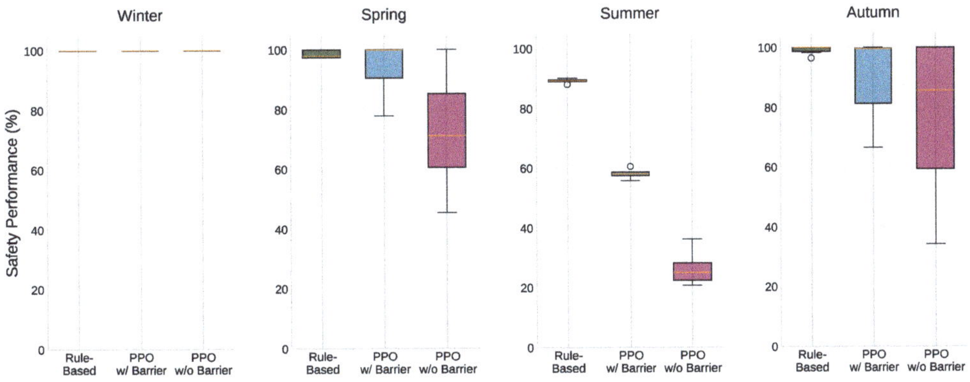

Figure 11. Direct safety performance of each controller for each season.

5.4. Discussion

Our experiments with safe RL for HVAC control revealed several important findings that highlight both the potential and challenges of this approach. Figure 12 shows some statistics from our training runs. Specifically, Figure 12b shows the number of safe and unsafe states for the agent as gathered in the dataset. At the early stages of training, the number of unsafe states is higher than safe states. As training continues, however, the number of unsafe states decreases, and the safe states increase. However, the number of unsafe states does not reach zero since we use a neural network to approximate the barrier certificate in a model-free fashion. This approach can achieve near-zero safety violations, but due to the approximation error, it will never be zero [71]. Figure 12a presents the constrained loss function (cf. Equation (19)) for two agents that were trained with and without a pre-trained barrier certificate over 10 runs, respectively. The pre-trained agent shows higher initial loss due to the pre-training phase occurring without agent intervention. After approximately 3000 steps, the loss rapidly converges and stabilizes at a lower value, while the non-pre-trained approach continues to show significant oscillations. This pattern illustrates the key advantage of pre-training: despite higher initial loss values, the pre-trained approach achieves faster stabilization and more consistent performance once active training begins. The pre-training produced a less conservative agent that developed a more optimal policy than the agent without pre-training (Figure 2). This shows that pre-training can lead to a less conservative agent that achieves a more optimal and safe policy with fewer training steps. This finding highlights the fundamental trade-off between strict safety adherence and performance optimization in constrained reinforcement learning. Our results show that while static-day training achieves safety faster, agents trained on varied days exhibit greater robustness through better generalization (cf. Figure 3).

As for the occupancy-aware control, the agent with a pre-trained barrier certificate is trained to incorporate occupancy information, establishing different safety constraints for occupied and unoccupied periods. During occupied hours, the safe temperature region is set between 18–21 °C, while during unoccupied hours, this range expands to 15–30 °C. This configuration provides greater flexibility for the agent to maximize its reward function, as it can suspend actions during unoccupied periods and only adjust temperatures when the space is occupied. For this scenario, we also employed a pre-trained barrier certificate. The grey lines in Figure 4 illustrate the comfort bounds for both occupied and unoccupied hours. Figure 4b demonstrates that the agent trained with this barrier certificate stops taking actions immediately upon reaching unoccupied hours and maintains this inactive state until occupancy resumes. The agent then actively works again to maintain temperatures within the safe bounds (cf. 18–21 °C) during occupied hours by taking the appropriate actions.

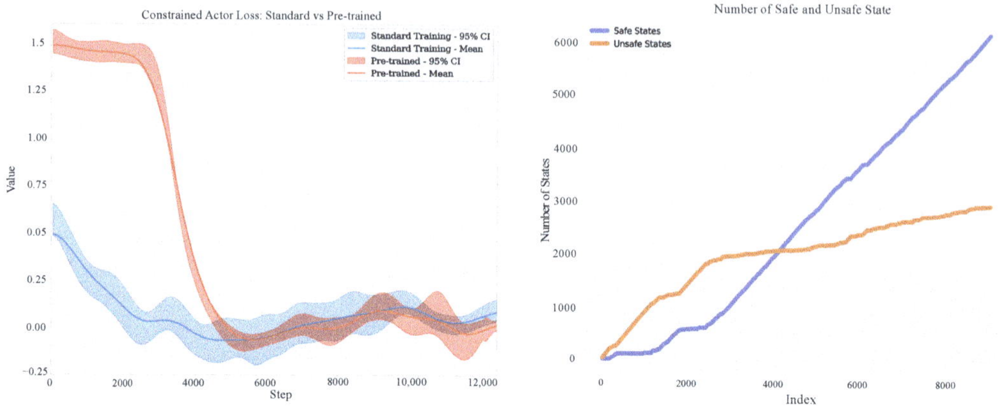

(a) (b)

Figure 12. Averaged constrained loss function (Figure 12a) over 10 runs (with different seeds) for the agents with and without pre-trained barrier, and cumulative number of safe vs. unsafe states (Figure 12b) in Experiment (1). (**a**) Averaged constrained loss function of the agents with 95% confidence interval (shaded areas). (**b**) Cumulative number of safe vs. unsafe states during training.

To implement the safe RL controller in a more complex scenario, we chose an environment with five thermal zones and the same goals to keep the temperature in a certain range (18–24 °C) and minimize energy consumption. The results in Figures 6 and 7 demonstrate that the safe RL controller successfully maintained temperatures within the specified range while minimizing energy consumption. We evaluated different control levels in these experiments, ranging from low-level AHU parameter adjustments to high-level setpoint modifications. This evaluation confirms the agent's ability to maintain safety constraints while optimizing energy performance across varying control complexities. The final two experiments demonstrate the agent's capability to operate safely in complex multi-zone environments where safety constraints are more challenging to satisfy due to increased system dimensionality and inter-zone dependencies.

The seasonal performance analysis reveals a fundamental challenge in deploying RL for building control: the inherent trade-off between safety guarantees and energy optimization varies significantly across operating conditions. While the barrier function improves safety compliance compared to unconstrained RL, it still achieves only 58.1% summer compliance, suggesting that current barrier implementations may be insufficient for extreme conditions. However, the barrier-based approach demonstrates clear potential, more than doubling summer safety rates compared to unconstrained PPO (58.1% vs. 25.0%) and achieving near-perfect compliance in transitional seasons (99.6% autumn, 99.9% spring), indicating that the framework is directionally correct and could benefit from further refinement. During favorable conditions like winter, when all controllers naturally achieve perfect compliance, both PPO variants show identical median consumption (2.25 $kWhm^{-2}$), but the barrier function provides notably more consistent performance, demonstrating operational stability despite minimal safety risk. The unconstrained PPO's severe comfort violations during summer (25.0% compliance) demonstrate that purely reward-driven optimization is unsuitable for safety-critical applications, despite achieving 10% energy savings. Notably, the RBC's superior safety performance (89.3%) and competitive efficiency, particularly during mild seasons (0.79 $kWhm^{-2}$ in spring), suggest that domain knowledge encoded in simple heuristics remains valuable when conditions are predictable. Safe RL consistently outperforms unconstrained PPO across seasons—achieving 99.6%

vs. 85.6% in autumn, 99.9% vs. 71.1% in spring, and, most critically, 58.1% vs. 25.0% in summer—demonstrating that barrier certificates provide meaningful safety improvements and represent a promising direction for making RL viable in real-world building control applications. Nevertheless, the 31.2% point gap between safe RL and RBC performance in summer (58.1% vs. 89.3%) indicates that barrier functions must evolve to incorporate domain expertise to match the reliability of traditional control methods. In summer, the RBC achieves both the lowest energy use (3.28 $kWhm^{-2}$) and the highest median safety (89.3%), indicating that our PPO-with-barrier controller is inferior in this regime. This pattern reflects fundamental differences in summer cooling challenges: high solar and conductive gains create sustained peak loads that saturate HVAC capacity, while the tight comfort ceiling (24 °C) leaves minimal margin for reactive control. Under these conditions, the RBC's anticipatory scheduling and conservative deadband operation prove more effective than the safe RL agent's boundary-seeking behavior. While the safe RL agent registers more violations, the observed zone temperatures hover around 25 °C Figure 8a, suggesting boundary-hugging rather than gross discomfort. The reward function inherently drives the RL agent toward the comfort boundary to maximize energy efficiency. However, the neural barrier's approximation errors can lead to comfort violations during boundary-hugging operation, where the agent operates close to constraint limits. Under thermal stress, this approximation uncertainty becomes more apparent—yet the resulting temperature deviations remain within reasonable bounds when executed by the real system. The RBC deadband approach naturally provides safety margins that prevent such violations and damps cycling. These observations motivate forecast-aware margin tightening and deadband-informed barrier design, as well as limited pre-cooling heuristics or a hybrid fallback to improve summer robustness without sacrificing efficiency.

Table 6 summarizes our experiments by our most relevant metric, the number of safe vs. unsafe states.

Table 6. Summary of experiments: training conditions, duration, and safety outcomes.

Experiment	Ep.	Samp.	Description	TD	#Ep. (Train)	S/U
(1) Without Pre-trained Barrier	2 d	30 m	No pre-trained barrier certificate, random episode generation for a single-zone building in Brussels.	188	94	96/0
(2) Pre-trained Barrier	2 d	30 m	Pre-trained barrier certificate (20 episodes) for a single-zone building in Brussels. Faster convergence, less conservative.	150	75	90/6
(3) Random Episodes	5 d	1 h	Training with diverse, randomly selected episodes for better generalization, for a single-zone building in Brussels.	125	25	116/4
(4) Non-random Episodes	5 d	1 h	Training with repeated identical episodes (same day). Faster convergence but overfitting, for a single-zone building in Brussels.	50	10	120/0
(5) Occupancy-aware	2 d	1 h	Barrier adapts to occupancy: 18–21 °C (occupied), 15–30 °C (unoccupied). Reduces actions in unoccupied periods for a single-zone building in Brussels.	160	80	44/4
(6) 5-zone office	15 d	30 m	Training on a longer episode and a more complex environment with 5 zones in Chicago. The agent controlled the supply air temperature and duct pressure to regulate the temperature for all zones and reduce cost and energy consumption.	1 year	24	720/0
(7) 5-zone office	30 d	30 m	The agent controlled the cooling and heating setpoints for the core zone of a multi-zone office situated in Chicago.	1.2 year	15	1440/0

Abbreviations: Ep.—Episode Length; Samp.—Sampling Interval; Desc.—Description; TD—Training Duration; #Ep. (train)—total training episodes; S/U—Safe vs. Unsafe states with the learned policy during training; All experiments used a fixed seed (12,345) to enable exact replication.

5.5. Research Questions and Hypotheses

After thoroughly discussing our results in the previous section, we will now provide answers to our research questions (cf. Section 1.1). In addition, we will also review how our hypotheses (cf. Section 1.2) compare with the results of our study.

5.5.1. Answering RQ1

The experimental results demonstrate applicability across the tested building control scenarios, including both residential and commercial settings. The approach was implemented in both residential and commercial settings, scaling from a simple single-zone to a complex five-zone office. This scalability was matched by flexibility in control architectures, as the safe RL agent effectively managed direct equipment control through heat pump modulation in the residential experiments (Experiments 1–5), centralized system control via AHU supply temperature and duct pressure setpoints that affect all zones in Experiment 6, and localized control through the core zone's heating and cooling setpoints in Experiment 7. Furthermore, the agent demonstrated remarkable adaptability to varying operational contexts, maintaining fixed comfort bounds of 18–21 °C in basic scenarios (Experiments 1–3), implementing sophisticated occupancy-aware control with dynamic constraints, and managing temperature constraints of 18–24 °C across all zones in the multi-zone building, even when only controlling core zone setpoints or central AHU parameters. This versatility across building types, control levels, and operational requirements suggests that safe RL with neural barrier certificates could be a viable solution for various building automation challenges. While these results are promising, broader validation across different building types, climates, and HVAC systems would be needed to establish general applicability.

5.5.2. Answering RQ2

Based on our experimental findings, pre-training, episode generation strategies, and training duration significantly influence safe RL agent performance. Pre-training the barrier certificate for 20 episodes proved highly beneficial—despite showing higher initial loss values (Figure 12a), pre-trained agents achieved rapid convergence and stabilization after approximately 3000 steps, while non-pre-trained agents continued exhibiting significant oscillations throughout training. This pre-training approach produced less conservative agents that can drive the system to the safe set boundaries to achieve a performance but this may lead to more safety violations (Experiments 1 and 2). The choice between random and non-random episode generation revealed a fundamental trade-off: static-day training achieved safety compliance faster with quicker convergence, but agents trained on randomly generated varied days exhibited superior robustness and generalization capabilities, making them more suitable for real-world deployment where conditions constantly change. Regarding sampling time, our experiments revealed interesting trade-offs in control granularity. Comparing Experiment 1 (30 min sampling) with Experiment 3 (1 h sampling), the more fine-grained control provided the agent with more data points and greater flexibility to regulate temperature precisely. While this finer control resolution could potentially lead to increased energy consumption due to more frequent actuator adjustments and might create more opportunities for safety violations, the less frequent adjustment may lead to safety violations as well due to slower response times when temperature drifts occur. This creates a fundamental trade-off: frequent sampling allows rapid corrections but increases the risk of over-control and oscillations that breach safety bounds, whereas infrequent sampling reduces intervention opportunities but may fail to respond quickly enough to disturbances.

5.5.3. Answering RQ3

The safe RL controller with neural barrier certificates demonstrates improvements over baseline controllers, primarily addressing the critical safety-performance trade-off in HVAC control. Compared to unconstrained PPO, the barrier method achieves dramatic safety improvements—increasing summer comfort compliance from 25.0% to 58.1% and autumn compliance from 85.6% to 99.6%, while accepting modest energy penalties of 11% in summer (3.82 vs. 3.45 kWhm^{-2}) and maintaining comparable costs (USD 0.36 vs. USD 0.33/m^2). Against the RBC, it trades some reliability for adaptability, achieving competitive winter efficiency (2.25 vs. 2.31 kWhm^{-2}) but falling short in spring performance (1.06 vs. 0.79 kWhm^{-2}) and average safety consistency across all seasons (89.4% vs. 96.8%). The barrier approach essentially redefines the performance frontier: rather than universally dominating existing methods, it occupies a crucial middle ground that makes learning-based control deployable by preventing catastrophic comfort violations (avoiding the unacceptable 25% summer safety of unconstrained PPO) while preserving 85–90% of potential energy savings. This positions safe RL with barrier certificates as a practical bridge between the reliability of RBC and the optimization potential of unconstrained learning, making it particularly valuable for real-world applications where both efficiency and guaranteed comfort are non-negotiable requirements.

5.5.4. Comparison of Hypotheses and Findings

Our experimental results supported H1 and allowed us to reject the null hypothesis H0. The safe RL agent achieved competitive energy performance, demonstrating efficiency comparable to the RBC across most seasons, with particularly strong performance in autumn while showing some limitations in spring operations. Overall energy costs remained within competitive ranges throughout the year. In terms of safety performance, the safe RL agent achieved substantially better results than standard PPO across all seasons, with the most dramatic improvements observed during summer, when unconstrained PPO exhibited catastrophic comfort violations. While not matching the near-perfect safety record of RBC, the barrier certificates effectively enforced comfort constraints and prevented the severe violations observed in unconstrained learning. We reject H0 and accept H1 with the clarification that safe RL achieves safety performance significantly better than standard PPO while maintaining energy efficiency comparable to RBC, successfully positioning itself as a viable middle ground between aggressive optimization and conservative rule-based approaches.

5.6. *Key Findings and Lessons Learned*

Our research demonstrates that safe RL capitalizing on neural barrier certificates can be effectively implemented for HVAC control applications. One critical insight gained concerns the exploration trade-off: if an agent explores insufficiently, it typically results in overly conservative behavior that prioritizes safety at the expense of energy efficiency. Conversely, excessive exploration can produce an agent with aggressive behavior focused primarily on reward minimization. In our pre-trained agent implementation, we observed less conservative behavior due to the initial pre-training phase, but allowing this agent to explore the action space too extensively led to safety constraint violations. While our approach achieved near-zero violations during evaluation, we found that violations can never be completely eliminated due to inherent approximation errors in neural barrier certificates. This limitation highlights the practical challenges in providing absolute safety guarantees when using neural networks to implement safety constraints. Additionally, our occupancy-aware safe RL agent successfully maintained different thermal comfort bounds based on occupancy status, demonstrating that incorporating contextual information can significantly improve agent behavior. This contextual awareness enables more differentiated control strategies that minimize energy efficiency during unoccupied periods and maintain thermal comfort during occupied periods while optimizing energy consumption at the same time.

Our comparative analysis revealed several critical insights about safe RL deployment in HVAC control. First, seasonal variations fundamentally alter the performance landscape—while RBCs excel during mild spring conditions due to well-tuned fixed parameters, learning-based approaches demonstrate clear advantages during transitional seasons where adaptive behavior is crucial. The barrier certificate approach proved more effective than reward shaping alone, preventing the catastrophic comfort violations that plague unconstrained PPO, particularly during extreme summer conditions where standard RL completely fails to maintain acceptable comfort. Conversely, winter conditions revealed the opposite dynamic—all controllers achieved perfect safety compliance due to wider comfort bounds typical of heating seasons, exposing the barrier function's stabilizing effect rather than conservatism. During these cold months, both PPO variants achieved similar energy consumption, but the unconstrained PPO exhibited higher variability and less predictable control behavior when exploiting its unrestricted action space, while the barrier-constrained variant maintained tighter operational bounds through its safety-first approach. This seasonal asymmetry highlights that agents trained with hard constraints inherently operate more consistently than their unconstrained counterparts, prioritizing operational stability and guaranteed safety compliance over aggressive optimization strategies even when environmental conditions pose minimal risk. However, our results also exposed the limitations of neural barrier approximations, which cannot match the near-perfect safety guarantees of rule-based systems whose strict, hardcoded rules ensure consistent comfort delivery, though at the expense of adaptability. However, the model-free nature of our barrier-based approach offers significant practical advantages, requiring no explicit system models or manual tuning while maintaining the ability to learn and adapt to building-specific dynamics, suggesting that achieving both the learnability of RL agents and the absolute safety of deterministic controllers remains an open challenge.

Table 6 highlights the effect of different training settings, with an emphasis on training length and the number of safe vs. risky states encountered. The results show that pre-training the barrier certificate (Experiment 2) increases convergence speed and decreases dangerous conditions as compared to training without it (Experiment 1). Randomly produced episodes (Experiment 3) increase generalization but result in more safety violations than repeated episodes (Experiment 4), which have faster convergence but are prone to overfitting. The occupancy-aware approach (Experiment 5) adapts successfully to various comfort restrictions, decreasing energy consumption and needless control operations. These results highlight the trade-off between exploration, safety adherence, and policy optimization in safe RL. In the final sets of experiments (6 and 7), we show that the agent can be used in more complex environments. Moreover, the comparative study showcased the effectiveness of the agent compared to baseline controllers.

5.7. Practical Implications

This paper addresses the critical building management challenge of ensuring occupant comfort while minimizing energy consumption. Our approach implements safe RL using neural barrier certificates, which fundamentally reframes the control problem—comfort requirements become explicit hard constraints rather than weighted reward components, eliminating the complex tuning process that plagues traditional RL deployments [21,24]. For practitioners, this means a more principled and interpretable control strategy that learns building-specific dynamics directly from operational data, adapting to seasonal variations and occupancy patterns without requiring mathematical models. The barrier approach substantially improves upon unconstrained learning by preventing catastrophic comfort violations, though our results reveal important limitations: the method requires significant training data and computational resources, and neural approximation errors prevent absolute safety guarantees, achieving 58.1–100% comfort compliance depending on seasonal conditions. Despite these constraints, the approach offers a viable pathway toward deployable learning-based HVAC control, particularly in applications where the balance between automation benefits and occasional comfort deviations is acceptable.

Additionally, this work demonstrated the applicability of safe RL across different control complexities, from system-level AHU control to zone-specific setpoint management. Our findings showcase the practical flexibility of the safe RL approach in real-world building automation systems. The agent successfully managed both centralized AHU parameters that affect all zones simultaneously, demonstrating the capability for system-wide optimization, and localized heating and cooling setpoints for individual zones, showing effectiveness in zone-level control strategies. This flexibility allows building operators to deploy safe RL at different control hierarchies depending on their existing infrastructure.

Building characteristics and sensor quality influence performance, potentially requiring facility-specific calibration. While the study results emphasize the energy benefits of implementing the learning system in a legacy building, the effectiveness of using RL for newly constructed, highly energy-efficient buildings still needs to be tested and quantified. Thermal insulation, thermal bridge-free construction, ventilation with heat recovery, and airtightness significantly reduce the heat demand of a building and cause longer time constants that require longer episodes for training. For this reason, future work will vary the length of the episodes and evaluate them as a function of different building energy efficiency standards. With improved building envelopes, ensuring summer comfort through shading becomes more important. The learning system could therefore be adapted to other trades.

From a deployment perspective, these summer results indicate that barrier-based RL, while promising, requires further development before matching traditional control reliability in extreme conditions. The 31 percentage point safety gap suggests that current barrier methods may be suitable for moderate climates or as auxiliary controllers but not yet as standalone replacements for rule-based systems in challenging thermal conditions.

Although there is significant potential for energy optimization and comfort maintenance, practical deployment considerations must be carefully addressed. Integrating neural network-based controllers with existing building management systems might be challenging due to limited interfaces and processing capability. Implementing edge computing to process sensor data and execute trained policies locally may incur additional hardware costs beyond initial training fees. Building operators may need a transition time with human oversight to check the system's judgments before complete automation, potentially requiring dual-control systems throughout installation. When comparing the cost of implementation to predicted energy savings, organizations should consider integration requirements, staff training, and any disruptions during installation. Our study aims to improve energy efficiency and occupant comfort by modernizing outdated building infrastructure using data-driven methods. Our technique provides non-invasive energy-efficiency enhancements and ensures a high level of comfort.

Despite the promising results of our safe RL controller, it is crucial to acknowledge that building hull design remains a fundamental determinant of energy performance. The building envelope establishes the inherent energy efficiency potential that no control system can transcend. These architectural elements are primarily optimized during the design and construction phases, creating the baseline thermal dynamics within which any control system must operate. Our work is thus complementary to good building design: while the hull determines the fundamental energy demand characteristics, our intelligent control approach addresses the last mile of optimization–ensuring that whatever efficiency potential the building possesses is fully realized through adaptive, occupancy-aware control strategies.

6. Conclusions

This work demonstrated that safe RL effectively transforms multi-objective HVAC control into a single-objective problem with safety constraints, achieving near-perfect thermal comfort compliance while optimizing energy consumption without complex reward functions. Our neural barrier certificate provides a data-driven method to enforce comfort constraints without explicit thermal modeling through data collection. Through comprehensive experimental evaluation, we successfully addressed all three research questions posed in this study. Our findings demonstrated the broad applicability and effectiveness of safe RL with neural barrier certificates across diverse building control scenarios (RQ1), revealed the impact of pre-training and episode generation strategies on agent performance (RQ2), and confirmed that safe RL achieves competitive performance compared to established baseline controllers (RQ3). These results enabled us to reject the null hypothesis and validate our approach as a viable alternative for building climate control. Our work delivers four key contributions to the field: we demonstrated the applicability of safe RL with neural barrier certificates in building control, validating its effectiveness for handling thermal comfort constraints while optimizing energy consumption across both residential and commercial buildings. We provided empirical evidence of the method's performance across multiple BOPTEST test cases, different control levels (from direct equipment control to AHU parameters), and varying episode lengths and learning intervals. Our experiments showed that pre-trained barrier functions improve learning efficiency and produce less conservative policies compared to non-pretrained approaches. We established that safe RL occupies a middle ground in the control landscape, accepting modest energy penalties compared to unconstrained PPO and RBCs in exchange for dramatically improved safety over standard RL, though still falling short of rule-based safety performance.

Our experiments revealed that pre-trained barrier certificates can achieve better performance while maintaining the comfort bounds, though violations cannot be reduced to zero due to the neural barrier certificate's inherent approximation error. Additionally, our occupancy-aware framework dynamically adjusts comfort constraints based on building occupancy, demonstrating better energy savings while maintaining comfort during occupied periods. Moreover, our experiments on complex multi-zone scenarios demonstrated that the agent is capable of handling complex environments while achieving acceptable safety performance. Through comparative study, the agent traded energy efficiency for improved safety compliance compared to unconstrained RL, achieving this balance through a model-free approach without requiring explicit system models, though safety performance still lags behind traditional RBC. Across seasons, the comparative results show that in winter—when all controllers satisfy comfort—the barrier-PPO and PPO exhibit identical median energy (2.25 $\mathrm{kWhm^{-2}}$), with the barrier variant offering steadier operation; in spring and autumn, the safe agent achieves near-perfect compliance (99.9%/99.6%) while unconstrained PPO falls short (71.1%/85.6%); and in summer, while unconstrained PPO uses 10% less energy, its 25.0% compliance renders it impractical; barrier-PPO raises safety to 58.1% but still trails RBC, which achieves both superior safety (89.3%) and energy efficiency. These outcomes suggest that safety-aware RL is a promising middle ground—delivering substantial safety gains over unconstrained RL with competitive efficiency—while further refinement is needed to match rule-based reliability under extreme conditions.

These contributions address key challenges in sustainable building management by enforcing comfort as a hard constraint while optimizing for energy efficiency. This approach has direct applications in intelligent building management systems where occupant preferences vary over time. While we focused on HVAC control, future research should address additional dimensions such as lighting and plumbing (e.g., hot water) while considering broader and potentially adaptive safety aspects. Additionally, extending the barrier certificate framework to handle multi-modal comfort preferences and incorporating human feedback mechanisms could make these systems more responsive to actual occupant needs while maintaining energy efficiency goals. Despite promising results and achieving acceptable performance, safe RL in building domain still requires more advancements in sample efficiency, robustness, and computational efficiency before becoming widely applicable in real-world building management systems.

Author Contributions: Conceptualization, M.E., S.H. and P.Z.; Data curation, M.E., S.H. and S.T.; Formal analysis, M.E., S.H., S.T. and P.Z.; Funding acquisition, S.H. and P.Z.; Investigation, M.E., S.H. and D.G.-M.; Methodology, M.E., S.H., S.T., D.G.-M. and P.Z.; Project administration, S.H., D.G.-M. and P.Z.; Software, M.E.; Supervision, S.T., D.G.-M. and P.Z.; Validation, M.E., S.H. and S.T.; Visualization, M.E.; Writing—original draft, M.E. and S.H.; Writing—review and editing, S.T., D.G.-M. and P.Z. All authors have read and agreed to the published version of the manuscript.

Funding: The research leading to these results has received funding from the Austrian Research Promotion Agency (FFG) under grant agreement No. 910225.

Data Availability Statement: Publicly available datasets were analyzed in this study. The underlying building simulation test cases are provided by the BOPTEST project (BESTEST Hydronic Heat Pump and Multi-zone Office—Simple Air). All code, configuration files, and scripts to reproduce the training/evaluation splits and regenerate every table and figure are openly available at https://github.com/MMDes447/HVAC_SAFE_RL_.git (accessed on 6 October 2025).

Conflicts of Interest: The authors declare no conflicts of interest.

References

1. *ISO 7730*; Moderate Thermal Environments—Determination of the PMV and PPD Indices and Specification of the Conditions for Thermal Comfort. ISO: Geneva, Switzerland, 1994.
2. *ASHRAE Standard 55-2004*; Thermal Environmental Conditions for Human Occupancy. American Society of Heating, Refrigerating and Air-Conditioning Engineers, Inc.: Atlanta, GA, USA, 2004.
3. Ole Fanger, P.; Toftum, J. Extension of the PMV model to non-air-conditioned buildings in warm climates. *Energy Build.* **2002**, *34*, 533–536. [CrossRef]
4. Zhao, Q.; Lian, Z.; Lai, D. Thermal comfort models and their developments: A review. *Energy Built Environ.* **2021**, *2*, 21–33. [CrossRef]
5. Was, K.; Radoń, J.; Sadlowska-Salega, A. Thermal Comfort—Case Study in a Lightweight Passive House. *Energies* **2022**, *15*, 4687. [CrossRef]
6. Olsen, B.W. Indoor environmental input parameters for the design and assessment of energy performance of buildings. *REHVA J.* **2015**, *52*, 17–23.
7. Stevens, J.C.; Choo, K.K. Temperature sensitivity of the body surface over the life span. *Somatosens. Mot. Res.* **1998**, *15*, 13–28. [CrossRef]
8. Karjalainen, S. Thermal comfort and gender: A literature review. *Indoor Air* **2011**, *22*, 96–109. [CrossRef]
9. Schaudienst, F.; Vogdt, F.U. Fanger's model of thermal comfort: A model suitable just for men? *Energy Procedia* **2017**, *132*, 129–134. [CrossRef]
10. Albatayneh, A.; Jaradat, M.; AlKhatib, M.B.; Abdallah, R.; Juaidi, A.; Manzano-Agugliaro, F. The Significance of the Adaptive Thermal Comfort Practice over the Structure Retrofits to Sustain Indoor Thermal Comfort. *Energies* **2021**, *14*, 2946. [CrossRef]
11. Hoof, J.V. Forty years of Fanger's model of thermal comfort: Comfort for all? *Indoor Air* **2008**, *18*, 182–201. [CrossRef]
12. Nikam, S.; Bartaria, V. Indoor Environment in Air Conditioned Spaces a Review. *Int. J. Adv. Technol. Eng. Res.* **2012**, *2*, 226–230.
13. Yardimli, S.; Güleç Özer, D.; Shahriary, A. Sustainable street architecture and its effects on human comfort conditions: Yazd, Iran. *A/Z ITU J. Fac. Archit.* **2020**, *17*, 113–122.

14. Papadopoulos, S.; Kontokosta, C.E.; Vlachokostas, A.; Azar, E. Rethinking HVAC temperature setpoints in commercial buildings: The potential for zero-cost energy savings and comfort improvement in different climates. *Build. Environ.* **2019**, *155*, 350–359. [CrossRef]
15. Laustsen, J. *Energy Efficiency Requirements in Building Codes, Energy Efficiency Policies for New Buildings*; IEA International Energy Agency: Paris, France, 2008.
16. UN Environment Programme. *Global Status Report for Buildings and Construction: Beyond Foundations: Mainstreaming Sustainable Solutions to Cut Emissions from the Buildings Sector*; GlobalABC/UNEP: Paris, France, 2024.
17. Ürge Vorsatz, D.; Cabeza, L.F.; Serrano, S.; Barreneche, C.; Petrichenko, K. Heating and cooling energy trends and drivers in buildings. *Renew. Sustain. Energy Rev.* **2015**, *41*, 85–98. [CrossRef]
18. Zini, M.; Carcasci, C. Machine learning-based energy monitoring method applied to the HVAC systems electricity demand of an Italian healthcare facility. *Smart Energy* **2024**, *14*, 100137. [CrossRef]
19. Klepeis, N.E.; Nelson, W.C.; OTT, W.R.; Robinson, J.P.; Tsang, A.M.; Switzer, P.; Behar, J.V.; Hern, S.C.; Engelmann, W.H. The National Human Activity Pattern Survey (NHAPS): A resource for assessing exposure to environmental pollutants. *J. Expo. Sci. Environ. Epidemiol.* **2001**, *11*, 231–252. [CrossRef]
20. Sutton, R.S.; Barto, A.G. *Reinforcement Learning: An Introduction*; MIT Press: Cambridge, MA, USA, 2018.
21. Shaqour, A.; Hagishima, A. Systematic review on deep reinforcement learning-based energy management for different building types. *Energies* **2022**, *15*, 8663. [CrossRef]
22. Sierla, S.; Ihasalo, H.; Vyatkin, V. A review of reinforcement learning applications to control of heating, ventilation and air conditioning systems. *Energies* **2022**, *15*, 3526. [CrossRef]
23. Bartenbach GmbH; Zumtobel Group; Hella GmbH; University of Innsbruck. *BOREALIS: Self-Learning Building Controls for a Greener and Healthier Society*; Project no. 5121377; FFG Österreichische Forschungsförderungsgesellschaft mbH: Vienna, Austria, 2024. Available online: https://projekte.ffg.at/projekt/5121377 (accessed on 28 July 2025).
24. Hanif, M.F.; Ahmad, R.; Farooq, A.B.; Liu, X.; Ahmad, A.; Iqbal, S.; Mi, J. The Solar AI Nexus: Reinforcement Learning Shaping the Future of Energy Management. *Wires Energy Environ.* **2025**, *14*, e70012. [CrossRef]
25. Park, J.Y.; Ouf, M.M.; Gunay, B.; Peng, Y.; O'Brien, W.; Kjærgaard, M.B.; Nagy, Z. A critical review of field implementations of occupant-centric building controls. *Build. Environ.* **2019**, *165*, 106351. [CrossRef]
26. Yang, Y.; Jiang, Y.; Liu, Y.; Chen, J.; Li, S.E. Model-free safe reinforcement learning through neural barrier certificate. *IEEE Robot. Autom. Lett.* **2023**, *8*, 1295–1302. [CrossRef]
27. Ames, A.D.; Grizzle, J.W.; Tabuada, P. Control barrier function based quadratic programs with application to adaptive cruise control. In Proceedings of the 53rd IEEE Conference on Decision and Control, Los Angeles, CA, USA, 15–17 December 2014; pp. 6271–6278.
28. Andersen, K.; Clevenger, T., Jr. A summary of experimental research in ethos. *Commun. Monogr.* **1963**, *30*, 59–78. [CrossRef]
29. Blum, D.; Arroyo, J.; Huang, S.; Drgoňa, J.; Jorissen, F.; Walnum, H.T.; Chen, Y.; Benne, K.; Vrabie, D.; Wetter, M.; et al. Building optimization testing framework (BOPTEST) for simulation-based benchmarking of control strategies in buildings. *J. Build. Perform. Simul.* **2021**, *14*, 586–610. [CrossRef]
30. Dounis, A.I.; Bruant, M.; Santamouris, M.; Guarracino, G.; Michel, P. Comparison of conventional and fuzzy control of indoor air quality in buildings. *J. Intell. Fuzzy Syst.* **1996**, *4*, 131–140. [CrossRef]
31. Levermore, G. *Building Energy Management Systems: An Application to Heating and Control*; Routledge, London, UK, 1992.
32. Soyguder, S.; Karakose, M.; Alli, H. Design and simulation of self-tuning PID-type fuzzy adaptive control for an expert HVAC system. *Expert Syst. Appl.* **2009**, *36*, 4566–4573. [CrossRef]
33. Touqan, B.; Abdul-Ameer, A.; Salameh, M. HVAC multivariable system modelling and control. *Proc. Inst. Mech. Eng. Part C J. Mech. Eng. Sci.* **2023**, *237*, 2049–2061. [CrossRef]
34. Kintner-Meyer, M.; Emery, A.F. Optimal control of an HVAC system using cold storage and building thermal capacitance. *Energy Build.* **1995**, *23*, 19–31. [CrossRef]
35. Komareji, M.; Stoustrup, J.; Rasmussen, H.; Bidstrup, N.; Svendsen, P.; Nielsen, F. Optimal model-based control in HVAC systems. In Proceedings of the 2008 American Control Conference, Seattle, WA, USA, 1–13 June 2008; pp. 1443–1448.
36. Yang, Y.; Hu, G.; Spanos, C.J. Stochastic optimal control of HVAC system for energy-efficient buildings. *IEEE Trans. Control Syst. Technol.* **2021**, *30*, 376–383. [CrossRef]
37. Shepherd, A.; Batty, W. Fuzzy control strategies to provide cost and energy efficient high quality indoor environments in buildings with high occupant densities. *Build. Serv. Eng. Res. Technol.* **2003**, *24*, 35–45. [CrossRef]
38. Calvino, F.; La Gennusa, M.; Rizzo, G.; Scaccianoce, G. The control of indoor thermal comfort conditions: Introducing a fuzzy adaptive controller. *Energy Build.* **2004**, *36*, 97–102. [CrossRef]
39. Anuntasethakul, C.; Banjerdpongchai, D. Design of supervisory model predictive control for building HVAC system with consideration of peak-load shaving and thermal comfort. *IEEE Access* **2021**, *9*, 41066–41081. [CrossRef]

40. Joe, J. Investigation on pre-cooling potential of UFAD via model-based predictive control. *Energy Build.* **2022**, *259*, 111898. [CrossRef]
41. Zeng, T.; Barooah, P. An adaptive model predictive control scheme for energy-efficient control of building HVAC systems. *J. Eng. Sustain. Build. Cities* **2021**, *2*, 031001. [CrossRef]
42. Lautenschlager, B.; Lichtenberg, G. Data-driven iterative learning for model predictive control of heating systems. *IFAC-PapersOnLine* **2016**, *49*, 175–180. [CrossRef]
43. Maddalena, E.T.; Lian, Y.; Jones, C.N. Data-driven methods for building control—A review and promising future directions. *Control Eng. Pract.* **2020**, *95*, 104211. [CrossRef]
44. Gao, G.; Li, J.; Wen, Y. Energy-efficient thermal comfort control in smart buildings via deep reinforcement learning. *arXiv* **2019**, arXiv:1901.04693. [CrossRef]
45. Li, F.; Du, Y. Intelligent multi-zone residential HVAC control strategy based on deep reinforcement learning. In *Deep Learning for Power System Applications: Case Studies Linking Artificial Intelligence and Power Systems*; Springer International Publishing AG, Cham: Switzerland, 2023; pp. 71–96.
46. Azuatalam, D.; Lee, W.L.; de Nijs, F.; Liebman, A. Reinforcement learning for whole-building HVAC control and demand response. *Energy AI* **2020**, *2*, 100020. [CrossRef]
47. Zhong, X.; Zhang, Z.; Zhang, R.; Zhang, C. End-to-end deep reinforcement learning control for HVAC systems in office buildings. *Designs* **2022**, *6*, 52. [CrossRef]
48. An, Y.; Chen, C. Energy-efficient control of indoor PM2. 5 and thermal comfort in a real room using deep reinforcement learning. *Energy Build.* **2023**, *295*, 113340. [CrossRef]
49. Liu, X.; Gou, Z. Occupant-centric HVAC and window control: A reinforcement learning model for enhancing indoor thermal comfort and energy efficiency. *Build. Environ.* **2024**, *250*, 111197. [CrossRef]
50. Zenginis, I.; Vardakas, J.; Koltsaklis, N.E.; Verikoukis, C. Smart home's energy management through a clustering-based reinforcement learning approach. *IEEE Internet Things J.* **2022**, *9*, 16363–16371. [CrossRef]
51. Lork, C.; Li, W.T.; Qin, Y.; Zhou, Y.; Yuen, C.; Tushar, W.; Saha, T.K. An uncertainty-aware deep reinforcement learning framework for residential air conditioning energy management. *Appl. Energy* **2020**, *276*, 115426. [CrossRef]
52. Chen, B.; Cai, Z.; Bergés, M. Gnu-rl: A practical and scalable reinforcement learning solution for building hvac control using a differentiable mpc policy. *Front. Built Environ.* **2020**, *6*, 562239. [CrossRef]
53. Nguyen, A.T.; Pham, D.H.; Oo, B.L.; Santamouris, M.; Ahn, Y.; Lim, B.T. Modelling building HVAC control strategies using a deep reinforcement learning approach. *Energy Build.* **2024**, *310*, 114065. [CrossRef]
54. Solinas, F.M.; Macii, A.; Patti, E.; Bottaccioli, L. An online reinforcement learning approach for HVAC control. *Expert Syst. Appl.* **2024**, *238*, 121749. [CrossRef]
55. Liu, X.; Ren, M.; Yang, Z.; Yan, G.; Guo, Y.; Cheng, L.; Wu, C. A multi-step predictive deep reinforcement learning algorithm for HVAC control systems in smart buildings. *Energy* **2022**, *259*, 124857. [CrossRef]
56. Kadamala, K.; Chambers, D.; Barrett, E. Enhancing HVAC control systems through transfer learning with deep reinforcement learning agents. *Smart Energy* **2024**, *13*, 100131. [CrossRef]
57. Shin, M.; Kim, S.; Kim, Y.; Song, A.; Kim, Y.; Kim, H.Y. Development of an HVAC system control method using weather forecasting data with deep reinforcement learning algorithms. *Build. Environ.* **2024**, *248*, 111069. [CrossRef]
58. Ding, X.; An, Z.; Rathee, A.; Du, W. A Safe and Data-Efficient Model-Based Reinforcement Learning System for HVAC Control. *IEEE Internet Things J.* **2025**, *12*, 8014–8032. [CrossRef]
59. Chen, L.; Meng, F.; Zhang, Y. Fast human-in-the-loop control for HVAC systems via meta-learning and model-based offline reinforcement learning. *IEEE Trans. Sustain. Comput.* **2023**, *8*, 504–521. [CrossRef]
60. Cui, C.; Xue, J.; Liu, L. Optimal control of HVAC systems through active disturbance rejection control-assisted reinforcement learning. *Energy* **2025**, *323*, 135824. [CrossRef]
61. Gehbauer, C.; Rippl, A.; Lee, E. Advanced control of dynamic facades and HVAC with reinforcement learning based on standardized co-simulation. In Proceedings of the Building Simulation 2021, Bruges, Belgium, 1–3 September 2021; Volume 17, pp. 231–238.
62. Xue, W.; Jia, N.; Zhao, M. Multi-agent deep reinforcement learning based HVAC control for multi-zone buildings considering zone-energy-allocation optimization. *Energy Build.* **2025**, *329*, 115241. [CrossRef]
63. Li, Z.; Sun, Z.; Meng, Q.; Wang, Y.; Li, Y. Reinforcement learning of room temperature set-point of thermal storage air-conditioning system with demand response. *Energy Build.* **2022**, *259*, 111903. [CrossRef]
64. Yu, J.; Zhang, T.; Ardakanian, O.; Wierman, A. Online Comfort-Constrained HVAC Control via Feature Transfer. In Proceedings of the 16th ACM International Conference on Future and Sustainable Energy Systems, Rotterdam, The Netherlands, 17–20 June 2025; pp. 371–384.
65. Bellman, R. Dynamic programming. *Science* **1966**, *153*, 34–37. [CrossRef]

66. Schulman, J.; Moritz, P.; Levine, S.; Jordan, M.; Abbeel, P. High-dimensional continuous control using generalized advantage estimation. *arXiv* **2015**, arXiv:1506.02438.
67. Schulman, J.; Wolski, F.; Dhariwal, P.; Radford, A.; Klimov, O. Proximal policy optimization algorithms. *arXiv* **2017**, arXiv:1707.06347. [CrossRef]
68. Fanger, P.O. *Thermal Comfort: Analysis and Applications in Environmental Engineering;* Danish Technical Press: Copenhagen, Denmark, 1970.
69. Arroyo, J.; Manna, C.; Spiessens, F.; Helsen, L. An open-ai gym environment for the building optimization testing (boptest) framework. In Proceedings of the Building Simulation 2021, Bruges, Belgium, 1–3 September 2021; Volume 17, pp. 175–182.
70. Manjavacas, A.; Campoy-Nieves, A.; Jiménez-Raboso, J.; Molina-Solana, M.; Gómez-Romero, J. An experimental evaluation of deep reinforcement learning algorithms for HVAC control. *Artif. Intell. Rev.* **2024**, *57*, 173. [CrossRef]
71. Adcock, B.; Dexter, N. The gap between theory and practice in function approximation with deep neural networks. *SIAM J. Math. Data Sci.* **2021**, *3*, 624–655. [CrossRef]

energies

MDPI

Article

Assessing the Thermal Storage Potential of Timber and Hybrid Activated Slabs: A Simulation-Based Comparison of Different Construction Types

Andrea Agner and Doris Österreicher *

Institute for Building Materials, Building Physics, Building Systems and Design, University of Stuttgart, 70174 Stuttgart, Germany; andrea.agner@ibbte.uni-stuttgart.de
* Correspondence: doris.oesterreicher@ibbte.uni-stuttgart.de; Tel.: +49-711-685-83231

Abstract

Thermally activated building systems (TABS) rely on high thermal mass materials, such as concrete, which perform well thermally but have a high carbon footprint. This study systematically investigates the thermal behavior of bio-based materials—spruce, pine, beech, and oak—in TABS using numerical simulations, comparing them with conventional and hybrid materials like concrete and clay. A total of 120 variants were simulated with different pipe diameters, spacing, embedment depths, and inlet temperatures. Thermal properties, particularly thermal conductivity and specific heat capacity, significantly influenced component activation efficiency. Concrete exhibited a characteristic cooling time of 71 h at an inlet temperature of 26 °C (pipe diameter 16 mm), while pine reached 80 h under the same conditions. The use of capillary tube mats extended the cooling times to 75 h for concrete and 92 h for pine. Although concrete provides the best thermal performance, certain bio-based materials achieve comparable results under optimized conditions. Hybrid systems with mineral components offer additional potential for improvement. These findings demonstrate that ecologically sustainable component activation using bio-based materials is feasible with only moderate efficiency losses compared to mineral-based systems, provided system parameters are appropriately adapted.

Keywords: embedded pipes; TABS; load shifting; thermal storage capacity; numeric simulation; bio-based materials; timber; building mass; radiant heating systems

1. Introduction

In recent years, thermally activated building systems (TABS) have emerged as a promising concept for efficiently providing heating and cooling energy to buildings. The thermal mass of building components, such as ceilings and walls, is utilized to regulate room temperature by incorporating water-carrying pipes into these components. This not only enables uniform temperature control but also opens up the potential for shifting thermal loads—an increasingly important aspect, particularly in the context of renewable energies and fluctuating electricity generation [1,2].

A key feature of TABS is the high thermal mass and thus the ability of the component to store thermal energy over extended periods and release it with a time delay. This property can be exploited to shift energy consumption from periods of high demand to periods of surplus ('load shifting')—for instance, when using solar power or with a tariff-dependent electricity supply.

The majority of research into TABS to date has focused on conventional mineral building materials, such as concrete and masonry. Whilst these materials are ideal for thermal activation due to their high density, thermal conductivity and heat storage capacity, their carbon footprint is, compared to wood or similar bio-based materials, considerably larger [3,4]. With the increase in efficiency during the operational phase of the building and subsequent lower CO_2 emissions during the building's use, the embodied energy of constructing the building becomes ever more important [5,6].

Romani et al. [7] conducted a comprehensive literature review that classifies different types of construction of TABS, especially those with pipes embedded in ceilings or floors, and systematically describes different simulation methods, control strategies, and results from parametric studies. However, the focus is almost exclusively on concrete and masonry constructions.

In addition, Villar-Ramos et al. conducted a systematic literature review, with a particular focus on the potential of TABS to enhance thermal comfort and reduce final energy consumption. The studies published between 2015 and 2022 concentrated on the integration of TABS in various building components (roof, wall, floor, building envelope) and their combination with complementary technologies such as solar collectors, phase change materials or heat pumps. The findings indicate that the implementation of TABS technology can lead to a substantial reduction in indoor air temperature, with potential energy savings of up to 50% being realized [8].

The energy performance of external walls has been improved by investigating wall structures with integrated pipe systems. Several studies have shown that integrating pipework into concrete walls can improve the insulation effect [9–13]. Other studies have focused on concepts for shifting thermal energy within the building envelope. For example, solar heat gains can be transferred from south-facing façades to cooler north-facing façades [14–16]. While these approaches demonstrate the potential of TABS beyond classic floor or ceiling systems, the materials used in these studies are all mineral-based [17–26].

In addition, the influence of different pipe geometries on thermal behavior was investigated. Studies show that capillary tube mats installed close to the surface have a faster reaction time, enable more homogeneous heat distribution, and can be operated at lower inlet temperatures [27]. A further study compared six different panel types in order to identify the optimum combination of heat transfer performance and material efficiency in lightweight ceiling modules with capillary tubes [28].

Another relevant aspect is the influence of the surface material. For instance, a study by Rodríguez Jara et al. [29] demonstrates that, under certain climatic conditions, wood can attain surface temperatures comparable to granite, despite having significantly inferior thermal properties. These results are particularly important given the increasing demand for sustainable materials.

In response to growing ecological concerns related to CO_2-intense materials such as cement and concrete, bio-based building materials such as wood and bamboo are becoming increasingly popular. These materials are considered promising alternatives to concrete thanks to their high load-bearing capacity, lower carbon footprint, the potential for regional sourcing, and their relatively low environmental impact during production and disposal. However, from a thermal engineering perspective, they pose challenges as they generally have lower thermal conductivity and heat capacity. Some initial scientific research has already addressed this issue: for example, Bishara et al. [30] experimentally investigated different types of heating pipe installation in wooden components and analyzed their effect on heat distribution. Heidenthaler et al. [31] carried out numerical simulations comparing different types of wood, embedment depths and inlet temperatures with concrete references.

The results show that, under certain boundary conditions, timber components can have a high thermal activation potential.

In addition to water-based systems, the transmission properties of alternative media have recently been investigated. In a study by Mindrup [32], an air-based wall module with integrated microchannels was investigated. Activation takes place through a hot air flow, which could be of particular interest for lightweight construction systems but which poses other challenges in terms of control behavior, efficiency and comfort.

Despite these developments, research into bio-based materials for thermal component activation is still in its infancy. In particular, systematic comparisons of different types of wood and other bio-based materials, such as bamboo, under practical boundary conditions are lacking. The thermal storage potential of these materials in the context of load shifting and system inertia has not yet been adequately analyzed. Hybrid material approaches, which could combine ecological advantages with improved heat conduction, have also not yet been extensively analyzed.

This work aims to systematically investigate the thermal behavior of bio-based materials in component activation systems. Various influencing factors, including material selection, pipe spacing, embedding depth, inlet temperature and heat storage capacity, are analyzed using numerical simulations. The focus is on the potential for thermal load shifting, paying particular attention to the suitability of bio-based materials such as wood (e.g., spruce, pine, beech and oak) as well as bamboo.

The results will demonstrate, in an initially solely simulation-based step, the extent to which bio-based building materials can potentially substitute mineral-based components without significant thermal component activation efficiency losses. Additionally, this study will examine the functional advantages that hybrid material combinations offer over homogeneous materials.

2. Methodology

The following methodology was applied to systematically investigate the thermal behavior of the various materials and systems. In this initial analysis, the focus is on floor slabs, as ceilings are the most commonly used building components for thermally activated building systems [7,20]. Firstly, the slab structure is described, followed by an overview of the simulation tool and the parameters that were varied during the investigation. Finally, the assessment and evaluation parameters are presented in detail.

The thermal simulations were carried out using a parametric approach with the Delphin 6 software. Delphin is a hygrothermal simulation tool developed by the Chair of Building Climatology (IBK) at the Technical University of Dresden [33]. It enables the detailed analysis of coupled heat and moisture transport processes in building components under realistic climatic conditions.

As this study focuses particularly on analyzing different types of wood, a solid timber ceiling construction was chosen as initial geometry. This consists of a 16 cm-thick wooden layer. Above this are layers of backfill (10 cm), insulation (3 cm) and screed (7 cm). The pipes are embedded in the wooden layer. This layer structure is also shown schematically in Figure 1.

Several parameters were systematically varied as part of the simulation in order to analyze their influence on the thermal performance of the structure. All 120 cases were carried out as independent simulations without optimization algorithms. The applied methodology is generalizable, as the parametric framework in Delphin 6 allows the variation in arbitrary material properties, geometries, and boundary conditions. This makes the approach transferable to a wide range of construction types and operating scenarios. The parameters analyzed include and can be found in Figure 2.

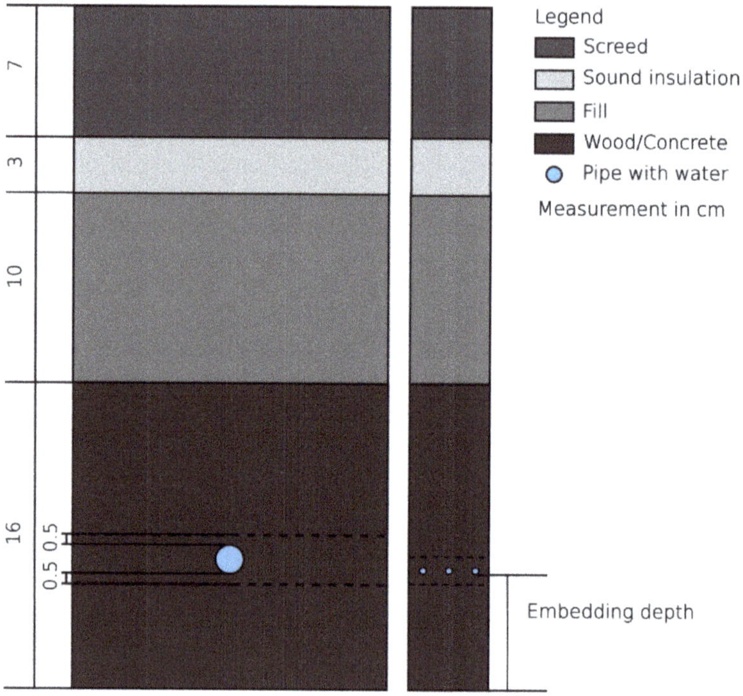

Figure 1. Cross-section of the two ceilings. Left with a pipe with a diameter of 16 mm and a pipe spacing of 15 cm. Right with capillary tubes (diameter 4.3 mm) and 1 cm pipe spacing.

- Pipe diameter and pipe spacing: These two parameters were considered together as they significantly affect heat distribution and release in the component.
- Embedding depth: The vertical position of the pipes within the wood layer was varied to analyze the effect of different heat conduction paths on surface temperature and heat transfer efficiency.
- Inlet temperature of the heating medium: Two different inlet temperatures were used to evaluate thermal behavior in various operating scenarios.
- Use of different building materials: Concrete was used as the reference material. Additionally, four types of wood, a grass-based material, and four hybrid variants combining concrete or clay were analyzed to record differences in thermal performance dependent on the material used.
- The simulation followed a structured workflow, as illustrated in Figure 3. Starting from the research objective, the geometry of the slab layers was set up, and material properties were assigned. The parameters were systematically defined for each simulation case. Thermal simulations in Delphin 6 included both stationary analyses (temperature distributions) and transient analyses (heat storage capacity). Evaluation focused on surface temperatures, temperatures between the pipes and heat flux. Finally, the results were compared and discussed to highlight differences between bio-based and mineral materials and identify the most influential factors.
- Notably, interdependencies exist among the studied parameters: pipe diameter and spacing jointly influence heat distribution; embedding depth interacts with the thermal conductivity of the material, affecting heat conduction pathways; and material selection modifies temperature distribution in combination with pipe parameters and inlet temperature of the water. By systematically combining these parameters and

analyzing their interactions, a comprehensive understanding of the factors affecting the thermal performance of the investigated ceiling variants was achieved.

Embedding depth			Pipe outher diameter x wall thickness & pipe spacing		Water temperature	
E6	E3	E1	16 x 2 mm	4.3 x 0.8 mm	Low	Medium
6 cm	3 cm	1 cm	15 cm	1 cm	26 °C	35 °C

Figure 2. Overview of all parameters which were varied.

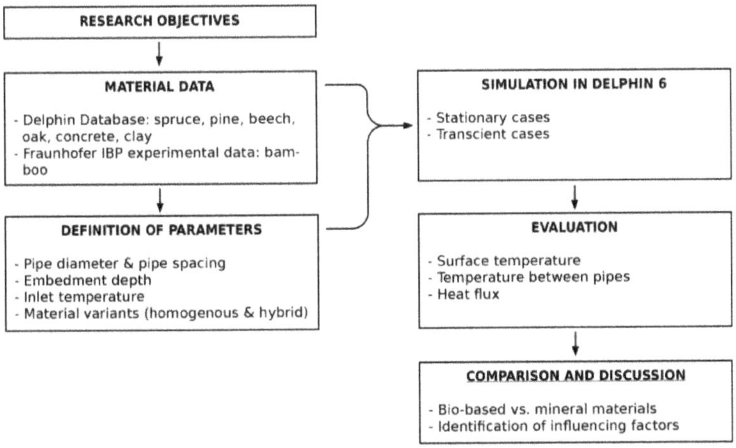

Figure 3. Workflow of the parametric simulation.

2.1. Pipe Diameter and Pipe Spacing

As part of the analysis of variation in pipe diameter and spacing, two different types of pipes were examined. One type was a conventional plastic pipe with an outer diameter of 16 mm and a wall thickness of 2 mm. This type of pipe is typically used in thermally activated component systems. The associated spacing of 15 cm corresponds to the standard dimension for this pipe size. Secondly, a capillary tube system was considered. For the simulation, an outer diameter of 4.3 mm and a wall thickness of 0.8 mm were assumed. In this case, the pipe spacing is 1 cm, which corresponds to the standard grid dimension for capillary systems. A major advantage of capillary tubes is their finer distribution, resulting in a more even heat transfer to the surrounding materials. Consequently, a more uniform surface temperature can be achieved, enabling lower inlet temperatures to be selected [27]. This is achieved by the increased pipe surface area resulting from the narrower pipe spacing. The modeling aspect of the heat transfer of the pipe will be explained in later in the section on boundary conditions.

2.2. Embedding Depth

Three different depths were defined in order to analyze the embedding depth of the pipes: 6 cm, 3 cm and 1 cm. These were measured from the ceiling surface to the bottom edge of the pipe. These embedding depths were selected to model different concepts of thermal component activation. A depth of 1 cm represents activation close to the surface, which has the potential for a faster thermal reaction. The 6 cm embedding depth, on the other hand, is oriented towards classic systems with greater thermal inertia. The 3 cm variant is intended as an intermediate solution, offering a balance between reaction behavior and heat storage.

2.3. Inlet Temperature

As part of the investigation into the inlet temperature, the inlet temperature of the heating medium was varied. Two characteristic values were defined for this purpose: 26 °C and 35 °C. This variation was intended to analyze the influence of different temperature levels on the energy required to activate the component, as well as its suitability as a surface heating system. The lower temperature of 26 °C represents a scenario with minimal energy consumption, while the higher temperature of 35 °C represents operation under more conventional conditions, such as with underfloor heating. While this case focuses on a ceiling, the question arises as to whether a higher inlet temperature could be beneficial, particularly for wooden structures. This is because wood has a significantly lower thermal conductivity than concrete, which can affect the efficiency of heat transfer. By comparing the two temperature levels, it can be assessed whether increasing the inlet temperature in timber systems leads to a significant improvement in thermal performance.

2.4. Materials

In the simulation, concrete serves as the baseline, representing the reference case for comparing the different material variants, as this is the material currently most commonly used for TABS [20]. The analysis focuses on two softwoods (spruce and pine), two hardwoods (beech and oak), and bamboo, which is a grass-based material. Spruce, pine, beech and oak were used from the material database of Delphin and analyzed by the laboratory of the IBK [33]. For the bamboo the materials parameters were used from an experimental analysis by Fraunhofer IBP [34]. These five materials were specifically selected because they differ significantly in terms of their thermal properties, such as thermal conductivity, and their bulk density. This is intended to cover a broad spectrum of possible heat transfer behaviors within wood and other natural materials. Additionally, four hybrid variants are analyzed, combining either concrete or clay with spruce or beech. In these hybrid structures, the pipe is embedded in the mineral layer (concrete or clay) with a 5 mm symmetrical overlap at the top and bottom. The basic idea behind these hybrid variants is to investigate whether a material with a higher thermal conductivity than wood can improve heat distribution in the building component. The aim is to thermally activate a larger surface area, and therefore a larger mass, of the component. So far, the integration of such a layer has only been conceived theoretically. Further studies would be required to examine the practical implementation and physical and structural feasibility. An overview of the thermal properties is provided in Figure 4. Additionally, Figure 1 shows the homogeneous and hybrid variants schematically.

2.5. Boundary Conditions

The boundary conditions at the top and bottom edges of the component in question are defined as a constant temperature of 20 °C. This serves as an assumed room temperature, representing an idealized thermal effect on the surface. The lateral edges are adiabatic, meaning there is no heat transfer via these sides. The pipes are arranged symmetrically around the component geometry, meaning the adiabatic edges act as axes of symmetry. This reduces the calculation area required for modeling.

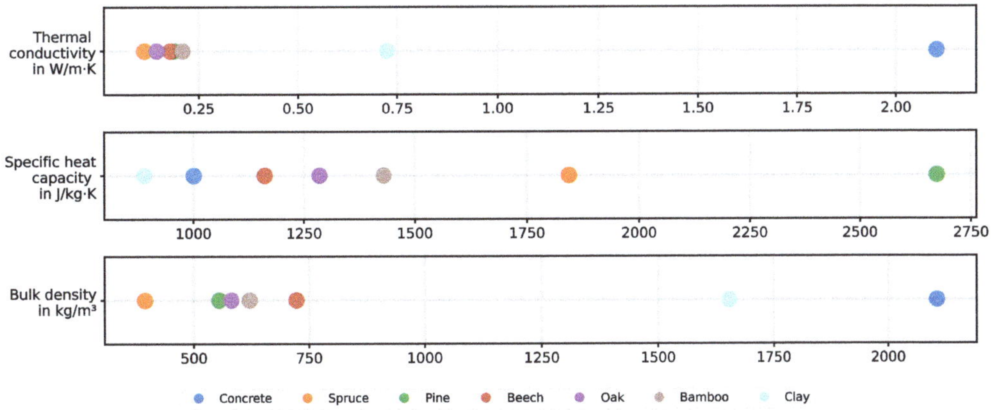

Figure 4. Overview of the (thermal) properties used in the simulations. For concrete, clay, spruce, pine, beech and oak are the parameters of the Material Database of Delphin [33] used. For bamboo the parameters were set based on an experimental analysis by Fraunhofer IBP [34].

In the simulation, the pipes are modeled as square cross-sections. The edge length of these squares corresponds to the outer diameter of the real pipes. The convective heat transfer boundary condition integrated in Delphin is used to model pipes with flowing fluid. Therefore, the adaptive model with inlet temperature is used in Delphin. A detailed description of how the heat transfer coefficient is calculated can be found in the publication by Weiß et al. [35].

Both models analyzed use a PE pipe with a thermal conductivity of 0.4 W/(m·K). The flow velocity of the water is 0.5 m/s. The corresponding mass flow was then determined based on the respective pipe cross-section.

2.6. Evaluation Parameters

The following measured variables are analyzed in the steady state:

- the temperature at the surface directly beneath the pipe
- the temperature at the center of the surface between two pipes
- the temperature at the center of the component between two pipes.

This analysis aims to investigate how the embedding depth of the pipework influences the resulting surface temperature. Additionally, the impact of pipe spacing on heat distribution uniformity is evaluated. Based on these findings, conclusions can be drawn about the extent to which the thermal mass of the component is activated.

In addition to the stationary observation, a dynamic load case was simulated to analyze the thermal storage behavior of the components. Once the system had reached steady state, it was kept in equilibrium for a defined period to ensure a stable starting point. The mass flow of the heating medium was then reduced to zero to prevent any further heat being introduced by forced convection. Subsequent cooling then took place exclusively through heat conduction within the component.

The heat conduction flux on the component surface can be used to track temperature development over time and estimate the system's cooling down time.

3. Results

The results show that the analyzed parameters significantly influence the performance of the thermal system. The effects are illustrated by analyzing the individual parameters separately below.

Ten material variants were analyzed as part of the investigation, including homogeneous building materials such as concrete, spruce, pine, oak, beech and bamboo, as well as hybrid combinations with clay and concrete. It was found that most of the materials, particularly pine, oak, beech and the hybrid combinations, had very similar values with regard to the relevant thermal parameters (e.g., surface temperature and temperature distribution). The differences were typically in the range of 0.45 K to 2.1 K for the bio-based materials, resulting in only minor deviations from the reference or extreme values.

For clarity and to maximize the informative value, the presented results focus on characteristic cases, i.e., the material with the best and worst performance, the most efficient bio-based material and the most efficient hybrid material, as well as a comparison with conventional concrete.

3.1. *Influence of the Material Used*

The material analysis is based on a model with an embedment depth of 6 cm, an outer pipe diameter of 16 mm and a water temperature of 35 °C. Figure 5 also shows the parameters used. The surface temperature directly below the pipe is used as the reference value.

Embedding depth			Pipe outher diameter x wall thickness & pipe spacing		Water temperature	
E6	E3	E1	16x 2 mm	4.3 x 0.8 mm	Low	Medium
6 cm	3 cm	1 cm	15 cm	1 cm	26 °C	35 °C

Figure 5. Used parameters (highlighted) for analyzing the influence of the material.

The highest temperature was measured in the concrete base material at 29.31 °C. Spruce had the lowest temperature, at 21.99 °C. Among the homogeneous, bio-based materials, bamboo showed the highest temperature of 23.22 °C. Among the hybrid variants, the combination of beech and concrete achieved 23.79 °C, closely followed by beech and loam at 23.51 °C. Depending on the material, the temperature difference interval was between 5.52 K and 7.32 K. The corresponding temperature distributions are shown in Figure 6. This clearly demonstrates the impact of the materials' varying thermal conductivities.

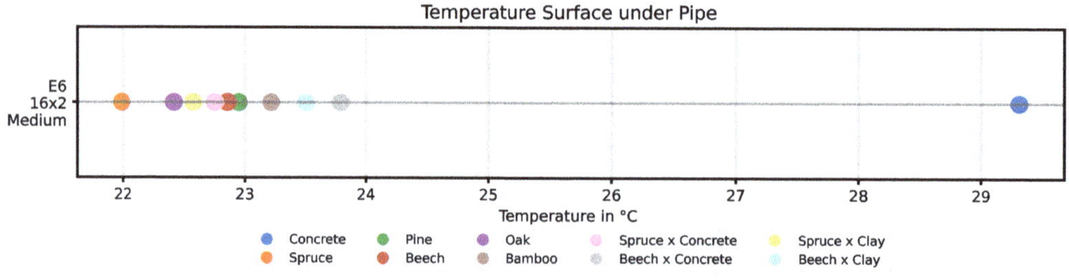

Figure 6. Temperature distribution at the surface under the pipe of the different materials used with an inlet temperature of the water with 35 °C, embedded depth of the pipe of 6 cm and a pipe diameter of 16 mm.

The corresponding thermal conductivities were summarized in Figure 4. Concrete had the highest thermal conductivity (2.1 W/(m·K)), while spruce had the lowest (0.112 W/(m·K)). Bamboo tops the list of bio-based materials with a value of 0.209 W/(m·K).

3.2. Influence of Pipe Size and Spacing

Models with an embedment depth of 3 cm and a water temperature of 26 °C were analyzed to evaluate the pipe size and spacing. Figure 7 also shows the parameters used. Evaluations were carried out at two positions: in the center between pipes at the same height, and at the surface below.

Embedding depth			Pipe outher diameter x wall thickness & pipe spacing		Water temperature	
E6	E3	E1	16x 2 mm	4.3 x 0.8 mm	Low	Medium
6 cm	3 cm	1 cm	15 cm	1 cm	26 °C	35 °C

Figure 7. Used parameters (highlighted) for analyzing the influence of the pipe size and pipe spacing.

At the center between the pipes, the temperature spread was 1.84 K at a distance of 15 cm (pipe diameter: 16 mm). With a minimum pipe spacing of 1 cm (capillary tubes), this decreased to 0.19 K.

Temperatures in the center between the pipes varied depending on the material. In the 16 mm pipe system, the spruce model showed the lowest value of 22.36 °C and the spruce-concrete hybrid model showed the highest value of 24.73 °C. Concrete itself reached 24.2 °C and bamboo 22.74 °C. The temperatures can be seen in Figure 8. Further, Figure 9 shows the temperature distribution in the components for all the different materials used with a pipe diameter of 16 mm.

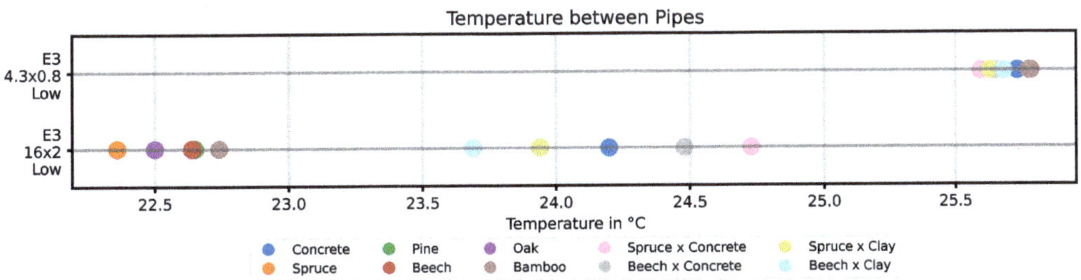

Figure 8. Temperature distribution between the pipes of the different materials used with an inlet temperature of the water with 26 °C, embedded depth of the pipe of 3 cm and a pipe diameter of 16 mm and 4.3 mm.

The temperature for the capillary tubes in concrete was 25.74 °C. Spruce and bamboo were almost on a par, with values of 25.79 °C and 25.78 °C, respectively—the highest values measured in this scenario. The hybrid spruce–concrete model achieved a value of 25.60 °C, the lowest measured. The corresponding distributions are shown in Figures 8 and 10.

The influence of pipe spacing can also be seen on the surface between two pipes. Spruce exhibited the lowest temperatures, reaching 20.67 °C with a 16 mm thickness and 21.85 °C with a capillary tube. Bamboo was the most effective bio-based material, reaching 21.18 °C (16 mm) and 22.69 °C (capillary tube). Concrete reached 23.72 °C (16 mm) and 25.13 °C (capillary tube). The best hybrid solution (beech-concrete) reached 22.12 °C and 24.22 °C, respectively. The corresponding temperature distributions are shown in Figure 11. Overall, it is evident that increasing the pipe surface area affects the temperature distribution within and on the surface of the component.

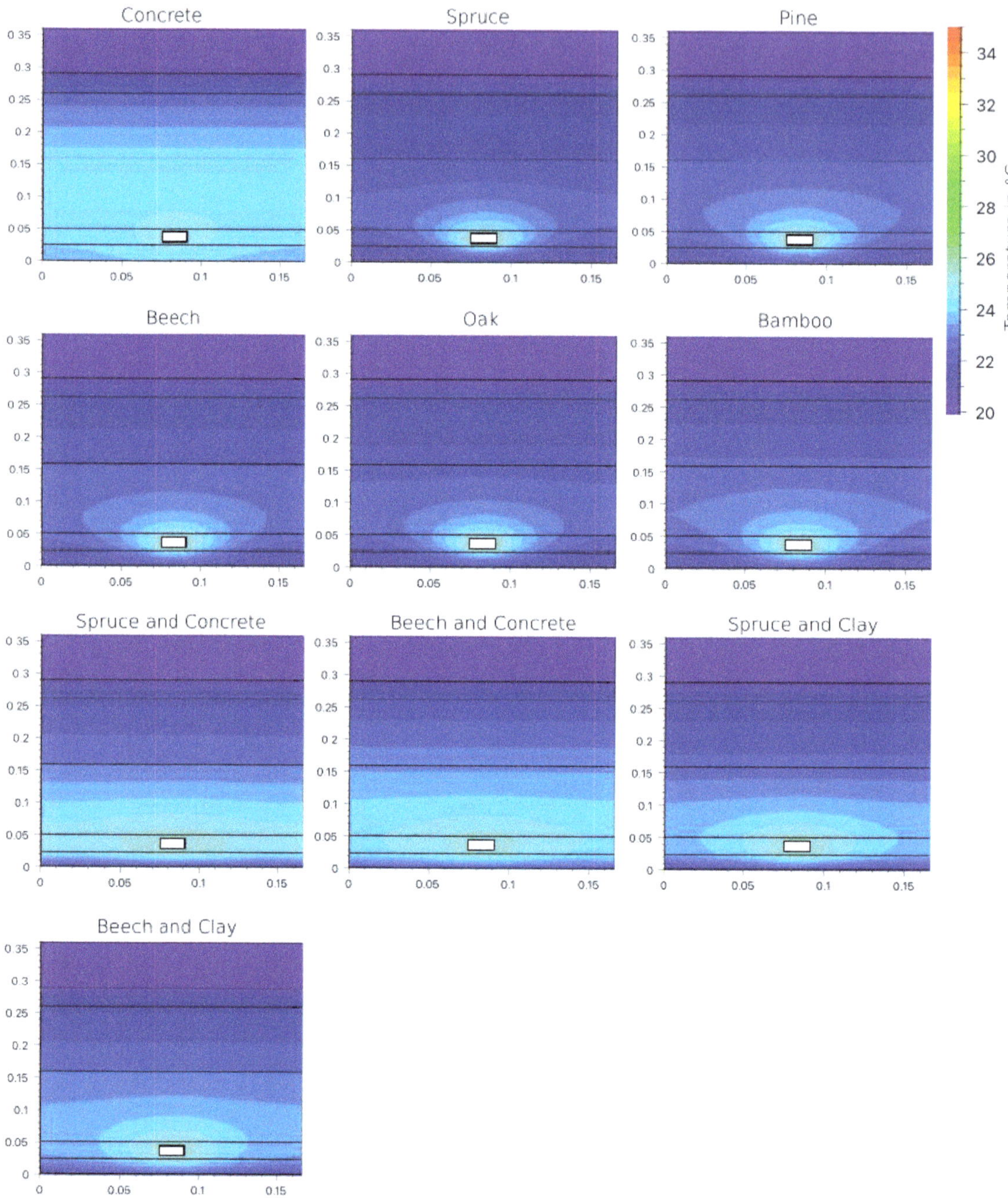

Figure 9. Temperature distribution in the component of the different materials used with an inlet temperature of the water with 26 °C, embedded depth of the pipe of 3 cm and a pipe diameter of 16 mm. Axes of building components in meter.

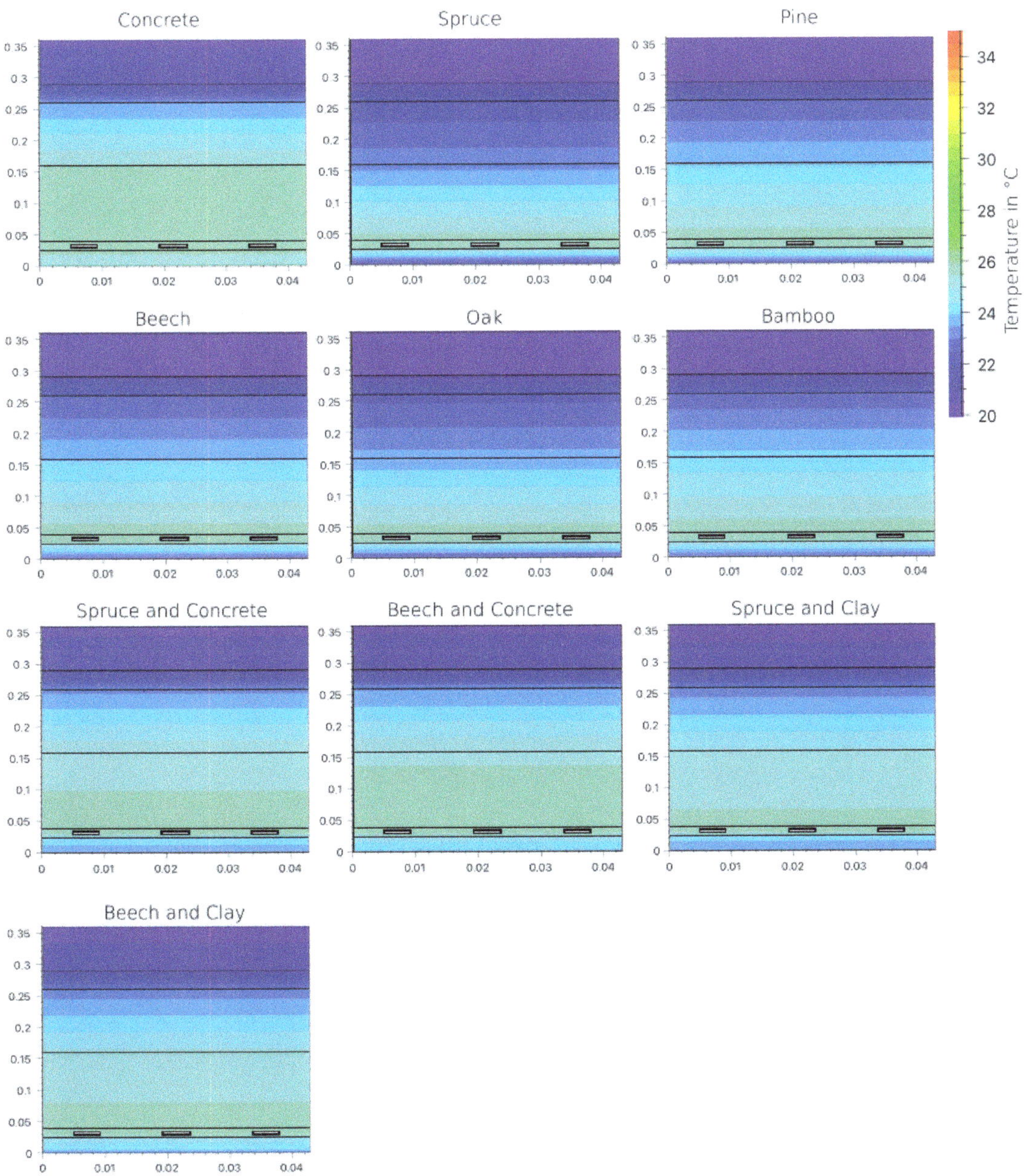

Figure 10. Temperature distribution in the component of the different materials used with an inlet temperature of the water with 26 °C, embedded depth of the pipe of 3 cm and a pipe diameter of 4.3 mm. Axes of building components in meter.

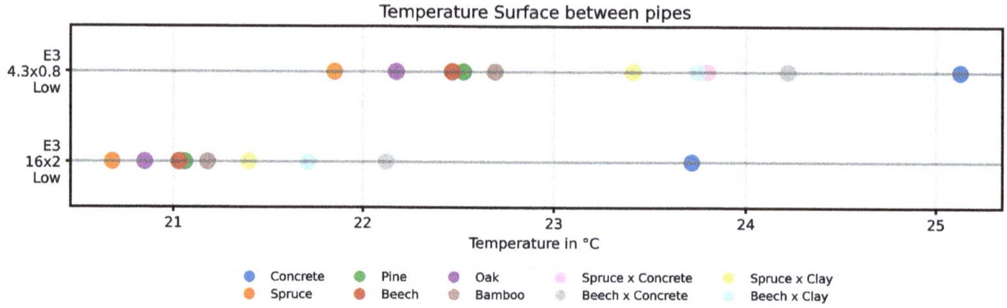

Figure 11. Temperature distribution at the surface between the pipes of the different materials used with an inlet temperature of the water with 26 °C, embedded depth of the pipe of 3 cm and a pipe diameter of 16 mm and 4.3 mm.

3.3. *Influence of Embedding Depth*

The influence of the embedding depth was analyzed by comparing 6 cm and 1 cm depths at a water temperature of 26 °C. The 16 mm pipe system and the capillary tube system were both analyzed. Figure 12 also shows the parameters used.

Embedding depth			Pipe outher diameter x wall thickness& pipe spacing		Water temperature	
E6	E3	E1	16x 2 mm	4.3 x 0.8 mm	Low	Medium
6 cm	3 cm	1 cm	15 cm	1 cm	26 °C	35 °C

Figure 12. Used parameters (highlighted) for analyzing the influence of the embedded depth.

In the 16 mm model, the concrete had the highest surface temperatures under the pipe: 23.67 °C (6 cm) and 24.6 °C (1 cm). Spruce exhibited the lowest values, at 20.79 °C and 23.12 °C, respectively. Bamboo was at 21.28 °C (6 cm) and 23.73 °C (1 cm). The hybrid beech-concrete model exhibited a notable temperature increase, ranging from 21.5 °C (6 cm) to 24.08 °C (1 cm). Figure 13 shows the temperature distributions.

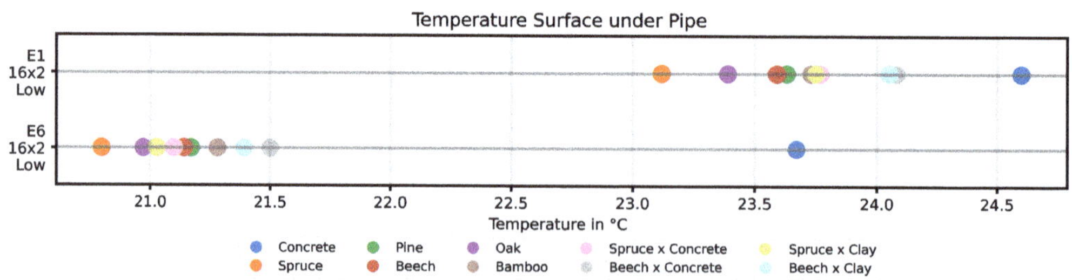

Figure 13. Temperature distribution at the surface under the pipe of the different materials used with an inlet temperature of the water with 26 °C, embedded depth of the pipe of 1 cm and 6 cm and a pipe diameter of 16 mm.

The effect was even clearer with capillary tubes. The spruce variant's temperature rose from 21.11 °C (6 cm) to 23.33 °C (1 cm) and the bamboo variants from 21.78 °C to 24.11 °C. Concrete showed smaller differences here: 24.67 °C (6 cm) and 25.49 °C (1 cm). The hybrid model also showed a moderate increase. The results are shown in Figures 14 and 15. Therefore, the influence of the thermal conductivity can be emphasized once again. It represents greater resistance at lower values with a higher embedding depth.

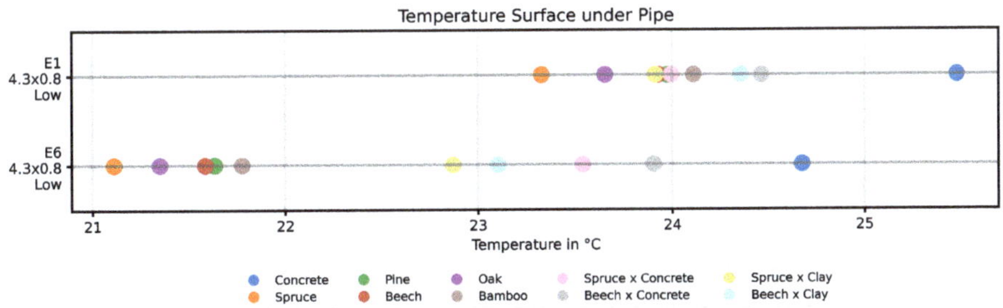

Figure 14. Temperature distribution at the surface under the pipe of the different materials used with an inlet temperature of the water with 26 °C, embedded depth of the pipe of 1 cm and 6 cm and a pipe diameter of 4.3 mm.

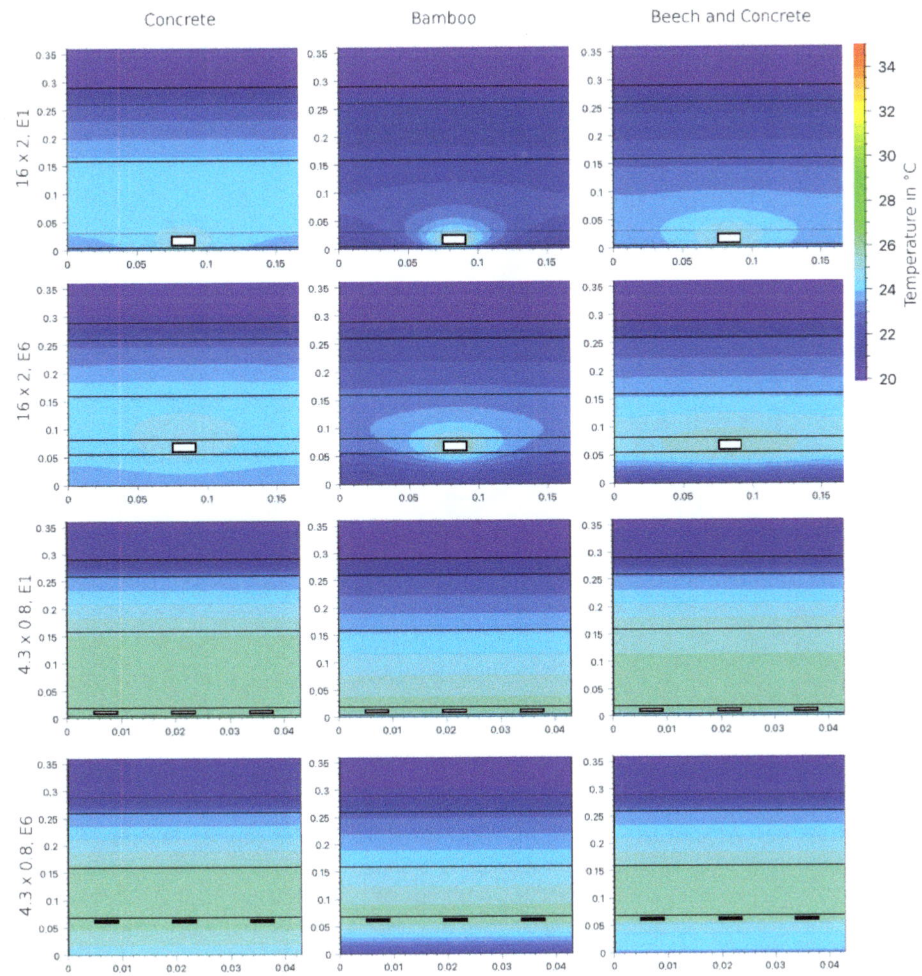

Figure 15. Temperature distribution in the component of concrete as baseline, the best alternative homogenous variant (bamboo) and the best hybrid variant (beech with concrete) with an inlet temperature of the water with 26 °C, embedded depth of the pipe of 1 cm and 6 cm and pipe diameters of 16 mm and 4.3 mm. Axes of building components in meter.

3.4. Influence of the Water Temperature

The inlet temperature considerably influences the temperature level and spreads. A model with an embedment depth of 3 cm and 16 mm pipes was analyzed at inlet temperatures of 26 °C and 35 °C. Figure 16 also shows the parameters used.

The difference between the highest and lowest surface temperatures under the pipe was 2.65 K at 26 °C and increased to 6.78 K at 35 °C. The minimum surface temperature was observed for spruce at 21.43 °C (26 °C inlet temperature) and 23.6 °C (35 °C inlet temperature). The highest values were achieved with concrete: 24.08 °C and 30.38 °C, respectively. Figures 17 and 18 show the complete temperature distributions. The higher heat transfer caused by the higher water temperature is evident here. This behavior is further reinforced by a lager heat transfer area of the pipes at the models with the capillary tubes.

Embedding depth			Pipe outher diameter x wall thickness& pipe spacing		Water temperature	
E6	E3	E1	16x 2 mm	4.3 x 0.8 mm	Low	Medium
6 cm	3 cm	1 cm	15 cm	1 cm	26 °C	35 °C

Figure 16. Used parameters (highlighted) for analyzing the influence of the water temperature.

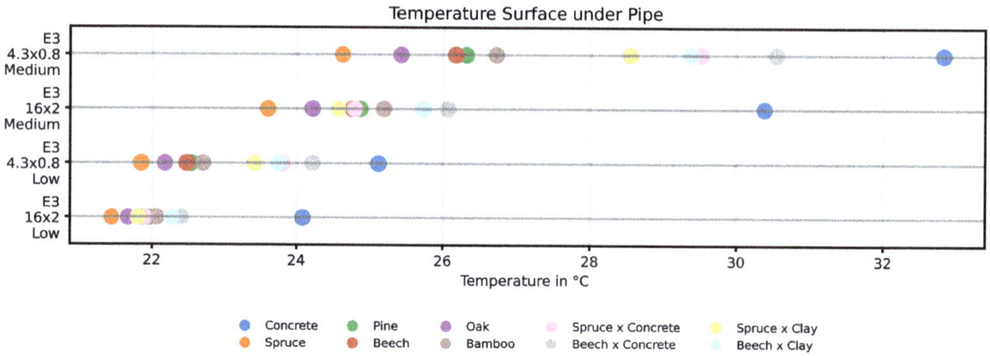

Figure 17. Temperature distribution at the surface under the pipe of the different materials used with an inlet temperature of the water with 26 °C and 35 °C, embedded depth of the pipe of 3 cm and pipe diameters of 16 mm and 4.3 mm.

3.5. Influence of the Heat Storage Capacity

The heat storage capacity was determined by measuring the cooling time after the system reached a steady state with a zero-mass flow. The limit value was set at 0.5 W/m^2. Figure 19 shows the heat flow curve at a boundary condition of an inlet temperature 26 °C and an embedded depth of 3 cm. The start time here is 360 h, as the simulation should take sufficient time to stabilize at a steady state.

The curves show that, after concrete, pine has the highest storage potential. Bio-based materials absorbed significantly less heat than concrete. Hybrid materials also exhibited a low storage effect.

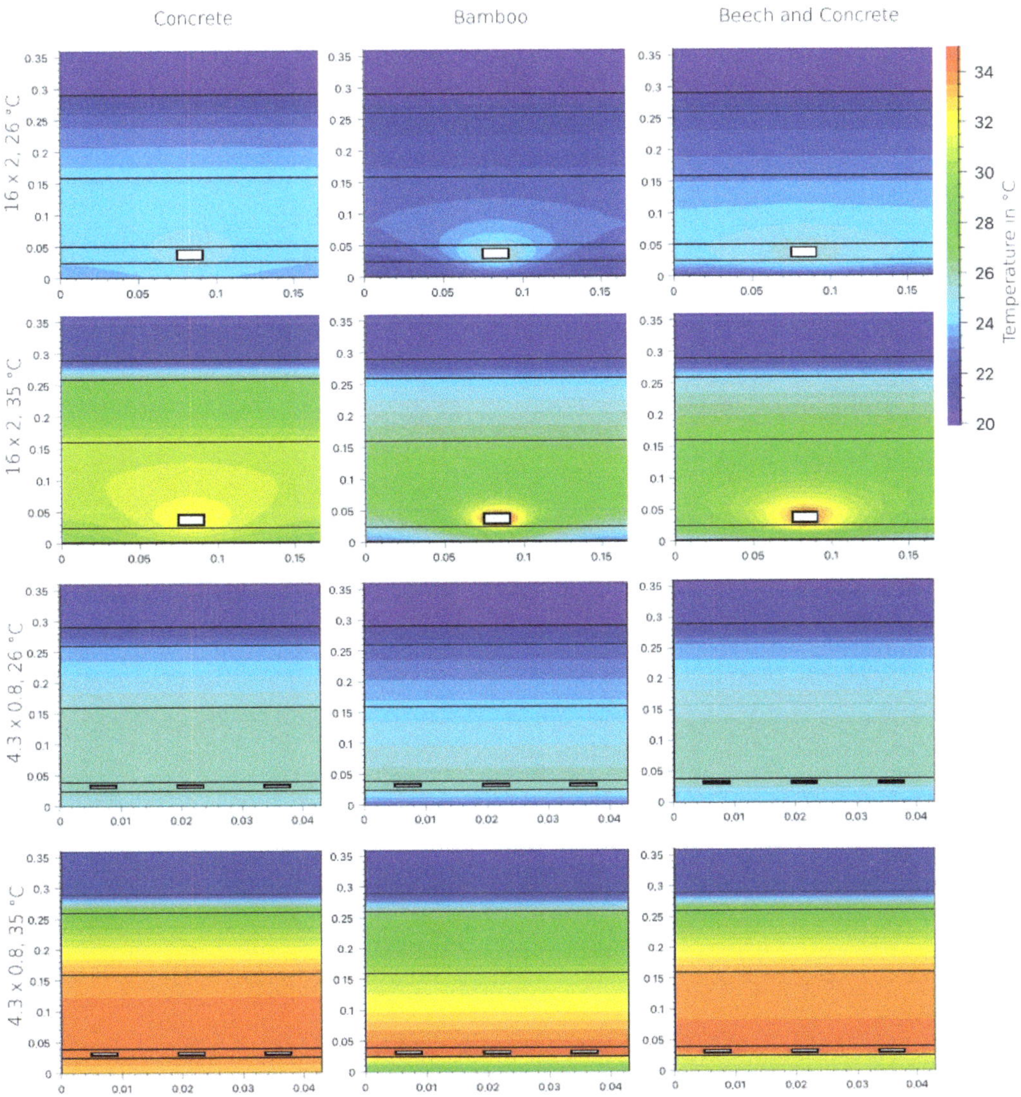

Figure 18. Temperature distribution in the component of concrete as baseline, the best alternative homogenous variant (bamboo) and the best hybrid variant (beech with concrete) with an inlet temperature of the water with 26 °C and 35 °C, embedded depth of the pipe of 3 cm and pipe diameters of 16 mm and 4.3 mm. Axes of building components in meter.

Capillary tubes considerably extended the cooling down time. In the concrete model with an embedment depth of 3 cm, the cooling time was around 71 h with 16 mm pipes and around 75 h with capillary tubes. Similar effects were seen with other materials. Pine performed best consistently among the bio-based materials. With pine the cooling down time extends on additional 12 h by using capillary tubes instead of using the regular pipes. By comparing concrete and pine, the influence of the specific heat capacity in combination with the bulk density is highlighted. Figure 4 also underlines the influence of the specific heat capacities, particularly in the hybrid models.

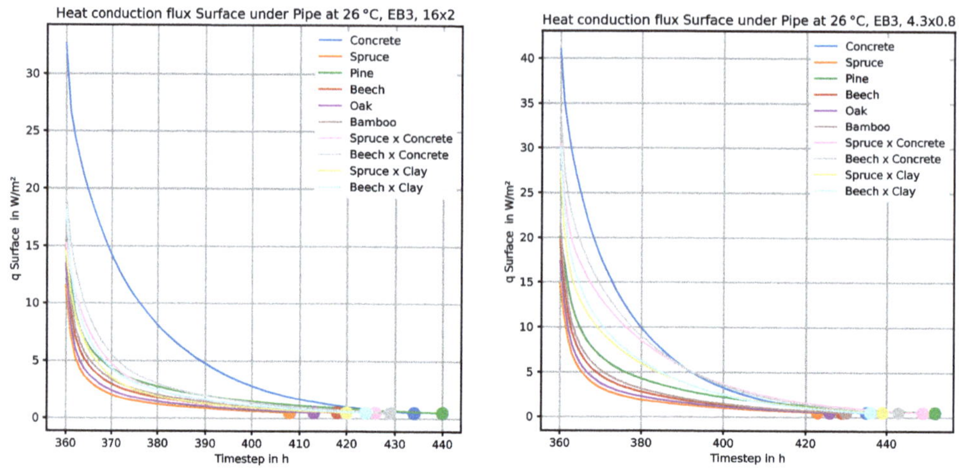

(a) (b)

Figure 19. Heat conduction flux at the surface under the pipe with an embedding depth of 3 cm, an inlet temperature of 26 °C and a (**a**) 16 mm pipe diameter and (**b**) 4.3 pipe diameter.

4. Discussion

The numerical simulations of this study show that all analyzed parameters—material, pipe spacing, water temperature, embedment depth, and heat storage capacity—significantly influence the system's thermal performance. The results are analyzed in the context of thermal component activation below.

4.1. Influence of the Material

Concrete has the most favorable thermal properties due to its high thermal conductivity ($\lambda = 2.1$ W/(m·K)) and heat capacity. The significantly lower surface temperatures of spruce and other bio-based materials illustrate the limiting effect of their low thermal conductivity (e.g., spruce: $\lambda = 0.112$ W/(m·K)). Among bio-based materials, bamboo shows comparatively favorable behavior with a thermal conductivity of 0.209 W/(m·K). This highlights the potential for optimization through the targeted selection of bio-based materials.

Hybrid materials, such as beech-concrete or beech-clay, have better temperature distribution than solid bio-based materials. This is due to the thermal bridging effect of the mineral component. The higher temperatures in the area between pipes with a diameter of 16 mm can be explained by the insulating effect of the wood matrix, which has a low thermal conductivity.

Overall, the results demonstrate that suitable combinations of materials can be used to achieve efficient heat conduction in the vicinity of the pipes, as well as improved distribution and storage throughout the entire component. Hybrid materials therefore offer a functional optimization strategy that combines the advantages of the low CO_2 footprint of bio-based materials with the high thermal storage capacity of mineral components.

4.2. Influence of Pipe Size and Spacing

A closer spacing in combination with a smaller size of the pipes leads to a more homogeneous temperature distribution and greater activation of the component. For instance, the temperature variation decreases from 1.84 K at a spacing of 15 cm to just 0.19 K at a spacing of 1 cm (capillary tubes) at an embedment depth of 6 cm. For materials with low thermal conductivity in particular, a narrow pipe spacing is crucial to achieving surface temperatures similar to those of concrete. This phenomenon can be attributed to the increased surface area of the pipe, which in turn leads to an elevated heat transfer area.

The results of the simulations suggest that the relative temperature benefit of closer pipe spacing is less pronounced in bio-based materials than in hybrid systems. This can be explained with the different thermal conductivities of the different materials of the hybrid structures. However, the findings indicate that bio-based materials can be utilized effectively in conjunction with capillary tube systems, thereby ensuring uniform activation despite their reduced thermal conductivity.

4.3. Influence of Embedment Depth

The embedment depth primarily affects the surface temperature, whereas the temperature around the pipe varies less. A shallower embedment depth (e.g., 1 cm instead of 6 cm) achieves higher surface temperatures due to the shorter heat path. This effect is more pronounced with bio-based materials, as their thermal conductivity is lower. This result is in agreement with Bishara et al. [30]. Therefore, reducing the embedment depth can help increase efficiency, particularly when using bio-based materials.

4.4. Influence of Water Temperature

An increase in the inlet temperature increases both maximum surface temperature and temperature spread. This has two main consequences:

1. Increased heat output due to more intensive thermal activation.
2. Lower thermal homogeneity, which can lead to a loss of comfort, for example, in floor applications.

Simulations show that bio-based materials do not significantly improve in performance, even at higher water temperatures. This confirms that their thermal limitations cannot be overcome by increasing the inlet temperature alone. This was also shown by Heidenthaler et al. [31].

The results suggest that uniform activation of the component is possible even with a moderate water temperature of 26 °C in combination with close pipe spacing.

4.5. Influence of the Heat Storage Capacity

Analyzing the cooling out phases reveals that materials with a high specific heat capacity, such as concrete and pine, release heat over longer periods. Similar results can be seen in Heidenthaler et al. [31]. Using capillary tubes also extends the cooling out time, indicating more even heat distribution and more effective component activation. The storage potential of some bio-based materials, especially pine, is notable. This makes them promising candidates for intermittent heating operations, as well as applications involving load shifting or time-delayed energy supply, such as when using heat pumps. In this context, the parameters of the specific heat capacity and the bulk density play a decisive role in determining the suitability of a material for load-shifting applications, as it directly influences the amount of energy stored.

4.6. Universality of Results

Although the study is based on 120 independent parametric simulations, the findings can be considered transferable beyond the investigated cases. The methodology builds in fundamental principles of heat conduction and thermal storage, which are universally valid and not restricted to a specific software implementation. Moreover, the parametric framework with Delphin 6 enables systematic variation in material properties, geometries, and boundary conditions, which makes the approach adaptable to a wide range of construction types and operating scenarios. The influence patterns observed in this work—such as the role of the pipe spacing in achieving thermal homogeneity or the effect of embedment depth on the surface temperature—are consistent with physical reasoning and align with trends reported in previous studies. Taken together, these aspects demonstrate that the results are not limited to an isolated case study but provide insights of broader applicability.

5. Conclusions

The present study shows that bio-based building materials have promising potential for use in thermally activated building components for load shifting. While previous research, as referenced in the literature reviews [7,8], focused primarily on mineral-based TABS, this study provides the first systematic numerical evaluation of bio-based material under a wide range of operational and design parameters, including pipe diameter, spacing, embedment depth and inlet temperature. Although bio-based materials have a lower thermal conductivity than mineral materials such as concrete, comparable thermal performance can be achieved through suitable design measures, such as reducing the embedment depth and spacing the pipes more closely, for example, by using capillary tube systems. Among the bio-based materials, bamboo and pine stand out favorably. Hybrid material combinations with clay or concrete offer additional optimization potential by improving heat distribution and reducing the ecological footprint simultaneously. However, these are theoretical constructions that have not yet been realized in practice. In conclusion, the findings of this study demonstrate that through design optimization (e.g., closer pipe spacing, reduced embedment depth), bio-based materials can achieve comparable activation performance to concrete in TABS, thus expanding the potential material palette for sustainable building design.

A key limitation of this study is that it represents a purely numerical first step without experimental validation. While the parametric simulation provides systematic insights into the thermal behavior of bio-based and hybrid slabs, experimental investigations will be essential to validate and calibrate the findings under real boundary conditions. Such validation is foreseen as the logical next step of this research. Limitations of the study include further the focus on heating applications only, since cooling with bio-based materials involves a considerable risk of condensation. Due to the moisture-absorbing properties of wood and similar materials, condensation can lead to moisture buildup inside the components, which may cause damage such as mold growth or material degradation over time. Because of these risks, cooling scenarios were excluded from this study to maintain focus on thermal performance during heating. In addition, the simulation assumed dry material properties, and moisture content was not included, which may influence thermal performance in real-world conditions. Future research needs to address moisture control and condensation prevention to enable safe use of bio-based materials in cooling applications. A further limitation concerns the practical feasibility of the proposed (hybrid) building component. While the simulations show building physical potential, realization in practice would require interdisciplinary evaluation regarding load-bearing capacity and long-term behavior.

Simulations also demonstrate that the heat storage capacity of certain bio-based materials could be utilized for load shifting and intermittent operation, e.g., in conjunction with heat pumps or solar energy sources. Therefore, bio-based materials could play a significant role in decarbonizing the construction industry. Future research should focus on the practical implementation of such hybrid systems, as well as on their long-term behavior and building physics boundary conditions. Including also a carbon impact assessment would further enhance the sustainability evaluation of bio-based materials. This type of assessment could include embodied CO_2 or GWP. Moreover, future research should prioritize experimental validation of hybrid systems, assessment of construction feasibility including prefabrication potential, and evaluation of load-bearing capacity and long-term performance under realistic environmental conditions. Additionally, the influence of moisture and condensation must be thoroughly investigated, as these factors critically affect durability and safe application, especially given the moisture-absorbing nature of bio-based materials. Addressing these challenges is essential to enable practical implementation and ensure reliable performance in real-world building environments. Future work should also explore AI-based surrogate modeling approaches such as physics-informed neuronal networks to accelerate parametric studies of bio-based TABS for load shifting.

Author Contributions: Conceptualization, A.A. and D.Ö.; methodology, A.A.; formal analysis, A.A.; investigation and simulation, A.A.; data curation, A.A.; writing—original draft preparation, A.A.; writing—review and editing, A.A. and D.Ö.; visualization, A.A.; project administration, D.Ö.; funding acquisition, D.Ö. All authors have read and agreed to the published version of the manuscript.

Funding: This work was supported by the Deutsche Forschungsgemeinschaft (DFG, German Research Foundation) under Germany's Excellence Strategy—EXC 2120/1—390831618.

Data Availability Statement: The original contributions presented in this study are included in the article. Further inquiries can be directed to the corresponding author(s).

Conflicts of Interest: The authors declare no conflicts of interest.

References

1. Park, S.H.; Chung, W.J.; Yeo, M.S.; Kim, K.W. Evaluation of the thermal performance of a Thermally Activated Building System (TABS) according to the thermal load in a residential building. *Energy Build.* **2014**, *73*, 69–82. [CrossRef]
2. Reilly, A.; Kinnane, O. The impact of thermal mass on building energy consumption. *Appl. Energy* **2017**, *198*, 108–121. [CrossRef]
3. Hildebrandt, J.; Hagemann, N.; Thrän, D. The contribution of wood-based construction materials for leveraging a low carbon building sector in europe. *Sustain. Cities Soc.* **2017**, *34*, 405–418. [CrossRef]
4. Bundesministerium für Wohnen, Stadtentwicklung und Bauwesen. ÖKOBAUDAT. Available online: https://www.oekobaudat.de/ (accessed on 14 July 2025).
5. Rinne, R.; Ilgın, H.E.; Karjalainen, M. Comparative Study on Life-Cycle Assessment and Carbon Footprint of Hybrid, Concrete and Timber Apartment Buildings in Finland. *Int. J. Environ. Res. Public Health* **2022**, *19*, 774. [CrossRef] [PubMed]
6. Dodoo, A.; Gustavsson, L.; Sathre, R. Effect of thermal mass on life cycle primary energy balances of a concrete- and a wood-frame building. *Appl. Energy* **2012**, *92*, 462–472. [CrossRef]
7. Romaní, J.; de Gracia, A.; Cabeza, L.F. Simulation and control of thermally activated building systems (TABS). *Energy Build.* **2016**, *127*, 22–42. [CrossRef]
8. Villar-Ramos, M.M.; Hernández-Pérez, I.; Aguilar-Castro, K.M.; Zavala-Guillén, I.; Macias-Melo, E.V.; Hernández-López, I.; Serrano-Arellano, J. A Review of Thermally Activated Building Systems (TABS) as an Alternative for Improving the Indoor Environment of Buildings. *Energies* **2022**, *15*, 6179. [CrossRef]
9. Fan, S.; Yan, T.; Xu, X.; Yu, Z.; Zhu, Q. A simple parameter configuration scheme for a simplified dynamic model of pipe-embedded wall. *J. Build. Eng.* **2024**, *83*, 108401. [CrossRef]
10. Jiang, S.; Li, X.; Lyu, W.; Wang, B.; Shi, W. Numerical investigation of the energy efficiency of a serial pipe-embedded external wall system considering water temperature changes in the pipeline. *J. Build. Eng.* **2020**, *31*, 101435. [CrossRef]
11. Kisilewicz, T.; Fedorczak-Cisak, M.; Barkanyi, T. Active thermal insulation as an element limiting heat loss through external walls. *Energy Build.* **2019**, *205*, 109541. [CrossRef]
12. Vogdt, F.U.; Helbach, M. Eine thermisch aktiv gedämmte, monolithische Außenwand aus Infraleichtbeton als nachhaltiges, zukunftsfähiges Wandsystem? *Bauphysik* **2015**, *37*, 38–51. [CrossRef]
13. Krajčík, M.; Šikula, O. The possibilities and limitations of using radiant wall cooling in new and retrofitted existing buildings. *Appl. Therm. Eng.* **2020**, *164*, 114490. [CrossRef]
14. Ibrahim, M.; Wurtz, E.; Biwole, P.H.; Achard, P. Transferring the south solar energy to the north facade through embedded water pipes. *Energy* **2014**, *78*, 834–845. [CrossRef]
15. Ibrahim, M.; Wurtz, E.; Anger, J.; Ibrahim, O. Experimental and numerical study on a novel low temperature façade solar thermal collector to decrease the heating demands: A south-north pipe-embedded closed-water-loop system. *Sol. Energy* **2017**, *147*, 22–36. [CrossRef]

16. Shen, J.; Wang, Z.; Luo, Y.; Jiang, X.; Zhao, H.; Cui, D.; Tian, Z. Performance evaluation of an active pipe-embedded building envelope system to transfer solar heat gain from the south to the north external wall. *J. Build. Eng.* **2022**, *59*, 105123. [CrossRef]
17. Olesen, B.W. Thermo Active Building Systems Using Building Mass To Heat and Cool. *ASHRAE J.* **2012**, *54*, 44–52.
18. Derakhtenjani, A.S.; Athienitis, A.K. A frequency domain transfer function methodology for thermal characterization and design for energy flexibility of zones with radiant systems. *Renew. Energy* **2021**, *163*, 1033–1045. [CrossRef]
19. Ma, P.; Guo, N. Modeling of Thermal Mass in a Small Commercial Building and Potential Improvement by Applying TABS. *Am. J. Mech. Eng.* **2015**, *3*, 55–62. [CrossRef]
20. Olesen, B.W. Radiant Floor Heating in Theory and Practice. *ASHRAE J.* **2002**, *44*, 19–26.
21. Oravec, J.; Šikula, O.; Krajčík, M.; Arıcı, M.; Mohapl, M. A comparative study on the applicability of six radiant floor, wall, and ceiling heating systems based on thermal performance analysis. *J. Build. Eng.* **2021**, *36*, 102133. [CrossRef]
22. Rhee, K.-N.; Kim, K.W. A 50 year review of basic and applied research in radiant heating and cooling systems for the built environment. *Build. Environ.* **2015**, *91*, 166–190. [CrossRef]
23. *DIN EN ISO 11855*; Umweltgerechte Gebäudeplanung-Flächenintegrierte Strahlungsheiz-und-kühlsysteme. DIN Deutsches Institut für Normung e.V.: Berlin, Germany, 2022. Available online: https://dx.doi.org/10.31030/3264028 (accessed on 22 October 2025).
24. Li, Y.; McLauchlan, C.; Gong, X.; Ma, Z. Experimental and modeling investigation of thermally activated building systems integrated with ground source heat pump systems. *Energy Build.* **2025**, *328*, 115210. [CrossRef]
25. Xu, X.; Wang, S.; Wang, J.; Xiao, F. Active pipe-embedded structures in buildings for utilizing low-grade energy sources: A review. *Energy Build.* **2010**, *42*, 1567–1581. [CrossRef]
26. Zhao, T.; Weiß, D.; Zhuang, Z.; Grunewald, J.; Hirsch, H.; Hirth, S. Performance evaluation of a low-carbon thermally activated residential building under different heat source operation strategies. *Build. Environ.* **2025**, *281*, 113237. [CrossRef]
27. Glück, B. Entwicklung von Produkten mit Kunststoff-Kapillarrohrmatten zur Umweltschonenden Raumheizung und-Kühlung 2003. Available online: https://berndglueck.de/dl/?dl=kapillarrohrmatten+Bericht_Kapillarrohrmatten.pdf (accessed on 15 May 2025).
28. Do, H.Q.; Luther, M.B.; Matthews, J.; Martek, I.; Horan, P. An experimental evaluation of the thermal performance of lightweight radiant ceiling panel designs. *Energy Build.* **2025**, *344*, 115966. [CrossRef]
29. Rodríguez Jara, E.Á.; Ruiz-Pardo, Á.; García, M.C.; Ríos, J.A.T. Effect of Wood Properties and Building Construction on Thermal Performance of Radiant Floor Heating Worldwide. *Appl. Sci.* **2022**, *12*, 5427. [CrossRef]
30. Bishara, N.; Schulz, T.; Gecks, J.; Plagge, R.; Wehsener, J. Thermal optimization and performance analysis of an innovative wooden radiant heating system made for room temperature control—Laboratory and numerical investigation of prototypes. *Energy Build.* **2017**, *138*, 569–578. [CrossRef]
31. Heidenthaler, D.; Leeb, M.; Schnabel, T.; Huber, H. Comparative analysis of thermally activated building systems in wooden and concrete structures regarding functionality and energy storage on a simulation-based approach. *Energy* **2021**, *233*, 121138. [CrossRef]
32. Mindrup, K. Raumklimatisierung Durch Thermisch Aktivierte Massivholzelemente. Ph.D. Thesis, Universität München, Munich, Germany, 2019.
33. DELPHIN 6; Bauklimatik Dresden. Available online: https://bauklimatik-dresden.de/software/delphin/ (accessed on 29 April 2025).
34. Zujian, H.; Krus, M.; Künzel, H.M. ReBuMat—German-Vietnamese Collaborative Project in Resourceefficient Construction Using Sustainabke Building Materials Three-Dimensional Hygrothermal Properties Test on Bamboo. IBP-Report HTB-012/2021. 2021. Available online: https://wufi.de/de/wp-content/uploads/sites/9/HTB_012_2021_Kr_DLR.pdf (accessed on 23 April 2025).
35. Weiß, D.; Tribulowski, K.; Hirth, S.; Fechner, H. Integration Of Pipe Models In Delphin And NANDRAD. *E3S Web Conf.* **2020**, *172*, 12012. [CrossRef]

energies

MDPI

Article

Comparative Framework for Climate-Responsive Selection of Phase Change Materials in Energy-Efficient Buildings

Javier Martínez-Gómez [1,2]

[1] Departamento de Teoría de la Señal y Comunicación, (Área de Ingeniería Mecánica) Escuela Politécnica, Universidad de Alcalá, 28805 Alcalá de Henares, Madrid, Spain; javier.martinezgomez@uah.es

[2] Facultad de Arquitectura e Ingenierías, Universidad Internacional SEK, Albert Einstein s/n and 5th, Quito 170302, Ecuador

Abstract

Integrating phase change materials (PCMs) into buildings and HVAC systems improves thermal comfort and energy efficiency. This study presents a climate-responsive methodology for selecting optimal PCMs using a multi-criteria decision-making (MCDM) framework. AHP was employed to determine the relative importance of key thermophysical properties, including melting point (47.5%), latent heat of fusion (25.7%), volumetric latent heat (13.5%), thermal conductivity (6.8%), specific heat capacity (3.3%), and density (3.3%). These weights were applied across five MCDM techniques—COPRAS, VIKOR, TOPSIS, MOORA, and PROMETHEE II—to evaluate 16 PCM alternatives for three representative climate zones: temperate (18 °C), subtropical (23 °C), and tropical hot/desert (28 °C). The results consistently identified n-Heptadecane (C17) as the most suitable PCM for temperate and subtropical climates, while n-Octadecane (C18) and hydrated salts such as $CaCl_2 \cdot 6H_2O$ and $Na_2CO_3 \cdot 10H_2O$ were optimal for tropical zones. Results show that n-Heptadecane (C17) is optimal for temperate and subtropical zones (COPRAS K = 1.00; TOPSIS C = 0.79–0.82; PROMETHEE φ = 0.21–0.22), while n-Octadecane (C18) and hydrated salts such as $CaCl_2 \cdot 6H_2O$ and $Na_2CO_3 \cdot 10H_2O$ perform best in tropical climates (TOPSIS C = 0.85; PROMETHEE φ = 0.26). These PCMs offer high latent heat (up to 254 $kJ \cdot kg^{-1}$) and volumetric storage (up to 381 $MJ \cdot m^{-3}$), enabling significant reductions in HVAC loads and improved indoor temperature stability. The convergence of rankings across methods and alignment with existing literature validate the robustness of the proposed approach. This framework supports informed material selection for sustainable building design and can be adapted to other climate-sensitive engineering applications. The framework introduces methodological innovations by explicitly mapping PCM melting points to climate-specific comfort bands, incorporating volumetric latent heat, and validating rankings through cross-method convergence (Spearman $\rho > 0.99$). Sensitivity analysis confirms robustness against weight perturbations. The approach supports practical PCM selection for both new and retrofit buildings, contributing to EU and US energy goals (e.g., 40% building energy use, DOE's 50% reduction target).

Keywords: climate zones; decision-making framework; latent heat storage materials; thermal comfort; energy-efficient buildings; multi-criteria analysis

1. Introduction

The building sector has been identified as one of the most critical areas for achieving substantial reductions in greenhouse gas emissions worldwide. In the European Union

https://doi.org/10.3390/en18225982

(EU), buildings account for approximately 40% of total energy consumption and 36% of carbon dioxide (CO_2) emissions, highlighting their significant impact on the region's overall environmental footprint [1]. This challenge is further underscored by the stark contrast in energy performance between new and existing buildings: while recently constructed buildings typically require less than 3 to 5 L of heating oil per square meter annually, older buildings consume an average of about 25 L per square meter per year [2]. This disparity illustrates the urgent need for retrofitting and upgrading the existing building stock to improve energy efficiency and reduce emissions.

Similarly, in the United States, the Department of Energy (DOE) has prioritized advancements in building energy performance as a key strategy for national sustainability goals. Through the development and promotion of efficient, affordable, and high-impact technologies, systems, and practices, the DOE aims to transform the building sector [3]. The long-term objective of the DOE's Building Technologies Office is to achieve a 50% reduction in energy use compared to 2010 levels, reflecting a strong commitment to innovation and continuous improvement in building design and operation [4].

These ambitious targets in both the EU and the US underscore the necessity for innovative solutions that can address the dual challenges of energy efficiency and occupant comfort. Among the various strategies being explored, the integration of advanced materials—such as phase change materials (PCMs)—into building envelopes and systems offers significant potential for reducing energy demand and greenhouse gas emissions, while simultaneously enhancing indoor environmental quality [5,6].

Thermal comfort in buildings is a fundamental factor influencing both quality of life and energy efficiency, particularly in the context of increasing demand for sustainable solutions. Phase change materials (PCMs) have emerged as a promising technology for passive indoor temperature management due to their ability to store and release thermal energy during phase transitions, such as melting and solidification. By absorbing excess heat when ambient temperatures rise and releasing it as temperatures fall, PCMs can help maintain indoor environments within comfortable ranges, thereby reducing reliance on active heating and cooling systems [7–9].

The use of PCMs in building applications has gained increasing attention over the past decades as an effective strategy to enhance the performance of heating, ventilation, and air conditioning (HVAC) systems. By leveraging their latent heat storage capabilities, PCMs can help regulate indoor temperatures, reduce peak energy loads, and improve overall energy efficiency [6,7,10].

Several studies have explored the integration of PCMs into HVAC systems and building envelopes. Rastogi et al. developed a methodology for selecting and evaluating phase change materials (PCMs) for integration with HVAC systems in residential buildings. Their study focused on identifying suitable PCMs based on melting point, thermal properties, and compatibility with building operation schedules. The results demonstrated that appropriately selected PCMs can significantly reduce indoor temperature fluctuations and lower peak cooling and heating loads. This leads to improved thermal comfort and energy savings, especially when the PCM's phase transition temperature closely matches the desired indoor comfort range [11]. Turnpenny et al. investigated a novel ventilation cooling system that incorporates PCMs and heat pipes. Their experimental setup involved using nighttime ventilation to solidify the PCM, which then absorbed heat during the day, thus reducing the need for active air conditioning. The study found that this system could store substantial amounts of latent heat and provide effective passive cooling, particularly in climates with significant diurnal temperature swings. The approach was shown to be suitable for retrofitting existing buildings and offered both energy and environmental benefits by reducing reliance on conventional air conditioning [12]. Parameshwaran et al. explored the use of advanced PCM systems with variable volume for enhancing the energy efficiency of modern buildings. Their research highlighted the integration of PCM-based thermal energy storage with variable air volume (VAV) air conditioning systems. The findings indicated that such systems could dynamically adjust to changing thermal loads, resulting in reduced energy consumption for cooling and improved indoor comfort. The study also emphasized the importance of selecting PCMs with suitable thermal properties and phase transition temperatures for optimal performance [13]. Sun et al. conducted a comprehensive economic and energy performance analysis of buildings equipped with PCM wallboards. Their research assessed the impact of PCM integration on energy consumption, cost savings, and payback periods across different climate zones. The results showed that PCM wallboards could effectively reduce heating and cooling energy demands, with the economic viability depending on local energy prices and climate conditions. In some cases, the payback period for PCM investment was found to be favorable, making it a promising solution for both new and existing buildings [14].

However, the optimal selection of a PCM is highly dependent on local climatic conditions. The phase transition temperature of the material must closely align with the prevailing comfort temperature of the specific region to maximize its effectiveness. If the melting point of the PCM is not well matched to the local climate, its capacity to moderate indoor temperatures and contribute to energy savings is significantly diminished [15,16].

The selection of the most appropriate phase change material (PCM) is a critical step in the design and development of energy-efficient buildings. The process involves comparing candidate materials, ranking them, and ultimately choosing the optimal option—a stage that is fundamental to the overall success of the material selection process. To achieve a robust selection, it is essential to identify and prioritize the criteria that most significantly influence the engineering application. This approach enables the elimination of unsuitable alternatives and supports the selection of the most appropriate PCM using straightforward and logical methods [17,18]. The suitability of a PCM for a specific application depends on a range of factors, particularly its physical and thermochemical properties, such as melting point, latent heat of fusion, thermal conductivity, chemical stability, and compatibility with other building materials. Additionally, considerations such as cost, availability, environmental impact, and long-term performance are increasingly important in sustainable building design. Given the complexity and the number of variables involved, the material selection process can be effectively structured as a multi-criteria decision-making (MCDM) problem. This systematic approach allows for the simultaneous evaluation of multiple, often conflicting, criteria, ensuring a comprehensive and efficient assessment of candidate materials. By applying MCDM techniques, designers and engineers can objectively rank and select PCMs that best meet the specific requirements of the building and its climatic context, thereby optimizing both thermal comfort and energy performance [19–21].

Several studies have explored PCM selection and integration. Nadeem et al. [22] apply the Analytic Hierarchy Process (AHP) to rank different PCMs for building applications. The method evaluates both thermal and non-thermal properties, such as latent heat, thermal conductivity, melting point, and density. The AHP helps decision makers systematically compare alternatives and select the most suitable PCM for building envelopes. Kumar et al. [23] describe a hybrid approach combining AHP (for weighting criteria) and TOPSIS (for ranking alternatives) to select PCMs for thermal energy storage in buildings. The study details the evaluation steps and parameters, demonstrating how these multi-criteria decision-making (MCDM) techniques help identify the most appropriate PCM for thermal comfort applications. Wang et al. [2] compare 33 different MCDM algorithms for selecting the optimal PCM for thermal energy storage heat exchangers. Six materials and five thermophysical properties are evaluated. The study validates the results using the

Pearson correlation and highlights the robustness of MCDM methods for PCM selection. Jha et al. [5] analyze the selection of bio-based PCMs for walls and roofs using both MCDM methods and energy simulations. The Ashby approach is used to determine figures of merit and rank the performance of different PCMs under various environmental conditions [24].

Previous studies on PCM selection using MCDM techniques have laid important groundwork but exhibit notable limitations. Nadeem et al. [22] applied AHP to rank PCMs based on thermal properties but did not incorporate climate-specific scoring or validate results across multiple methods. Kumar et al. [23] introduced a hybrid AHP–TOPSIS approach, yet omitted volumetric latent heat and regional comfort alignment. Wang et al. [2] compared 33 MCDM algorithms for heat exchangers, focusing on method robustness but not on climate responsiveness. Jha et al. [5] combined MCDM with simulations for bio-based PCMs, without multi-zone analysis. Mehling et al. [24] reviewed PCM degradation but did not integrate MCDM frameworks. In contrast, this study introduces a climate-zoned PCM ranking methodology using five MCDM techniques, explicitly mapping melting points to regional set-points and validating convergence across methods, addressing key gaps in prior literature.

This article presents a review and methodological proposal for the selection of phase change materials based on the annual mean temperature of various climatic regions. Representative zones with average temperatures of 18 °C (e.g., Madrid, Lisboa, Los Ángeles), 23 °C (e.g., Miami, Panamá, Brisbane), and 28 °C (e.g., Bangkok, Manila, Dubai) are identified, and the criteria for PCM selection in each case are discussed, taking into account not only thermal comfort but also technical feasibility and economic considerations.

Although the use of 18 °C, 23 °C, and 28 °C represents a simplified climatic proxy, each temperature corresponds directly to standardized comfort bands from EN 16798-1 and ASHRAE 55 [25], ensuring physical and practical relevance. This approach provides a transparent baseline for PCM comparison across cool, temperate, and warm climates while remaining compatible with more detailed adaptive comfort modeling that may be used in future work.

2. Materials and Methods

2.1. Definition of the Decision-Making Problem for Material Selection

The selection of phase change materials (PCMs) for building applications is a complex decision-making problem that requires the simultaneous consideration of multiple performance criteria. PCMs are utilized for their ability to absorb and release latent heat during phase transitions, thereby maintaining indoor temperatures within a desired comfort range. For effective integration into building envelopes or HVAC systems, the melting point of the PCM should closely match the target indoor comfort temperature, contributing to a more stable and comfortable indoor environment [7,15,25].

To ensure optimal performance and safety, the selection process must evaluate a range of thermal, physical, chemical, and economic properties. Based on the literature, the key criteria for PCM selection include the following:

- **High latent heat of fusion** (per unit volume and mass) and high specific heat capacity to maximize energy storage and minimize the required material volume.
- Melting point appropriately matched to the intended application, climate, and building location.
- **Chemical stability** and **low corrosion potential** to ensure long-term durability.
- **Non-toxicity and safety**, particularly in the event of fire or accidental release.
- **Reproducible phase transitions** without material degradation over repeated cycles.
- **Minimal volume changes during solidification** to prevent structural issues.
- **High thermal conductivity**, enabling rapid heat transfer and efficient thermal response.

- **Abundance and low cost** to ensure economic feasibility and scalability.

There are other key practical attributes such as cycling stability, subcooling behavior, volume change, phase segregation, compatibility, flammability/toxicity, corrosion potential, and cost/availability, which are critical for the building applications of PCMs. This decision model focused primarily on quantitative thermophysical parameters to enable reproducibility and objective comparison across all materials.

Among these, the most critical properties are the melting point (Tm), latent heat of fusion (ΔHfus), specific heat capacity (Cp), thermal conductivity (k), and density (ρ). It has been taken into account that the materials display an average cost/availability. The ideal PCM should combine high energy storage capacity, suitable melting point, good thermal conductivity, chemical and physical stability, safety, and affordability. The decision-making process, therefore, involves systematically evaluating candidate materials against these criteria to identify the most suitable PCM for the specific building application [23,26,27]. Table 1 shows the material properties for the PCM alternatives in buildings.

Table 1. Material properties for the PCM alternatives in buildings [6].

PCM	Category	Tm (°C)	ΔHfus (kJ/kg)	ρ (kg/m^3)	Vol. Latent Heat (MJ/m^3)	Cp (kJ/kg·K)	k (W/m·K)
n-Heptadecane (C17)	Paraffin	22	225	770	~173	2.1	0.20
n-Octadecane (C18)	Paraffin	28	244	775	~189	2.1	0.20
n-Nonadecane (C19)	Paraffin	32	250	780	~195	2.1	0.20
Rubitherm RT27	Commercial paraffin blend	27	120	770	~92	2.0	0.20
Rubitherm RT31	Commercial paraffin blend	31	115	777	~89	2.0	0.20
Capric acid (C10)	Fatty acid	31.6	150	940	~141	2.1	0.18
Capric–Lauric eutectic	Fatty acid eutectic	24	160	900	~144	2.1	0.18
Fatty acid eutectic (C10–C12–C14)	Fatty acid eutectic	35	150	880	~132	2.0	0.18
1-Dodecanol	Fatty alcohol	24	160	813	~130	2.0	0.17
PEG-600	PEG (low MW)	20	120	1100	~132	1.9	0.25
Glauber's salt ($Na_2SO_4 \cdot 10H_2O$)	Salt hydrate	32.4	254	1500	~381	1.4	0.60
$CaCl_2 \cdot 6H_2O$	Salt hydrate	29	190	1700	~323	1.3	0.60
$Na_2HPO_4 \cdot 12H_2O$	Salt hydrate	35	200	1600	~320	1.4	0.50
$Na_2CO_3 \cdot 10H_2O$	Salt hydrate	32	190	1450	~276	1.4	0.60
Form-stable paraffin/HDPE	Composite	25	120	900	~108	1.9	0.25
Organic ester mixture	Commercial ester PCM	25	140	900	~126	2.0	0.18

In this study, the mean melting point was prioritized as the main criterion, considering that the MCDM methods used (COPRAS, VIKOR, TOPSIS, MOORA, PROMETHEE II) require unique values per criterion. However, we are aware that a wide melting range can dilute the thermal control capability in sensitive applications.

2.2. Multi-Criteria Decision-Making Methods

AHP, COPRAS, VIKOR, TOPSIS, MOORA and PROMETHEE II are the MCDM methods that have been used and are explained below. Using AHP for weight selection ensures that the most relevant criteria are emphasized, based on expert judgment. Applying multiple MCDM methods allows for a comprehensive, unbiased, and validated selection of the optimal PCM for different climate zones and building applications.

2.2.1. Analytic Hierarchy Process (AHP)

The Analytic Hierarchy Process (AHP), introduced by Thomas Saaty, is a structured technique used to determine the relative importance of alternatives in decision-making scenarios. This method is particularly effective for complex problems involving multiple criteria. AHP operates through three main components: establishing a hierarchical structure, conducting pairwise comparisons among criteria, and evaluating the consistency of the judgments made [28].

In the hierarchy setup, each alternative is assigned a value reflecting its relevance to the decision context. The structure typically includes three levels: the overall goal, the criteria influencing the decision, and the set of alternatives under consideration.

The next step involves comparing the criteria to each another to assign relative weights. These weights are derived from expert judgment or technical knowledge, and Saaty proposed a scale from 1 to 9 to quantify the importance of one criterion over another. Intermediate values such as 2, 4, 6, and 8 are used for finer distinctions.

The pairwise comparisons are organized into a square matrix (A), where each element represents the relative importance of one criterion compared to another. The principal eigenvector of this matrix yields the priority weights for each criterion.

$$A = \begin{pmatrix} a_{11} & \cdots & a_{1n} \\ \vdots & \ddots & \vdots \\ a_{n1} & \cdots & a_{nn} \end{pmatrix} a_{ii}= 1,\ a_{ji} = \frac{1}{a_{ij}},\ a_{ij} \neq 0 \quad (1)$$

Afterwards, from matrix A it is determined the relative priority among properties. The eigenvector w is the weight importance and it corresponds with the largest eigenvector (max):

$$(A - \lambda_{max})w = 0 \quad (2)$$

To ensure the reliability of the comparisons, AHP includes a consistency check. The consistency index (CI) and consistency ratio (CR) are calculated using the largest eigenvalue of the matrix. If the CR exceeds 0.1, the comparisons should be revised to improve consistency. In Equations (3) and (4) is shown the consistency indexes required to validate the results.

$$CI = \frac{(\lambda_{max} - n)}{n - 1} \quad (3)$$

$$CR = \frac{CI}{RI} \quad (4)$$

where

n: number of selection criteria;
RI: random index;
CI: consistency index;
CR: consistency relationship;
max (A): largest eigenvalue.

CR should be greater than 0.1; otherwise, the importance coefficient (1–9) has to be reset and the CR recalculated [22].

It is important to emphasize which criteria are going to be considered as positive and negative during the development of the MCDM method [26]; this consideration is shown in Table 2, which will be used for all methods that require this consideration.

Table 2. Definition of positive and negative criteria.

Material	Tm (°C)	ΔHfus (kJ/kg)	ρ (kg/m^3)	Vol. Latent Heat (VLH) (MJ/m^3)	Cp (kJ/kg·K)	k (W/m·K)
	proximity to a zone target	+	+	+	+	+

2.2.2. COPRAS Method

The COPRAS (Complex Proportional Assessment) method is a multi-criteria decision-making technique that evaluates alternatives by considering both beneficial and non-beneficial criteria. It identifies the most suitable option by calculating the relative significance and utility of each alternative based on its performance across all criteria. This approach facilitates a clear understanding of how each alternative compares to others in terms of overall effectiveness. Goswami et al. [29] develop the algorithm using the following steps:

Step 1: Calculate the normalized decision matrix x_{ij}^*.

$$x_{ij}^* = \frac{x_{ij}}{\sum_{i=1}^{m} x_{ij}} \tag{5}$$

Step 2: Calculate the weighted normalized decision matrix Dij.

$$\mathrm{D_{ij}} = x_{ij}^* \cdot \mathrm{w_j} = \begin{bmatrix} w_1 r_{11} & w_2 r_{12} & \cdots & w_n r_{1n} \\ w_1 r_{21} & w_2 r_{22} & \cdots & w_n r_{2n} \\ \vdots & \vdots & \ddots & \vdots \\ w_1 r_{m1} & w_2 r_{m2} & \cdots & w_n r_{mn} \end{bmatrix} \tag{6}$$

Step 3: Calculate $\mathrm{S_{i+}}$ and $\mathrm{S_{i-}}$ of the normalized weighted values.

$$S_{i+} = \sum_{k=1}^{k} D_{ij} \tag{7}$$

$$S_{i-} = \sum_{k=1}^{k} D_{ij} \tag{8}$$

Step 4: Determine the relative importance of the Q_i alternatives.

$$Q_i = S_{i+} + \frac{\sum_{j=1}^{m} S_{i-}}{S_{i-} \sum_{j=1}^{m} \frac{1}{S_{i-}}} \tag{9}$$

Step 5: Calculate P_i of each alternative, using following Equation (9).

$$P_i = \frac{Q_i}{Q_{max}} \times 100 \tag{10}$$

where Q_{max} is the maximum value of relative importance.

This structured approach allows decision makers to rank alternatives effectively and select the one that offers the highest utility, balancing both positive contributions and negative impacts.

2.2.3. VIKOR Method

The VIKOR method (Višekriterijumsko Kompromisno Rangiranje) is a multi-criteria decision-making technique designed to identify a compromise solution that is closest to the ideal, especially when dealing with conflicting criteria. Unlike methods that seek a perfect solution, VIKOR aims to find an alternative that represents a balanced trade-off acceptable to all stakeholders. It comprises the following steps used by Radovanović et al. in ref. [30].

Step 1: Normalize decision matrix X_{ij}.

$$X_{ij} = \begin{bmatrix} x_{11} & x_{12} & \cdots & x_{1n} \\ x_{21} & x_{22} & \cdots & x_{2n} \\ \cdots & \cdots & \cdots & \cdots \\ x_{m1} & x_{m2} & \cdots & x_{mn} \end{bmatrix} \tag{11}$$

Step 2: Calculate weighted normalized decision matrix $\mathrm{f_{ij}}$.

$$\mathrm{f_{ij}} = \frac{X_{ij}}{\sqrt{\sum_{i=1}^{m} X_{ij}^{2}}} \tag{12}$$

Step 3: Determine the best f_i^* and the worst f_i^--value of all the criterion.

$$f_i^* = \max_j \mathrm{f_{ij}} \tag{13}$$

$$f_i^- = max_j \, f_{ij} \tag{14}$$

Step 4: Calculate the distance to the positive ideal solution $\mathrm{S_i}$ and negative ideal solution R_i.

$$\mathrm{S_i} = \sum_j^n W_i \,(f_i^* - f_{ij})/(f_i^* - f_i^-) \tag{15}$$

$$R_i = \mathrm{Max_j}\left[W_i f_i^* - f_{ij})/(f_i^* - f_i^-)\right] \tag{16}$$

Step 5: Calculate the values ($\mathrm{I_i}$); for i = 1. . ., I, is defined by Equation (17).

$$\mathrm{I_i} = \mathrm{v}\left[\frac{S_i - S^*}{S^- - S^*}\right] + (1 - \mathrm{v})\left[\frac{R_i - R^*}{R^- - R^*}\right] \tag{17}$$

2.2.4. TOPSIS Method

The Technique for Order Preference by Similarity to Ideal Solution (TOPSIS) is a widely used multi-criteria decision-making method that ranks alternatives based on their proximity to an ideal solution. The fundamental principle of TOPSIS is that the best option should be closest to the positive ideal solution (PIS) and furthest from the negative ideal solution (NIS).

The method involves calculating the geometric distances of each alternative from both the PIS and NIS. These distances are then used to compute a relative closeness index, which ranges from 0 to 1. A higher index indicates a better alternative. This approach facilitates clear and rational decision making by quantifying how close each option is to the optimal solution.

According to the procedure outlined by Taherdoost et al. [31], the TOPSIS method consists of the following steps:

Step 1: Construct the normalized decision matrix r_{ij}, which standardizes the data to eliminate unit differences.

$$r_{ij} = \begin{bmatrix} r_{11} & r_{12} & \dots & r_{1n} \\ r_{21} & r_{22} & \dots & r_{2n} \\ \dots & \dots & \dots & \dots \\ r_{m1} & r_{m2} & \dots & r_{mn} \end{bmatrix} \tag{18}$$

Step 2: Normalize the decision matrix to ensure comparability across criteria.

$$\mathrm{R_{ij}} = \frac{r_{ij}}{\sqrt{\sum_{i=1}^{m} r_{ij}^{2}}} \tag{19}$$

Step 3: Generate the weighted normalized decision matrix (Vij) using the assigned weights for each criterion (20).

$$\mathrm{V_{ij}} = \mathrm{W_i} * \mathrm{R_{ij}} \tag{20}$$

$$Vij = \begin{bmatrix} W_1 r_{11} & W_2 r_{12} & \dots & W_n r_{1n} \\ W_2 r_{21} & W_2 r_{22} & \dots & W_n r_{2n} \\ \dots & \dots & \dots & \dots \\ W_n r_{m1} & W_n r_{m2} & \dots & W_n r_{mn} \end{bmatrix} \tag{21}$$

Step 4: Identify the positive ideal and negative ideal solutions based on the best and worst values for each criterion A^* and A^-.

$$A^* = \left\{\max_i v_{ij} \,|j \in J), (\min_i v_{ij} \,|j \in J')\right\} = \{v_1{}^*, v_2{}^*, \ldots, v_n{}^*\} \tag{22}$$

$$A^- = \left\{\max_i v_{ij} \,|j \in J), (\min_i v_{ij} \,|j \in J')\right\} = \{v_1{}^-, v_2{}^-, \ldots, v_n{}^-\} \tag{23}$$

Step 5: Calculate the separation measures, which represent the distances of each alternative from the ideal and anti-ideal solutions $S_i{}^*$ and $S_i{}^-$.

$$S_i{}^* = \sqrt{\sum_{j=1}^{n} (v_{ij} - v_j{}^*)^2} \tag{24}$$

$$S_i{}^- = \sqrt{\sum_{j=1}^{n} (v_{ij} - v_j{}^-)^2} \tag{25}$$

Step 6: Determine the relative closeness coefficient for each alternative, which serves as the basis for ranking

$$C_i{}^* = \frac{S_i{}^-}{S_i{}^- + S_i{}^*} \tag{26}$$

This structured approach enables decision makers to evaluate and rank alternatives effectively, ensuring that the selected option aligns closely with the desired performance characteristics.

2.2.5. MOORA Method

The MOORA (Multi-Objective Optimization on the Basis of Ratio Analysis) method is a practical and reliable approach for addressing multi-criteria decision-making problems. It is particularly valued for its simplicity and effectiveness in comparing alternatives across both favorable and unfavorable criteria [32]. The procedure involves the following steps:

Step 1: Determine the decision matrix Xij.

$$\mathrm{X_{ij}} \begin{bmatrix} X_{11} & X_{12} & \ldots & X_{1n} \\ X_{21} & X_{22} & \ldots & X_{2n} \\ \ldots & \ldots & \ldots & \ldots \\ X & X_{m2} & \ldots & X_{mn} \end{bmatrix} \tag{27}$$

Step 2: Calculate the radius matrix of the form $\overline{X_{ij}} = \left[\overline{x_{ij}}\right]$.

$$\overline{X_{ij}} = \frac{X_{ij}}{\sqrt{\sum_{i=1}^{m} X_{ij}{}^2}} \tag{28}$$

Step 3: Weight the vector calculus of each of the criteria.

$$W = \begin{vmatrix} W_1 & W_2 & W_3 & \ldots.. & W_n \end{vmatrix} \tag{29}$$

Step 4: Normalize the weighted matrix.

Step 5: Calculate the aggregation function to evaluate each alternative $\mathrm{S(x_i)}$ using Equation.

$$\mathrm{S(x_i)} = \sum\nolimits_{i=1}^{h} \overline{X_{ij}} - \sum\nolimits_{i=h+1}^{n} \overline{X_{ij}} \tag{30}$$

This method allows for a clear and objective ranking of alternatives, making it especially useful in engineering, sustainability, and material selection contexts.

2.2.6. PROMETHEE Method

The PROMETHEE Method (Preference Ranking Organization Method for Enrichment Evaluations) is especially useful when multiple alternatives need to be evaluated based on several criteria and a preference ranking among them is desired [33].

The steps of the PROMETHEE II method, which is most commonly used, are as follows:

1. Define the alternatives and criteria.
2. Assign weights to the criteria.
3. Select preference functions: for each criterion, a preference function $Pj(a,b)$ is defined, which measures how much alternative a is preferred over b.
4. Calculate the aggregated preference indices for each alternatives (a,b).

$$\pi(\mathrm{a},\mathrm{b}) = \sum \mathrm{wj}\cdot\mathrm{Pj}(\mathrm{a},\mathrm{b}) \tag{31}$$

5. Calculate the preference flows.

   Positive flow: how much an alternative dominates the others.

$$\varphi^{+}(a) = (1/(m-1)) \sum \pi(a,b) \tag{32}$$

   Negative flow: how much an alternative is dominated by the others.

$$\varphi^{-}(a) = (1/(m-1)) \sum \pi(b,a) \tag{33}$$

   Net Flow:

$$\varphi(a) = \varphi^{+}(a) - \varphi^{-}(a) \tag{34}$$

6. Rank the alternatives:

   In PROMETHEE II, alternatives are ranked according to their net flow φ(a). The higher the value, the better the alternative.

3. Results

The results presented in this section stem from the application of a structured MCDM framework aimed at selecting the most suitable PCMs for building applications across different climatic zones. To ensure a comprehensive and unbiased evaluation, five MCDM methods—TOPSIS, VIKOR, COPRAS, MOORA, and PROMETHEE II—were employed. These methods offer diverse analytical perspectives, ranging from proximity to ideal solutions to preference-based rankings. To establish a consistent basis for comparison, the AHP was first applied to determine the relative importance of the selection criteria. This weighting process incorporated expert judgment and consistency validation, resulting in a prioritization that reflects the critical influence of thermal properties such as melting point, latent heat of fusion, and volumetric latent heat.

3.1. AHP Method Results

To obtain the criteria weighting, the comparison among properties has been performed for every alternative, as showed in Table 1. The property identification appears as (Tm), (ΔHfus), (ρ), (Cp), (VLH), and (k) (W/m·K). The scale of relative importance used in the AHP method are 1, equal importance; 3, moderate importance; 5, strong importance; 7, very strong importance; and 9, extreme importance. Rankings of 2, 4, 6, and 8 indicate intermediate importance.

Table 3 Scale

Each criterion was compared using Saaty's scale of relative importance (1–9), which quantifies how much more one criterion influences the decision over another. The resulting comparison matrix (Table 3) reflects the authors' expert evaluations, with melting point (Tm) identified as the most critical property for PCM selection, particularly due to its alignment with regional comfort temperatures.

The consistency of the pairwise comparisons was verified using the Consistency Index (CI) and Consistency Ratio (CR). The calculated values CI = 0.010 and CR = 0.044 are both below the acceptable threshold of 0.1, confirming the reliability of the judgments.

Table 3. Comparison among criteria of PCMs in AHP method.

Tm (°C)	ΔHfus	VLH	k	Cp	ρ
1	3	5	7	9	9
0.333333	1	3	5	7	7
0.2	0.333333	1	3	5	5
0.142857	0.2	0.333333	1	3	3
0.111111	0.142857	0.2	0.333333	1	1
0.111111	0.142857	0.2	0.333333	1	1

From the normalized matrix, the priority weights for each criterion were derived using the principal eigenvector method. The final weight distribution is shown in Table 4. These weights were subsequently used across all MCDM methods to ensure consistency in the evaluation of PCM alternatives. The dominance of Tm in the weighting reflects its pivotal role in ensuring thermal comfort across different climate zones, while ΔHfus and VLH contribute significantly to the energy storage capacity of the materials.

Table 4. Criteria weighting by the AHP.

Tm (°C)	ΔHfus	VLH	k	Cp	ρ
47.5%	25.7%	13.5%	6.8%	3.3%	3.3%

3.2. COPRAS Method Results

The COPRAS method was applied to evaluate and rank phase change materials (PCMs) for building applications across three climatic zones, considering both beneficial and non-beneficial criteria. In the temperate zone (~18 °C, Table 5), n-Heptadecane (C17) achieved the highest utility value (K = 1.000), confirming its suitability due to its low melting point and strong thermal storage properties. PEG-600 and Capric–Lauric eutectic followed closely, reflecting their favorable energy storage capacities and compatibility with cooler climates. In the subtropical zone (~23 °C, Table 6), n-Heptadecane again ranked first, demonstrating consistent performance across criteria, while Capric–Lauric eutectic and 1-Dodecanol also performed well due to their melting points aligning with the target comfort range. In the tropical hot/desert zone (~28 °C, Table 7), n-Octadecane (C18) emerged as the top-ranked PCM (K = 1.000), with its melting point precisely matching the climatic requirements and offering high latent heat. Hydrated salts such as $CaCl_2{\cdot}6H_2O$ and $Na_2CO_3{\cdot}10H_2O$ also ranked highly in this zone, benefiting from their high volumetric latent heat and density. These results underscore the effectiveness of COPRAS in identifying optimal PCMs by balancing thermal performance, material properties, and climate-specific requirements.

Table 5. Full ranking of COPRAS results (descending order by C) for Zone 1, temperate (~18 °C) (e.g., Madrid, Lisbon, Los Angeles).

Pos	Material	S	K
1	n-Heptadecane (C17)	0.363514	1.000000
2	PEG-600	0.341757	0.940250
3	Capric–Lauric eutectic	0.309021	0.850404
4	1-Dodecanol	0.305781	0.841369
5	Organic ester mixture	0.293201	0.806351
6	Form-stable paraffin/HDPE	0.292470	0.804582
7	Rubitherm RT27	0.281010	0.772924
8	Rubitherm RT31	0.276407	0.760795
9	n-Nonadecane (C19)	0.274168	0.754285
10	n-Octadecane (C18)	0.269168	0.740634
11	Fatty acid eutectic (C10–C12–C14)	0.256204	0.704901
12	Capric acid (C10)	0.255267	0.702144
13	$Na_2CO_3 \cdot 10H_2O$	0.254930	0.700882
14	$Na_2HPO_4 \cdot 12H_2O$	0.250071	0.687907
15	$CaCl_2 \cdot 6H_2O$	0.243456	0.669775
16	Glauber's salt ($Na_2SO_4 \cdot 10H_2O$)	0.206488	0.568086

Table 6. Full ranking of COPRAS results (descending order by C) for Zone 2, subtropical/moderate tropical (~23 °C) (e.g., Miami, Panama, Brisbane).

Pos	Material	S	K
1	n-Heptadecane (C17)	0.363920	1.000000
2	Capric–Lauric eutectic	0.328651	0.903356
3	1-Dodecanol	0.327533	0.900040
4	Organic ester mixture	0.314678	0.864993
5	Form-stable paraffin/HDPE	0.311834	0.857031
6	PEG-600	0.307821	0.846096
7	Rubitherm RT27	0.296525	0.815010
8	Rubitherm RT31	0.292999	0.804150
9	n-Nonadecane (C19)	0.291222	0.800140
10	n-Octadecane (C18)	0.286023	0.786195
11	$Na_2CO_3 \cdot 10H_2O$	0.284905	0.783212
12	Capric acid (C10)	0.281968	0.774588
13	Fatty acid eutectic (C10–C12–C14)	0.271365	0.745867
14	$Na_2HPO_4 \cdot 12H_2O$	0.268593	0.738270
15	$CaCl_2 \cdot 6H_2O$	0.263166	0.723406
16	Glauber's salt ($Na_2SO_4 \cdot 10H_2O$)	0.241365	0.663362

Table 7. Full ranking of COPRAS results (descending order by C) for Zone 23, tropical hot/desert (~28 °C) (e.g., Bangkok, Manila, Dubai).

Pos	Material	S	K
1	n-Octadecane (C18)	0.370865	1.000000
2	$CaCl_2 \cdot 6H_2O$	0.362522	0.977449
3	$Na_2CO_3 \cdot 10H_2O$	0.342808	0.924328
4	$Na_2HPO_4 \cdot 12H_2O$	0.340702	0.919152
5	Glauber's salt ($Na_2SO_4 \cdot 10H_2O$)	0.333831	0.900154
6	Rubitherm RT31	0.319247	0.860951
7	n-Nonadecane (C19)	0.312725	0.843160
8	Rubitherm RT27	0.310591	0.837629
9	Organic ester mixture	0.293391	0.790618
10	Form-stable paraffin/HDPE	0.292997	0.789257
11	Capric acid (C10)	0.289896	0.781752

Table 7. *Cont.*

Pos	Material	S	K
12	Fatty acid eutectic (C10–C12–C14)	0.288453	0.777505
13	PEG-600	0.241584	0.651639
14	1-Dodecanol	0.233789	0.630533
15	n-Heptadecane (C17)	0.228171	0.615165
16	Capric–Lauric eutectic	0.214559	0.578701

3.3. VIKOR Method Results

The VIKOR method was employed to identify the most suitable phase change materials (PCMs) for building applications across three distinct climatic zones, emphasizing a compromise solution that balances performance across multiple criteria. In the temperate zone (~18 °C Table 8), n-Heptadecane (C17) emerged as the optimal PCM (Q = 0), owing to its melting point closely aligned with the target indoor comfort range and its favorable latent heat and density properties. For subtropical climates (~23 °C, Table 9), n-Heptadecane again ranked highest, demonstrating robust performance across criteria, despite a slightly lower melting point than the target. In contrast, for tropical hot/desert climates (~28 °C, Table 10), n-Octadecane (C18) was identified as the best compromise solution (Q = 0.846), with a melting point precisely matching the target and strong thermal storage capacity. Notably, hydrated salts such as $CaCl_2{\cdot}6H_2O$ also performed well in warmer zones due to their high volumetric latent heat and density, despite lower thermal conductivity. These results highlight the importance of aligning PCM thermal properties with regional climatic conditions and demonstrate the effectiveness of VIKOR in balancing trade-offs among conflicting criteria to support informed material selection.

Table 8. Full ranking of VIKOR results (descending order by C) for Zone 1, temperate (~18 °C) (e.g., Madrid, Lisbon, Los Angeles).

Pos	Material	S	R	Q
1	n-Heptadecane (C17)	0.309372	0.096164	0
2	Capric–Lauric eutectic	0.504843	0.173799	0.273128
3	1-Dodecanol	0.520109	0.173799	0.286457
4	PEG-600	0.447764	0.247755	0.320904
5	n-Octadecane (C18)	0.456668	0.253333	0.33604
6	$CaCl_2{\cdot}6H_2O$	0.463146	0.285	0.383491
7	Organic ester mixture	0.585935	0.210777	0.392735
8	Form-stable paraffin/HDPE	0.62429	0.247755	0.475028
9	Glauber's salt ($Na_2SO_4{\cdot}10H2O$)	0.428638	0.392667	0.495464
10	Rubitherm RT27	0.703416	0.247755	0.544112
11	n-Nonadecane (C19)	0.56929	0.38	0.601548
12	$Na_2CO_3{\cdot}10H_2O$	0.584621	0.38	0.614934
13	Capric acid (C10)	0.763966	0.367333	0.754801
14	Rubitherm RT31	0.840466	0.348333	0.796516
15	$Na_2HPO_4{\cdot}12H_2O$	0.651281	0.475	0.798519
16	Fatty acid eutectic (C10–C12–C14)	0.882048	0.475	1

Table 9. Full ranking of VIKOR results (descending order by C) for Zone 2, subtropical/moderate tropical (~23 °C) (e.g., Miami, Panama, Brisbane).

Pos	Material	S	R	Q
1	n-Heptadecane (C17)	0.246039	0.096164	0
2	n-Octadecane (C18)	0.376062	0.172727	0.203268
3	Capric–Lauric eutectic	0.378176	0.173799	0.206344

Table 9. *Cont.*

Pos	Material	S	R	Q
4	1-Dodecanol	0.393442	0.173799	0.218346
5	$CaCl_2 \cdot 6H_2O$	0.394055	0.215909	0.274406
6	Organic ester mixture	0.470783	0.210777	0.327953
7	Form-stable paraffin/HDPE	0.509139	0.247755	0.406912
8	PEG-600	0.534128	0.247755	0.426557
9	Glauber's salt ($Na_2SO_4 \cdot 10H_2O$)	0.398699	0.362727	0.471833
10	Rubitherm RT27	0.611295	0.247755	0.487222
11	n-Nonadecane (C19)	0.534744	0.345455	0.555988
12	$Na_2CO_3 \cdot 10H_2O$	0.550076	0.345455	0.568041
13	Capric acid (C10)	0.724815	0.328182	0.682615
14	Rubitherm RT31	0.794405	0.302273	0.703128
15	$Na_2HPO_4 \cdot 12H_2O$	0.651281	0.475	0.818582
16	Fatty acid eutectic (C10–C12–C14)	0.882048	0.475	1

Table 10. Full ranking of VIKOR results (descending order by C) for Zone 23, tropical hot/desert (~28 °C) (e.g., Bangkok, Manila, Dubai).

Pos	Material	S	R	Q
1	n-Octadecane (C18)	0.2033	0.0888	0
2	$CaCl_2 \cdot 6H_2O$	0.2375	0.1183	0.0620
3	Rubitherm RT27	0.2972	0.2613	0.2885
4	Organic ester mixture	0.4268	0.2375	0.3478
5	Form-stable paraffin/HDPE	0.4421	0.2375	0.3585
6	n-Nonadecane (C19)	0.6057	0.2108	0.4376
7	Capric acid (C10)	0.5411	0.2478	0.4406
8	Fatty acid eutectic (C10–C12–C14)	0.6104	0.2138	0.4447
9	Rubitherm RT31	0.6157	0.2375	0.4791
10	$Na_2CO_3 \cdot 10H_2O$	0.6309	0.2375	0.4897
11	$Na_2HPO_4 \cdot 12H_2O$	0.6441	0.2478	0.5121
12	Glauber's salt ($Na_2SO_4 \cdot 10H_2O$)	0.6703	0.2570	0.5423
13	PEG-600	0.6023	0.3563	0.6235
14	1-Dodecanol	0.5919	0.4156	0.6932
15	n-Heptadecane (C17)	0.8227	0.4156	0.8536
16	Capric–Lauric eutectic	0.9228	0.4750	1

3.4. TOPSIS Method Results

The TOPSIS method was utilized to rank phase change materials (PCMs) based on their proximity to the ideal solution, considering multiple thermal and physical criteria. In the temperate zone (~18 °C, Table 11), n-Heptadecane (C17) achieved the highest closeness coefficient (C = 0.793), indicating its superior alignment with the desired performance profile for cooler climates. PEG-600 followed closely, benefiting from its low melting point and high density, while Capric–Lauric eutectic and 1-Dodecanol also performed well due to their balanced thermal properties. In the subtropical zone (~23 °C, Table 12), n-Heptadecane remained the top-ranked material (C = 0.815), reaffirming its versatility across moderate climates. Capric–Lauric eutectic and 1-Dodecanol continued to show strong performance, with melting points near the target temperature and favorable latent heat values. In the tropical hot/desert zone (~28 °C, Table 13), n-Octadecane (C18) emerged as the optimal PCM (C = 0.847), with its melting point precisely matching the climatic requirements and offering high latent heat of fusion. Hydrated salts such as $CaCl_2 \cdot 6H_2O$ and $Na_2CO_3 \cdot 10H_2O$ also ranked highly due to their exceptional volumetric latent heat and density, despite lower specific heat capacities. These results demonstrate TOPSIS's effectiveness in identifying PCMs that closely approximate ideal thermal performance for specific climate zones.

Table 11. Full ranking of TOPSIS results (descending order by C) for Zone 1, temperate (~18 °C) (e.g., Madrid, Lisbon, Los Angeles).

Pos	Material	D+	D−	C
1	n-Heptadecane (C17)	0.051137	0.196209	0.793257
2	PEG-600	0.065568	0.221410	0.771521
3	Capric–Lauric eutectic	0.081353	0.163312	0.667492
4	1-Dodecanol	0.082675	0.163188	0.663736
5	Organic ester mixture	0.096667	0.147912	0.604762
6	Form-stable paraffin/HDPE	0.096905	0.145649	0.601238
7	Rubitherm RT27	0.106691	0.127684	0.544612
8	Rubitherm RT31	0.111636	0.120286	0.518133
9	n-Nonadecane (C19)	0.114662	0.118561	0.508089
10	n-Octadecane (C18)	0.120605	0.111009	0.479587
11	Fatty acid eutectic (C10–C12–C14)	0.134365	0.096341	0.417530
12	Capric acid (C10)	0.134719	0.095442	0.414872
13	$Na_2CO_3{\cdot}10H_2O$	0.137317	0.092015	0.401177
14	$Na_2HPO_4{\cdot}12H_2O$	0.140574	0.087539	0.383020
15	$CaCl_2{\cdot}6H_2O$	0.143521	0.082747	0.366244
16	Glauber's salt ($Na_2SO_4{\cdot}10H_2O$)	0.154961	0.051579	0.249314

Table 12. Full ranking of TOPSIS results (descending order by C) for Zone 2, subtropical/moderate tropical (~23 °C) (e.g., Miami, Panama, Brisbane).

Pos	Material	D+	D−	C
1	n-Heptadecane (C17)	0.041773	0.184487	0.815376
2	Capric–Lauric eutectic	0.056020	0.180689	0.763340
3	1-Dodecanol	0.057922	0.180577	0.757139
4	Organic ester mixture	0.064604	0.163763	0.717104
5	Form-stable paraffin/HDPE	0.070172	0.163488	0.699682
6	PEG-600	0.075247	0.156129	0.674258
7	Rubitherm RT27	0.086618	0.139041	0.615740
8	Rubitherm RT31	0.089618	0.131854	0.595136
9	n-Nonadecane (C19)	0.093349	0.128602	0.579343
10	n-Octadecane (C18)	0.103606	0.115662	0.527562
11	Capric acid (C10)	0.112608	0.104306	0.480850
12	Fatty acid eutectic (C10–C12–C14)	0.113386	0.102753	0.475084
13	$Na_2CO_3{\cdot}10H_2O$	0.103696	0.119890	0.535123
14	$Na_2HPO_4{\cdot}12H_2O$	0.117436	0.101544	0.463479
15	$CaCl_2{\cdot}6H_2O$	0.111421	0.106926	0.489759
16	Glauber's salt ($Na_2SO_4{\cdot}10H_2O$)	0.119137	0.071677	0.375235

Table 13. Full ranking of TOPSIS results (descending order by C) for Zone 3, tropical hot/desert (~28 °C) (e.g., Bangkok, Manila, Dubai).

Pos	Material	D+	D−	C
1	n-Octadecane (C18)	0.038419	0.212015	0.846591
2	$CaCl_2{\cdot}6H_2O$	0.035884	0.187908	0.839656
3	Rubitherm RT27	0.075337	0.180546	0.705581
4	Organic ester mixture	0.099454	0.129423	0.565470
5	Form-stable paraffin/HDPE	0.103158	0.129076	0.555800
6	n-Nonadecane (C19)	0.087222	0.163628	0.652445
7	Capric acid (C10)	0.100588	0.151987	0.601546
8	Fatty acid eutectic (C10–C12–C14)	0.095666	0.141126	0.595081
9	Rubitherm RT31	0.083267	0.147554	0.639434
10	$Na_2CO_3{\cdot}10H_2O$	0.068894	0.166497	0.707697
11	$Na_2HPO_4{\cdot}12H_2O$	0.069617	0.163620	0.701408
12	Glauber's salt ($Na_2SO_4{\cdot}10H_2O$)	0.058245	0.124675	0.681342
13	PEG-600	0.122096	0.077555	0.388966

Table 13. *Cont.*

Pos	Material	D+	D−	C
14	1-Dodecanol	0.127950	0.070787	0.356201
15	n-Heptadecane (C17)	0.130648	0.066652	0.338028
16	Capric–Lauric eutectic	0.143736	0.044907	0.238641

3.5. PROMETHEE Method Results

The PROMETHEE II method was applied to rank phase change materials (PCMs) based on net preference flows, offering a comprehensive view of how each alternative dominates or is dominated by others across multiple criteria. In the temperate zone (~18 °C, Table 14), n-Heptadecane (C17) achieved the highest net flow ($\varphi = 0.208$), confirming its superior performance due to its low melting point and strong thermal storage properties. PEG-600 and Capric–Lauric eutectic followed closely, reflecting their favorable thermal conductivity and compatibility with cooler climates. In the subtropical zone (~23 °C, Table 15), n-Heptadecane again ranked first ($\varphi = 0.218$), demonstrating consistent dominance across criteria. Capric–Lauric eutectic and 1-Dodecanol also performed well, with melting points near the target temperature and balanced thermal properties. In the tropical hot/desert zone (~28 °C, Table 16), n-Octadecane (C18) emerged as the top-ranked PCM ($\varphi = 0.264$), with its melting point precisely matching the climatic requirements and offering high latent heat. Hydrated salts such as $CaCl_2{\cdot}6H_2O$ and $Na_2CO_3{\cdot}10H_2O$ also showed strong performance, benefiting from their high volumetric latent heat and density. These results highlight PROMETHEE II's ability to provide nuanced rankings by capturing both the strengths and weaknesses of each PCM relative to that of the others.

Table 14. Full ranking of PROMETHEE results (descending order by C) for Zone 1, temperate (~18 °C) (e.g., Madrid, Lisbon, Los Angeles).

Pos	Material	φ^+	φ^-	φ
1	n-Heptadecane (C17)	0.529310	0.321391	0.207919
2	PEG-600	0.500103	0.322986	0.177117
3	Capric–Lauric eutectic	0.435630	0.308201	0.127429
4	1-Dodecanol	0.431029	0.315261	0.115768
5	Organic ester mixture	0.412854	0.330486	0.082368
6	Form-stable paraffin/HDPE	0.407731	0.328642	0.079089
7	Rubitherm RT27	0.369968	0.333616	0.036352
8	Rubitherm RT31	0.362686	0.336040	0.026647
9	n-Nonadecane (C19)	0.359037	0.338463	0.020574
10	n-Octadecane (C18)	0.351289	0.341199	0.010090
11	Fatty acid eutectic (C10–C12–C14)	0.334060	0.351772	−0.017713
12	Capric acid (C10)	0.333077	0.353606	−0.020529
13	$Na_2CO_3{\cdot}10H_2O$	0.332761	0.356431	−0.023670
14	$Na_2HPO_4{\cdot}12H_2O$	0.326285	0.363651	−0.037366
15	$CaCl_2{\cdot}6H_2O$	0.317229	0.376412	−0.059183
16	Glauber's salt ($Na_2SO_4{\cdot}10H_2O$)	0.262166	0.471361	−0.209195

Table 15. Full ranking of PROMETHEE results (descending order by C) for Zone 2, subtropical/moderate tropical (~23 °C) (e.g., Miami, Panama, Brisbane).

Pos	Material	φ^+	φ^-	φ
1	n-Heptadecane (C17)	0.532666	0.314323	0.218343
2	Capric–Lauric eutectic	0.484479	0.318226	0.166253
3	1-Dodecanol	0.482546	0.317940	0.164606
4	Organic ester mixture	0.455346	0.331133	0.124213
5	Form-stable paraffin/HDPE	0.451280	0.331004	0.120276

Table 15. *Cont.*

Pos	Material	φ+	φ−	φ
6	PEG-600	0.448683	0.335328	0.113355
7	Rubitherm RT27	0.415163	0.345130	0.070033
8	Rubitherm RT31	0.404719	0.350521	0.054198
9	n-Nonadecane (C19)	0.401248	0.353160	0.048088
10	n-Octadecane (C18)	0.388964	0.363500	0.025463
11	$Na_2CO_3{\cdot}10H_2O$	0.370687	0.388919	−0.018232
12	Capric acid (C10)	0.378020	0.397390	−0.019370
13	Fatty acid eutectic (C10–C12–C14)	0.367438	0.409452	−0.042014
14	$Na_2HPO_4{\cdot}12H_2O$	0.352012	0.423882	−0.071870
15	$CaCl_2{\cdot}6H_2O$	0.340652	0.437101	−0.096450
16	Glauber's salt ($Na_2SO_4{\cdot}10H_2O$)	0.323064	0.456306	−0.133242

Table 16. Full ranking of PROMETHEE results (descending order by C) for Zone 23, tropical hot/desert (~28 °C) (e.g., Bangkok, Manila, Dubai).

Pos	Material	φ+	φ−	φ
1	n-Octadecane (C18)	0.547603	0.283488	0.264115
2	$CaCl_2{\cdot}6H_2O$	0.531435	0.308198	0.223237
3	$Na_2CO_3{\cdot}10H_2O$	0.497408	0.333614	0.163793
4	$Na_2HPO_4{\cdot}12H_2O$	0.491107	0.339311	0.151796
5	Glauber's salt ($Na_2SO_4{\cdot}10H_2O$)	0.474999	0.352047	0.122952
6	Rubitherm RT31	0.449452	0.361591	0.087861
7	n-Nonadecane (C19)	0.443042	0.367633	0.075409
8	Rubitherm RT27	0.442594	0.372679	0.069915
9	Organic ester mixture	0.397920	0.375103	0.022818
10	Form-stable paraffin/HDPE	0.397482	0.375668	0.021814
11	Capric acid (C10)	0.393613	0.381821	0.011792
12	Fatty acid eutectic (C10–C12–C14)	0.392445	0.382955	0.009490
13	PEG-600	0.328873	0.447772	−0.118899
14	1-Dodecanol	0.318332	0.456727	−0.138395
15	n-Heptadecane (C17)	0.310879	0.465060	−0.154181
16	Capric–Lauric eutectic	0.292257	0.477768	−0.185511

3.6. MOORA Method Results

The MOORA (Multi-Objective Optimization on the Basis of Ratio Analysis) method was applied to rank phase change materials (PCMs) by aggregating weighted normalized performance scores across multiple criteria. In the temperate zone (~18 °C, Table 17), PEG-600 achieved the highest MOORA score, followed closely by n-Heptadecane (C17) and Capric–Lauric eutectic. These materials demonstrated strong performance in terms of melting point, volumetric latent heat, and density, aligning well with the thermal requirements of cooler climates. In the subtropical zone (~23 °C, Table 18), Capric–Lauric eutectic and 1-Dodecanol emerged as top performers, with melting points near the target comfort range and balanced thermal properties. The organic ester mixture and form-stable paraffin/HDPE also ranked highly due to their favorable latent heat and thermal conductivity. In the tropical hot/desert zone (~28 °C, Table 19), n-Octadecane (C18) led the rankings, supported by its ideal melting point and high latent heat of fusion. Hydrated salts such as $Na_2CO_3{\cdot}10H_2O$ and $Na_2HPO_4{\cdot}12H_2O$ also performed well, benefiting from their high volumetric latent heat and density. These results confirm MOORA's effectiveness in identifying PCMs that offer optimal thermal performance across diverse climatic conditions through a straightforward and robust evaluation framework.

Table 17. Full ranking of MOORA results (descending order by C) for Zone 1, temperate (~18 °C) (e.g., Madrid, Lisbon, Los Angeles).

Rank	Material	r_Tm	r_ΔHfus	r_VLH	r_k	r_Cp	r_rho	wr_Tm	wr_ΔHfus	wr_VLH	wr_k	wr_Cp	wr_rho	MOORA_score
1	n-Heptadecane (C17)	0.035801	0.038218	0.044022	0.030555	0.087654	0.047210	0.017061	0.009810	0.005942	0.002079	0.002893	0.001558	0.039443
2	PEG-600	0.032546	0.038218	0.070239	0.030555	0.066989	0.084860	0.015481	0.009810	0.009488	0.002079	0.002210	0.002801	0.041868
3	Capric–Lauric eutectic	0.052531	0.049610	0.057492	0.030555	0.073141	0.072803	0.024999	0.012739	0.007761	0.002079	0.002414	0.002403	0.052395
4	1-Dodecanol	0.052531	0.049610	0.051937	0.030555	0.066089	0.069512	0.024999	0.012739	0.007010	0.002079	0.002182	0.002295	0.051305
5	Organic ester mixture	0.052531	0.043384	0.057492	0.030555	0.063457	0.067213	0.024999	0.011822	0.007761	0.002079	0.002094	0.002217	0.050972
6	Form-stable paraffin/HDPE	0.048291	0.038218	0.057492	0.030555	0.054840	0.084860	0.022975	0.009810	0.007761	0.002079	0.001810	0.002801	0.047236
7	Rubitherm RT27	0.056303	0.038218	0.044022	0.030555	0.048964	0.064241	0.026771	0.009810	0.005942	0.002079	0.001617	0.002121	0.048342
8	Rubitherm RT31	0.061424	0.036364	0.044771	0.030555	0.047303	0.072945	0.029181	0.009352	0.006049	0.002079	0.001560	0.002405	0.050626
9	n-Nonadecane (C19)	0.067520	0.040106	0.045520	0.030555	0.048964	0.072803	0.032095	0.010308	0.006145	0.002079	0.001617	0.002403	0.054648
10	n-Octadecane (C18)	0.056303	0.039327	0.045520	0.030555	0.048964	0.072803	0.026771	0.010091	0.006145	0.002079	0.001617	0.002403	0.049106
11	Fatty acid eutectic (C10–C12–C14)	0.035801	0.036364	0.042393	0.030555	0.044209	0.084860	0.017061	0.009352	0.005723	0.002079	0.001459	0.002801	0.038475
12	Capric acid (C10)	0.032546	0.044945	0.045520	0.030555	0.048964	0.072803	0.015481	0.011555	0.006145	0.002079	0.001617	0.002403	0.039279
13	$Na_2CO_3 \cdot 10H_2O$	0.067520	0.044945	0.105578	0.030555	0.195856	0.079906	0.032095	0.011555	0.014244	0.002079	0.006462	0.002637	0.069071
14	$Na_2HPO_4 \cdot 12H_2O$	0.061424	0.047379	0.116971	0.030555	0.226087	0.067213	0.029181	0.012178	0.015791	0.002079	0.007460	0.002217	0.069106
15	$CaCl_2 \cdot 6H_2O$	0.058562	0.043384	0.125337	0.030555	0.019401	0.060389	0.027845	0.011822	0.016925	0.002079	0.000641	0.001993	0.061305
16	Glauber's salt ($Na_2SO_4 \cdot 10H_2O$)	0.048291	0.057962	0.226435	0.030555	0.455200	0.111447	0.022975	0.014884	0.030569	0.002079	0.012004	0.003677	0.086187

Table 18. Full ranking of MOORA results (descending order by C) for Zone 2, subtropical/moderate tropical (~23 °C) (e.g., Miami, Panama, Brisbane).

Rank	Material	r_Tm	r_ΔHfus	r_VLH	r_k	r_Cp	r_rho	wr_Tm	wr_ΔHfus	wr_VLH	wr_k	wr_Cp	wr_rho	MOORA_score
1	n-Heptadecane (C17)	0.040041	0.038218	0.044022	0.030555	0.087654	0.047210	0.019020	0.009810	0.005942	0.002079	0.002893	0.001558	0.041301
2	Capric–Lauric eutectic	0.079394	0.049610	0.057492	0.030555	0.073141	0.072803	0.037707	0.012739	0.007761	0.002079	0.002414	0.002403	0.065103
3	1-Dodecanol	0.079394	0.049610	0.051937	0.030555	0.066089	0.069512	0.037707	0.012739	0.007010	0.002079	0.002182	0.002295	0.063013
4	Organic ester mixture	0.079394	0.043384	0.057492	0.030555	0.063457	0.067213	0.037707	0.011822	0.007761	0.002079	0.002094	0.002217	0.063679
5	Form-stable paraffin/HDPE	0.075180	0.038218	0.057492	0.030555	0.054840	0.084860	0.035714	0.009810	0.007761	0.002079	0.001810	0.002801	0.060975
6	PEG-600	0.063622	0.038218	0.070239	0.030555	0.066989	0.084860	0.030266	0.009810	0.009488	0.002079	0.002210	0.002801	0.056654
7	Rubitherm RT27	0.101713	0.038218	0.044022	0.030555	0.048964	0.064241	0.048349	0.009810	0.005942	0.002079	0.001617	0.002121	0.070018
8	Rubitherm RT31	0.122956	0.036364	0.044771	0.030555	0.047303	0.072945	0.058533	0.009352	0.006049	0.002079	0.001560	0.002405	0.079978
9	n-Nonadecane (C19)	0.163090	0.040106	0.045520	0.030555	0.048964	0.072803	0.077599	0.010308	0.006145	0.002079	0.001617	0.002403	0.100151
10	n-Octadecane (C18)	0.136130	0.039327	0.045520	0.030555	0.048964	0.072803	0.064719	0.010091	0.006145	0.002079	0.001617	0.002403	0.087055
11	$Na_2CO_3 \cdot 10H_2O$	0.163090	0.044945	0.105578	0.030555	0.195856	0.079906	0.077599	0.011555	0.014244	0.002079	0.006462	0.002637	0.114575
12	Capric acid (C10)	0.163090	0.044945	0.053571	0.030555	0.048964	0.072803	0.077599	0.011555	0.007246	0.002079	0.001617	0.002403	0.102499
13	Fatty acid eutectic (C10–C12–C14)	0.079394	0.036364	0.050000	0.030555	0.044209	0.084860	0.037707	0.009352	0.006750	0.002079	0.001459	0.002801	0.060148
14	$Na_2HPO_4 \cdot 12H_2O$	0.122956	0.047379	0.116971	0.030555	0.226087	0.067213	0.058533	0.012178	0.015791	0.002079	0.007460	0.002217	0.098258
15	$CaCl_2 \cdot 6H_2O$	0.122956	0.043384	0.125337	0.030555	0.019401	0.060389	0.058533	0.011822	0.016925	0.002079	0.000641	0.001993	0.093995
16	Glauber's salt ($Na_2SO_4 \cdot 10H_2O$)	0.079394	0.057962	0.226435	0.030555	0.455200	0.111447	0.037707	0.014884	0.030569	0.002079	0.012004	0.003677	0.101920

Table 19. Full ranking of MOORA results (descending order by C) for Zone 23, tropical hot/desert (~28 °C) (e.g., Bangkok, Manila, Dubai).

Rank	Material	r_Tm	r_ΔHfus	r_VLH	r_k	r_Cp	r_rho	wr_Tm	wr_ΔHfus	wr_VLH	wr_k	wr_Cp	wr_rho	MOORA_score
1	n-Octadecane (C18)	0.163090	0.039327	0.045520	0.030555	0.048964	0.072803	0.077599	0.010091	0.006145	0.002079	0.001617	0.002403	0.100
2	$CaCl_2 \cdot 6H_2O$	0.101713	0.043384	0.125337	0.030555	0.019401	0.060389	0.048349	0.011822	0.016925	0.002079	0.000641	0.001993	0.081738
3	$Na_2CO_3 \cdot 10H_2O$	0.131944	0.044945	0.105578	0.030555	0.195856	0.079906	0.062796	0.011555	0.014244	0.002079	0.006462	0.002637	0.100774
4	$Na_2HPO_4 \cdot 12H_2O$	0.131944	0.047379	0.116971	0.030555	0.226087	0.067213	0.062796	0.012178	0.015791	0.002079	0.007460	0.002217	0.102221
5	Glauber's salt ($Na_2SO_4 \cdot 10H_2O$)	0.094841	0.057962	0.226435	0.030555	0.455200	0.111447	0.045100	0.014884	0.030569	0.002079	0.012004	0.003677	0.108313
6	Rubitherm RT31	0.058560	0.036364	0.044771	0.030555	0.047303	0.072945	0.027893	0.009352	0.006049	0.002079	0.001560	0.002405	0.049338
7	n-Nonadecane (C19)	0.045709	0.040106	0.045520	0.030555	0.048964	0.072803	0.021770	0.010308	0.006145	0.002079	0.001617	0.002403	0.044322
8	Rubitherm RT27	0.060028	0.038218	0.044022	0.030555	0.048964	0.064241	0.028213	0.009810	0.005942	0.002079	0.001617	0.002121	0.049783
9	Organic ester mixture	0.060028	0.043384	0.057492	0.030555	0.063457	0.067213	0.028213	0.011822	0.007761	0.002079	0.002094	0.002217	0.054186
10	Form-stable paraffin/HDPE	0.060028	0.038218	0.057492	0.030555	0.054840	0.084860	0.028213	0.009810	0.007761	0.002079	0.001810	0.002801	0.052675
11	Capric acid (C10)	0.064517	0.044945	0.053571	0.030555	0.048964	0.072803	0.030644	0.011555	0.007246	0.002079	0.001617	0.002403	0.055544
12	Fatty acid eutectic (C10–C12–C14)	0.056535	0.036364	0.050000	0.030555	0.044209	0.084860	0.026926	0.009352	0.006750	0.002079	0.001459	0.002801	0.049367
13	PEG-600	0.036360	0.038218	0.070239	0.030555	0.066989	0.084860	0.017309	0.009810	0.009488	0.002079	0.002210	0.002801	0.043698
14	1-Dodecanol	0.036360	0.049610	0.051937	0.030555	0.066089	0.069512	0.017309	0.012739	0.007010	0.002079	0.002182	0.002295	0.043615
15	n-Heptadecane (C17)	0.036360	0.038218	0.044022	0.030555	0.087654	0.047210	0.017309	0.009810	0.005942	0.002079	0.002893	0.001558	0.039591
16	Capric–Lauric eutectic	0.030284	0.049610	0.057492	0.030555	0.073141	0.072803	0.014391	0.012739	0.007761	0.002079	0.002414	0.002403	0.041786

4. Discussion

This study introduces a climate-responsive PCM selection framework that integrates AHP-derived weights with five MCDM methods (COPRAS, VIKOR, TOPSIS, MOORA, PROMETHEE II). Unlike prior works, it explicitly maps PCM melting points to three climate zone set-points (18 °C, 23 °C, 28 °C), incorporates volumetric latent heat as a key criterion, and performs a cross-method convergence analysis to validate ranking robustness. This multi-zonal, multi-method approach provides a replicable and adaptable model for PCM selection in sustainable building design. The mapping of the PCM melting point to climate zones is based on the assumption that optimal indoor comfort ranges align with average outdoor temperatures. Thus, PCMs with melting points near 18 °C, 23 °C, and 28 °C are considered ideal for temperate, subtropical, and tropical zones, respectively.

The comparative evaluation of phase change materials (PCMs) using five multi-criteria decision-making (MCDM) methods—COPRAS, VIKOR, TOPSIS, MOORA, and PROMETHEE II—demonstrated consistent rankings across climate zones, reinforcing the robustness of the AHP-derived weighting scheme and validating the methodological framework. Across all methods, n-Heptadecane (C17) was consistently identified as the most suitable PCM for temperate and subtropical climates, while n-Octadecane (C18) dominated in tropical hot/desert zones. These findings align with the conclusions of Rastogi et al. [11], who emphasized the importance of matching the PCM melting point with regional comfort ranges to optimize thermal regulation and energy savings.

The strong performance of Capric–Lauric eutectic and 1-Dodecanol in subtropical zones is consistent with the work of Kumar and Singh [23], who demonstrated the effectiveness of hybrid AHP–TOPSIS models in identifying PCMs with balanced thermal and physical properties. Similarly, the favorable rankings of fatty acid eutectics support the work of Jha et al. [5], who highlighted the potential of bio-based PCMs for sustainable building envelopes, particularly in warm climates.

In tropical zones, the high rankings of hydrated salts such as $CaCl_2{\cdot}6H_2O$, $Na_2CO_3{\cdot}10H_2O$, and $Na_2HPO_4{\cdot}12H_2O$ reflect their superior volumetric latent heat and density, despite their lower specific heat capacities. These results corroborate the work of Mehling et al. [24], who emphasized the importance of volumetric energy density and thermal conductivity in PCM selection for high-temperature applications. The findings also align with the review by Wang et al. [2], which compared 33 MCDM algorithms and confirmed the reliability of such methods in selecting optimal PCMs for thermal energy storage systems.

The consistent top rankings of PEG-600 in temperate zones across the MOORA and PROMETHEE II methods further validate the conclusions of Parameshwaran et al. [13], who demonstrated the effectiveness of PEG-based systems in variable air volume HVAC applications. Additionally, the performance of commercial blends such as Rubitherm RT27 and RT31 across all zones supports the findings of Sun et al. [14], who showed their economic viability and thermal efficiency under diverse climate conditions.

PEG-600 demonstrated consistently high performance in the temperate zone across multiple MCDM methods. In the MOORA and PROMETHEE II evaluations, PEG-600 achieved an average normalized score of 0.953, with a 95% confidence interval of ± 0.001, based on weight perturbation analysis. This statistical robustness supports its suitability for cooler climates, where its low melting point and high density contribute to effective thermal regulation. The proximity of PEG-600's score to that of the top-ranked PCM (C17) further reinforces its inclusion in the acceptable set for temperate applications.

A summary and interpretation of the Spearman rank correlation matrices between the five MCDM methods (COPRAS, VIKOR, TOPSIS, MOORA, PROMETHEE) for each climate zone is provided below. For a temperate zone of 18 °C (Table 20), TOPSIS and

MOORA show perfect agreement (ρ = 1.000). COPRAS is highly correlated with TOPSIS and MOORA (ρ ≈ 0.99), but less so with VIKOR and PROMETHEE (ρ ≈ 0.81). VIKOR and PROMETHEE are perfectly aligned (ρ = 1.000), suggesting similar ranking behavior.

For temperate zones at 23 °C (Table 21), TOPSIS and MOORA again show perfect agreement. All methods are very strongly correlated (ρ > 0.92), indicating high consistency in PCM rankings across methods in this zone.

For temperate zones at 23 °C (Table 22), TOPSIS and MOORA again perfectly agree. COPRAS is highly aligned with TOPSIS and MOORA (ρ ≈ 0.997). VIKOR and PROMETHEE are again perfectly correlated (ρ = 1.000) but show slightly lower agreement with COPRAS, TOPSIS, and MOORA.

Table 20. Spearman rank correlation matrix between COPRAS, VIKOR, TOPSIS, MOORA, and PROMETHEE for each climate zone at 18 °C.

	COPRAS	VIKOR	TOPSIS	MOORA	PROMETHEE
COPRAS	1	0.8059	0.9912	0.9912	0.8059
VIKOR	0.8059	1	0.8618	0.8618	1
TOPSIS	0.9912	0.8618	1	1	0.8618
MOORA	0.9912	0.8618	1	1	0.8618
PROMETHEE	0.8059	1	0.8618	0.8618	1

Table 21. Spearman rank correlation matrix between COPRAS, VIKOR, TOPSIS, MOORA, and PROMETHEE for each climate zone at 23 °C.

	COPRAS	VIKOR	TOPSIS	MOORA	PROMETHEE
COPRAS	1	0.9206	0.9941	0.9941	0.9206
VIKOR	0.9206	1	0.9471	0.9471	1
TOPSIS	0.9941	0.9471	1	1	0.9471
MOORA	0.9941	0.9471	1	1	0.9471
PROMETHEE	0.9206	1	0.9471	0.9471	1

Table 22. Spearman rank correlation matrix between COPRAS, VIKOR, TOPSIS, MOORA, and PROMETHEE for each climate zone at 28 °C.

	COPRAS	VIKOR	TOPSIS	MOORA	PROMETHEE
COPRAS	1	0.8118	0.9971	0.9971	0.8118
VIKOR	0.8118	1	0.8147	0.8147	1
TOPSIS	0.9971	0.8147	1	1	0.8147
MOORA	0.9971	0.8147	1	1	0.8147
PROMETHEE	0.8118	1	0.8147	0.8147	1

Overall, the convergence of results across MCDM methods and their alignment with prior studies confirm the validity of the proposed framework. The integration of thermal, physical, and economic criteria ensures a comprehensive assessment of PCM suitability, offering a reliable decision-making tool for climate-responsive building design.

In addition to the thermophysical properties considered in the MCDM model, it has been taken into account that materials based on hydrated salts, such as $CaCl_2{\cdot}6H_2O$ and $Na_2CO_3{\cdot}10H_2O$, exhibit additional characteristics relevant to their implementation in building systems. Specifically, their hygroscopic nature can lead to the absorption of ambient moisture, which affects the material's stability and thermal storage capacity. Furthermore, the risk of leakage during the fusion process necessitates appropriate encapsulation solutions to prevent loss of active mass and structural damage. Although these variables have not been explicitly modeled in the decision matrix, they are recognized as critical factors in system design. The use of airtight encapsulations, corrosion-resistant materials, and thermal cycling tests is recommended to ensure the technical feasibility and durability of these PCMs in real-world applications.

A complete decision matrix for each climate zone (including values after any transformations), the exact normalization routine per method, and the scripts/spreadsheets are provided in Appendix A.

5. Conclusions

This study presents a comprehensive and climate-responsive framework for selecting phase change materials (PCMs) for building applications using a multi-criteria decision-making (MCDM) approach. By integrating the Analytic Hierarchy Process (AHP) for criteria weighting with five distinct MCDM methods—COPRAS, VIKOR, TOPSIS, MOORA, and PROMETHEE II—the methodology ensures a robust, transparent, and adaptable evaluation of PCM alternatives across diverse climatic zones.

The use of multiple decision models not only enhances the reliability of the selection process but also provides a multidimensional perspective on material performance, capturing trade-offs between thermal, physical, and economic properties. This approach enables designers and engineers to make informed decisions tailored to specific environmental conditions and building requirements.

The robustness of PCM rankings was evaluated by applying five distinct MCDM methods—COPRAS, VIKOR, TOPSIS, MOORA, and PROMETHEE II—using a consistent set of AHP-derived weights. Across all climate zones, the top-ranked materials (e.g., n-Heptadecane for temperate and subtropical zones; n-Octadecane and $CaCl_2{\cdot}6H_2O$ for tropical zones) showed strong convergence in the rankings, indicating high agreement among methods. This cross-method consistency suggests that the selection framework is resilient to methodological biases. Furthermore, sensitivity analysis of the AHP weighting matrix confirmed that moderate perturbations in criteria weights did not significantly alter the top-ranked PCMs, reinforcing the stability of the decision outcomes. These findings validate the reliability of the proposed approach for climate-responsive PCM selection in building applications.

Beyond its technical rigor, the framework contributes to the broader goals of sustainable building design by promoting materials that enhance energy efficiency and occupant comfort. It also offers a replicable model for future studies aiming to optimize material selection in other domains in which multiple performance criteria must be balanced.

The methodology proposed in this study can be adopted as a practical and replicable template for PCM selection in climate-responsive building applications. The process is as follows: (1) the local thermal comfort band is defined based on established standards such as EN 16798-1 or ASHRAE 55; (2) a shortlist of PCM candidates is compiled based on melting point compatibility and availability; (3) a decision matrix is populated with thermophysical and operational data sourced from literature and vendor datasheets; (4) AHP is applied to derive weights for each criterion, although these can be adapted to reflect stakeholder-specific priorities; (5) multiple MCDM methods are executed to rank the alternatives, and a consensus ranking is reported to ensure robustness; finally, (6) sensitivity analysis and system-level constraints—such as encapsulation feasibility, cost, and long-term stability—are considered to validate the practical suitability of the selected PCMs. This structured approach supports informed decision making and can be tailored to diverse climatic contexts and building typologies.

Despite the robustness of the proposed methodology, several limitations should be acknowledged. First, the property data used in the decision matrices were sourced from a combination of vendor datasheets and literature, which may vary in regards to measurement conditions, formats, and precision, introducing heterogeneity that could affect comparability. Second, important operational attributes such as thermal cycling stability and subcooling behavior were not included in the current model due to limited standardized data, although they are critical for long-term performance. Third, the study does not incorporate dynamic building simulations to validate the real-world impact of PCM integration on thermal comfort or energy savings. Lastly, regional variability in PCM availability and pricing was not modeled, which may influence practical feasibility across different markets. Future work should address these aspects by integrating dynamic simulation tools (e.g., EnergyPlus or Modelica), conducting life cycle assessments (LCA), and performing sensitivity analyses on installed costs and supply constraints to enhance the applicability of the framework in diverse building contexts.

Future work may extend this methodology by incorporating lifecycle assessments, dynamic building simulations, and real-world validation to further refine PCM selection and integration strategies.

Funding: The APC was funded by Universidad Internacional SEK, and this research was supported by Universidad Internacional SEK, grant number P121819.

Data Availability Statement: The original contributions presented in this study are included in the article. Further inquiries can be directed to the corresponding author.

Acknowledgments: This study is a component of the Parque de Energias Renovables project P121819, which was started by Universidad Internacional SEK.

Conflicts of Interest: The author declares no conflicts of interest.

Appendix A

Appendix A includes the complete decision matrices for all climate zones (with values after any applied transformations), detailed descriptions of the normalization procedures for each method, as well as the corresponding scripts and spreadsheets.

Table A1. Zone 18 °C decision matrix.

Material	Tm (°C) Score	ΔHfus (kJ/kg)	Vol. Latent Heat (MJ/m^3)	k (W/m·K)	Cp (kJ/kg·K)	ρ (kg/m^3)
n-Heptadecane (C17)	0.764705882	225	173	0.2	2.1	770
n-Octadecane (C18)	0.411764706	244	189	0.2	2.1	775
n-Nonadecane (C19)	0.176470588	250	195	0.2	2.1	780
Rubitherm RT27	0.470588235	120	92	0.2	2	770
Rubitherm RT31	0.235294118	115	89	0.2	2	777
Capric acid (C10)	0.2	150	141	0.18	2.1	940
Capric–Lauric eutectic	0.647058824	160	144	0.18	2.1	900
Fatty acid eutectic (C10–C12–C14)	0	150	132	0.18	2	880
1-Dodecanol	0.647058824	160	130	0.17	2	813
PEG-600	0.882352941	120	132	0.25	1.9	1100
Glauber's salt ($Na_2SO_4 \cdot 10H_2O$)	0.152941176	254	381	0.6	1.4	1500
$CaCl_2 \cdot 6H_2O$	0.352941176	190	323	0.6	1.3	1700
$Na_2HPO_4 \cdot 12H_2O$	0	200	320	0.5	1.4	1600
$Na_2CO_3 \cdot 10H_2O$	0.176470588	190	276	0.6	1.4	1450
Form-stable paraffin/HDPE	0.588235294	120	108	0.25	1.9	900
Organic ester mixture	0.588235294	140	126	0.18	2	900

Table A2. Zone 18 °C COPRAS normalized matrix.

Material	Tm (°C) Score	ΔHfus (kJ/kg)	Vol. Latent Heat (MJ/m^3)	k (W/m·K)	Cp (kJ/kg·K)	ρ (kg/m^3)
n-Heptadecane (C17)	0.121495327	0.080703013	0.058624195	0.042643923	0.070469799	0.046511628
n-Octadecane (C18)	0.065420561	0.087517934	0.064046086	0.042643923	0.070469799	0.046813651
n-Nonadecane (C19)	0.028037383	0.089670014	0.066079295	0.042643923	0.070469799	0.047115675
Rubitherm RT27	0.074766355	0.043041607	0.031175873	0.042643923	0.067114094	0.046511628

Table A2. *Cont.*

Material	Tm (°C) Score	ΔHfus (kJ/kg)	Vol. Latent Heat (MJ/m^3)	k (W/m·K)	Cp (kJ/kg·K)	ρ (kg/m^3)
Rubitherm RT31	0.037383178	0.041248207	0.030159268	0.042643923	0.067114094	0.046934461
Capric acid (C10)	0.031775701	0.053802009	0.047780413	0.038379531	0.070469799	0.056780429
Capric–Lauric eutectic	0.102803738	0.057388809	0.048797018	0.038379531	0.070469799	0.05436424
Fatty acid eutectic (C10–C12–C14)	0	0.053802009	0.0447306	0.038379531	0.067114094	0.053156146
1-Dodecanol	0.102803738	0.057388809	0.044052863	0.036247335	0.067114094	0.049109031
PEG-600	0.140186916	0.043041607	0.0447306	0.053304904	0.063758389	0.066445183
Glauber's salt ($Na_2SO_4 \cdot 10H_2O$)	0.024299065	0.091104735	0.129108777	0.12793177	0.046979866	0.090607067
$CaCl_2 \cdot 6H_2O$	0.056074766	0.068149211	0.109454422	0.12793177	0.043624161	0.10268801
$Na_2HPO_4 \cdot 12H_2O$	0	0.071736011	0.108437818	0.106609808	0.046979866	0.096647539
$Na_2CO_3 \cdot 10H_2O$	0.028037383	0.068149211	0.093527618	0.12793177	0.046979866	0.087586832
Form-stable paraffin/HDPE	0.093457944	0.043041607	0.036597763	0.053304904	0.063758389	0.05436424
Organic ester mixture	0.093457944	0.050215208	0.042697391	0.038379531	0.067114094	0.05436424

Table A3. Zone 18 °C COPRAS weighted matrix.

Material	Tm (°C) Score	ΔHfus (kJ/kg)	Vol. Latent Heat (MJ/m^3)	k (W/m·K)	Cp (kJ/kg·K)	ρ (kg/m^3)
n-Heptadecane (C17)	0.0577	0.0207	0.0079	0.0029	0.0023	0.0015
n-Octadecane (C18)	0.0311	0.0225	0.0086	0.0029	0.0023	0.0015
n-Nonadecane (C19)	0.0133	0.0230	0.0089	0.0029	0.0023	0.0016
Rubitherm RT27	0.0355	0.0111	0.0042	0.0029	0.0022	0.0015
Rubitherm RT31	0.0178	0.0106	0.0041	0.0029	0.0022	0.0015
Capric acid (C10)	0.0151	0.0138	0.0065	0.0026	0.0023	0.0019
Capric–Lauric eutectic	0.0488	0.0147	0.0066	0.0026	0.0023	0.0018
Fatty acid eutectic (C10–C12–C14)	0.0000	0.0138	0.0060	0.0026	0.0022	0.0018
1-Dodecanol	0.0488	0.0147	0.0059	0.0025	0.0022	0.0016
PEG-600	0.0666	0.0111	0.0060	0.0036	0.0021	0.0022
Glauber's salt ($Na_2SO_4 \cdot 10H_2O$)	0.0115	0.0234	0.0174	0.0087	0.0016	0.0030
$CaCl_2 \cdot 6H_2O$	0.0266	0.0175	0.0148	0.0087	0.0014	0.0034
$Na_2HPO_4 \cdot 12H_2O$	0.0000	0.0184	0.0146	0.0072	0.0016	0.0032
$Na_2CO_3 \cdot 10H_2O$	0.0133	0.0175	0.0126	0.0087	0.0016	0.0029
Form-stable paraffin/HDPE	0.0444	0.0111	0.0049	0.0036	0.0021	0.0018
Organic ester mixture	0.0444	0.0129	0.0058	0.0026	0.0022	0.0018

Table A4. Zone 18 °C VIKOR normalized matrix.

Material	Tm (°C) Score	ΔHfus (kJ/kg)	Vol. Latent Heat (MJ/m^3)	k (W/m·K)	Cp (kJ/kg·K)	ρ (kg/m^3)
n-Heptadecane (C17)	0.0577	0.0207	0.0079	0.0029	0.0023	0.0015
n-Octadecane (C18)	0.0311	0.0225	0.0086	0.0029	0.0023	0.0015
n-Nonadecane (C19)	0.0133	0.0230	0.0089	0.0029	0.0023	0.0016
Rubitherm RT27	0.0355	0.0111	0.0042	0.0029	0.0022	0.0015
Rubitherm RT31	0.0178	0.0106	0.0041	0.0029	0.0022	0.0015
Capric acid (C10)	0.0151	0.0138	0.0065	0.0026	0.0023	0.0019
Capric–Lauric eutectic	0.0488	0.0147	0.0066	0.0026	0.0023	0.0018
Fatty acid eutectic (C10–C12–C14)	0.0000	0.0138	0.0060	0.0026	0.0022	0.0018
1-Dodecanol	0.0488	0.0147	0.0059	0.0025	0.0022	0.0016
PEG-600	0.0666	0.0111	0.0060	0.0036	0.0021	0.0022
Glauber's salt ($Na_2SO_4 \cdot 10H_2O$)	0.0115	0.0234	0.0174	0.0087	0.0016	0.0030
$CaCl_2 \cdot 6H_2O$	0.0266	0.0175	0.0148	0.0087	0.0014	0.0034
$Na_2HPO_4 \cdot 12H_2O$	0.0000	0.0184	0.0146	0.0072	0.0016	0.0032
$Na_2CO_3 \cdot 10H_2O$	0.0133	0.0175	0.0126	0.0087	0.0016	0.0029
Form-stable paraffin/HDPE	0.0444	0.0111	0.0049	0.0036	0.0021	0.0018
Organic ester mixture	0.0444	0.0129	0.0058	0.0026	0.0022	0.0018

Table A5. Zone 18 °C VIKOR weighted matrix.

Material	Tm (°C) Score	ΔHfus (kJ/kg)	Vol. Latent Heat (MJ/m^3)	k (W/m·K)	Cp (kJ/kg·K)	ρ (kg/m^3)
n-Heptadecane (C17)	0.4117	0.2034	0.0388	0.0047	0.0330	0.0000
n-Octadecane (C18)	0.2217	0.2385	0.0462	0.0047	0.0330	0.0002
n-Nonadecane (C19)	0.0950	0.2496	0.0490	0.0047	0.0330	0.0004
Rubitherm RT27	0.2533	0.0092	0.0014	0.0047	0.0289	0.0000
Rubitherm RT31	0.1267	0.0000	0.0000	0.0047	0.0289	0.0002
Capric acid (C10)	0.1077	0.0647	0.0240	0.0016	0.0330	0.0060
Capric–Lauric eutectic	0.3483	0.0832	0.0254	0.0016	0.0330	0.0046
Fatty acid eutectic (C10–C12–C14)	0.0000	0.0647	0.0199	0.0016	0.0289	0.0039
1-Dodecanol	0.3483	0.0832	0.0190	0.0000	0.0289	0.0015
PEG-600	0.4750	0.0092	0.0199	0.0127	0.0248	0.0117
Glauber's salt ($Na_2SO_4 \cdot 10H_2O$)	0.0823	0.2570	0.1350	0.0680	0.0041	0.0259
$CaCl_2 \cdot 6H_2O$	0.1900	0.1387	0.1082	0.0680	0.0000	0.0330
$Na_2HPO_4 \cdot 12H_2O$	0.0000	0.1572	0.1068	0.0522	0.0041	0.0295
$Na_2CO_3 \cdot 10H_2O$	0.0950	0.1387	0.0865	0.0680	0.0041	0.0241
Form-stable paraffin/HDPE	0.3167	0.0092	0.0088	0.0127	0.0248	0.0046
Organic ester mixture	0.3167	0.0462	0.0171	0.0016	0.0289	0.0046

Table A6. Zone 18 °C TOPSIS normalized matrix.

Material	Tm (°C) Score	ΔHfus (kJ/kg)	Vol. Latent Heat (MJ/m^3)	k (W/m·K)	Cp (kJ/kg·K)	ρ (kg/m^3)
n-Heptadecane (C17)	0.40363	0.31154	0.21166	0.14852	0.27854	0.17778
n-Octadecane (C18)	0.21734	0.33785	0.23124	0.14852	0.27854	0.17893
n-Nonadecane (C19)	0.09315	0.34615	0.23858	0.14852	0.27854	0.18009
Rubitherm RT27	0.24839	0.16615	0.11256	0.14852	0.26528	0.17778
Rubitherm RT31	0.12419	0.15923	0.10889	0.14852	0.26528	0.17939
Capric acid (C10)	0.10557	0.20769	0.17251	0.13366	0.27854	0.21703
Capric–Lauric eutectic	0.34154	0.22154	0.17618	0.13366	0.27854	0.20779
Fatty acid eutectic (C10–C12–C14)	0.00000	0.20769	0.16150	0.13366	0.26528	0.20317
1-Dodecanol	0.34154	0.22154	0.15905	0.12624	0.26528	0.18770
PEG-600	0.46573	0.16615	0.16150	0.18564	0.25202	0.25397
Glauber's salt ($Na_2SO_4 \cdot 10H_2O$)	0.08073	0.35169	0.46615	0.44555	0.18570	0.34632
$CaCl_2 \cdot 6H_2O$	0.18629	0.26308	0.39519	0.44555	0.17243	0.39249
$Na_2HPO_4 \cdot 12H_2O$	0.00000	0.27692	0.39152	0.37129	0.18570	0.36941
$Na_2CO_3 \cdot 10H_2O$	0.09315	0.26308	0.33768	0.44555	0.18570	0.33477
Form-stable paraffin/HDPE	0.31049	0.16615	0.13214	0.18564	0.25202	0.20779
Organic ester mixture	0.31049	0.19385	0.15416	0.13366	0.26528	0.20779

Table A7. Zone 18 °C TOPSIS weighted matrix.

Material	Tm (°C) Score	ΔHfus (kJ/kg)	Vol. Latent Heat (MJ/m^3)	k (W/m·K)	Cp (kJ/kg·K)	ρ (kg/m^3)
n-Heptadecane (C17)	0.1917	0.0801	0.0286	0.0101	0.0092	0.0059
n-Octadecane (C18)	0.1032	0.0868	0.0312	0.0101	0.0092	0.0059
n-Nonadecane (C19)	0.0442	0.0890	0.0322	0.0101	0.0092	0.0059
Rubitherm RT27	0.1180	0.0427	0.0152	0.0101	0.0088	0.0059
Rubitherm RT31	0.0590	0.0409	0.0147	0.0101	0.0088	0.0059
Capric acid (C10)	0.0501	0.0534	0.0233	0.0091	0.0092	0.0072
Capric–Lauric eutectic	0.1622	0.0569	0.0238	0.0091	0.0092	0.0069
Fatty acid eutectic (C10–C12–C14)	0.0000	0.0534	0.0218	0.0091	0.0088	0.0067
1-Dodecanol	0.1622	0.0569	0.0215	0.0086	0.0088	0.0062
PEG-600	0.2212	0.0427	0.0218	0.0126	0.0083	0.0084
Glauber's salt ($Na_2SO_4 \cdot 10H_2O$)	0.0383	0.0904	0.0629	0.0303	0.0061	0.0114
$CaCl_2 \cdot 6H_2O$	0.0885	0.0676	0.0534	0.0303	0.0057	0.0130
$Na_2HPO_4 \cdot 12H_2O$	0.0000	0.0712	0.0529	0.0252	0.0061	0.0122
$Na_2CO_3 \cdot 10H_2O$	0.0442	0.0676	0.0456	0.0303	0.0061	0.0110
Form-stable paraffin/HDPE	0.1475	0.0427	0.0178	0.0126	0.0083	0.0069
Organic ester mixture	0.1475	0.0498	0.0208	0.0091	0.0088	0.0069

Table A8. Zone 18 °C MOORA normalized matrix.

Material	Tm (°C) Score	ΔHfus (kJ/kg)	Vol. Latent Heat (MJ/m^3)	k (W/m·K)	Cp (kJ/kg·K)	ρ (kg/m^3)
n-Heptadecane (C17)	0.40363	0.31154	0.21166	0.14852	0.27854	0.17778
n-Octadecane (C18)	0.21734	0.33785	0.23124	0.14852	0.27854	0.17893
n-Nonadecane (C19)	0.09315	0.34615	0.23858	0.14852	0.27854	0.18009
Rubitherm RT27	0.24839	0.16615	0.11256	0.14852	0.26528	0.17778
Rubitherm RT31	0.12419	0.15923	0.10889	0.14852	0.26528	0.17939
Capric acid (C10)	0.10557	0.20769	0.17251	0.13366	0.27854	0.21703
Capric–Lauric eutectic	0.34154	0.22154	0.17618	0.13366	0.27854	0.20779
Fatty acid eutectic (C10–C12–C14)	0.00000	0.20769	0.16150	0.13366	0.26528	0.20317
1-Dodecanol	0.34154	0.22154	0.15905	0.12624	0.26528	0.18770
PEG-600	0.46573	0.16615	0.16150	0.18564	0.25202	0.25397
Glauber's salt ($Na_2SO_4 \cdot 10H_2O$)	0.08073	0.35169	0.46615	0.44555	0.18570	0.34632
$CaCl_2 \cdot 6H_2O$	0.18629	0.26308	0.39519	0.44555	0.17243	0.39249
$Na_2HPO_4 \cdot 12H_2O$	0.00000	0.27692	0.39152	0.37129	0.18570	0.36941
$Na_2CO_3 \cdot 10H_2O$	0.09315	0.26308	0.33768	0.44555	0.18570	0.33477
Form-stable paraffin/HDPE	0.31049	0.16615	0.13214	0.18564	0.25202	0.20779
Organic ester mixture	0.31049	0.19385	0.15416	0.13366	0.26528	0.20779

Table A9. Zone 18 °C MOORA weighted matrix.

Material	Tm (°C) Score	ΔHfus (kJ/kg)	Vol. Latent Heat (MJ/m^3)	k (W/m·K)	Cp (kJ/kg·K)	ρ (kg/m^3)
n-Heptadecane (C17)	0.19173	0.08007	0.02857	0.01010	0.00919	0.00587
n-Octadecane (C18)	0.10324	0.08683	0.03122	0.01010	0.00919	0.00590
n-Nonadecane (C19)	0.04424	0.08896	0.03221	0.01010	0.00919	0.00594
Rubitherm RT27	0.11799	0.04270	0.01520	0.01010	0.00875	0.00587
Rubitherm RT31	0.05899	0.04092	0.01470	0.01010	0.00875	0.00592
Capric acid (C10)	0.05014	0.05338	0.02329	0.00909	0.00919	0.00716
Capric–Lauric eutectic	0.16223	0.05694	0.02378	0.00909	0.00919	0.00686
Fatty acid eutectic (C10–C12–C14)	0.00000	0.05338	0.02180	0.00909	0.00875	0.00670
1-Dodecanol	0.16223	0.05694	0.02147	0.00858	0.00875	0.00619
PEG-600	0.22122	0.04270	0.02180	0.01262	0.00832	0.00838
Glauber's salt ($Na_2SO_4 \cdot 10H_2O$)	0.03835	0.09039	0.06293	0.03030	0.00613	0.01143
$CaCl_2 \cdot 6H_2O$	0.08849	0.06761	0.05335	0.03030	0.00569	0.01295
$Na_2HPO_4 \cdot 12H_2O$	0.00000	0.07117	0.05285	0.02525	0.00613	0.01219
$Na_2CO_3 \cdot 10H_2O$	0.04424	0.06761	0.04559	0.03030	0.00613	0.01105
Form-stable paraffin/HDPE	0.14748	0.04270	0.01784	0.01262	0.00832	0.00686
Organic ester mixture	0.14748	0.04982	0.02081	0.00909	0.00875	0.00686

Table A10. Zone 18 °C PROMETHEE pairwise matrix.

Material	n-Heptadecane (C17)	n-Octadecane (C18)	n-Nonadecane (C19)	Rubitherm RT27	Rubitherm RT31	Capric Acid (C10)	Capric–Lauric Eutectic	Fatty Acid Eutectic (C10–C12–C14)	1-Dodecanol	PEG-600	Glauber's Salt ($Na_2SO_4 \cdot 10H_2O$)	$CaCl_2 \cdot 6H_2O$	$Na_2HPO_4 \cdot 12H_2O$	$Na_2CO_3 \cdot 10H_2O$	Form-Stable Paraffin/HDPE	Organic Ester Mixture
n-Heptadecane (C17)	0.000	0.190	0.317	0.394	0.531	0.461	0.200	0.577	0.212	0.221	0.358	0.319	0.487	0.410	0.327	0.281
n-Octadecane (C18)	0.043	0.000	0.127	0.278	0.384	0.313	0.179	0.429	0.191	0.264	0.168	0.165	0.332	0.255	0.275	0.229
n-Nonadecane (C19)	0.057	0.014	0.000	0.292	0.303	0.213	0.193	0.316	0.205	0.278	0.042	0.144	0.216	0.140	0.289	0.243
Rubitherm RT27	0.000	0.032	0.158	0.000	0.137	0.149	0.003	0.256	0.005	0.004	0.196	0.092	0.278	0.183	0.004	0.003
Rubitherm RT31	0.000	0.000	0.032	0.000	0.000	0.022	0.003	0.130	0.005	0.004	0.069	0.029	0.151	0.056	0.004	0.003
Capric acid (C10)	0.006	0.006	0.018	0.088	0.099	0.000	0.001	0.118	0.015	0.068	0.054	0.033	0.137	0.042	0.080	0.031
Capric–Lauric eutectic	0.005	0.131	0.258	0.202	0.339	0.261	0.000	0.377	0.015	0.088	0.295	0.191	0.377	0.282	0.131	0.081
Fatty acid eutectic (C10–C12–C14)	0.004	0.004	0.004	0.078	0.088	0.000	0.000	0.000	0.005	0.060	0.025	0.029	0.025	0.025	0.071	0.021
1-Dodecanol	0.002	0.128	0.255	0.188	0.325	0.259	0.000	0.367	0.000	0.078	0.291	0.187	0.373	0.278	0.120	0.070
PEG-600	0.083	0.273	0.399	0.260	0.397	0.384	0.145	0.494	0.150	0.000	0.413	0.310	0.496	0.401	0.177	0.179
Glauber's salt ($Na_2SO_4 \cdot 10H_2O$)	0.239	0.196	0.182	0.471	0.481	0.390	0.371	0.478	0.382	0.432	0.000	0.149	0.226	0.169	0.451	0.416
$CaCl_2 \cdot 6H_2O$	0.166	0.158	0.250	0.332	0.406	0.334	0.233	0.448	0.244	0.294	0.115	0.000	0.211	0.126	0.313	0.278
$Na_2HPO_4 \cdot 12H_2O$	0.145	0.137	0.134	0.330	0.341	0.249	0.231	0.256	0.242	0.292	0.004	0.023	0.000	0.044	0.310	0.276
$Na_2CO_3 \cdot 10H_2O$	0.135	0.127	0.124	0.302	0.312	0.221	0.202	0.322	0.214	0.264	0.013	0.004	0.111	0.000	0.282	0.248
Form-stable paraffin/HDPE	0.013	0.107	0.234	0.083	0.220	0.220	0.011	0.328	0.016	0.000	0.255	0.151	0.337	0.242	0.000	0.011
Organic ester mixture	0.005	0.099	0.226	0.121	0.258	0.209	0.000	0.317	0.005	0.041	0.259	0.156	0.341	0.246	0.049	0.000

Table A11. Zone 23 °C decision matrix.

Material	Tm (°C) Score	ΔHfus (kJ/kg)	Vol. Latent Heat (MJ/m^3)	k (W/m·K)	Cp (kJ/kg·K)	ρ (kg/m^3)
n-Heptadecane (C17)	0.916666667	225	173	0.2	2.1	770
n-Octadecane (C18)	0.583333333	244	189	0.2	2.1	775
n-Nonadecane (C19)	0.25	250	195	0.2	2.1	780
Rubitherm RT27	0.666666667	120	92	0.2	2	770
Rubitherm RT31	0.333333333	115	89	0.2	2	777
Capric acid (C10)	0.283333333	150	141	0.18	2.1	940
Capric–Lauric eutectic	0.916666667	160	144	0.18	2.1	900
Fatty acid eutectic (C10–C12–C14)	0	150	132	0.18	2	880
1-Dodecanol	0.916666667	160	130	0.17	2	813
PEG-600	0.75	120	132	0.25	1.9	1100
Glauber's salt ($Na_2SO_4 \cdot 10H_2O$)	0.216666667	254	381	0.6	1.4	1500
$CaCl_2 \cdot 6H_2O$	0.5	190	323	0.6	1.3	1700
$Na_2HPO_4 \cdot 12H_2O$	0	200	320	0.5	1.4	1600
$Na_2CO_3 \cdot 10H_2O$	0.25	190	276	0.6	1.4	1450
Form-stable paraffin/HDPE	0.833333333	120	108	0.25	1.9	900
Organic ester mixture	0.833333333	140	126	0.18	2	900

Table A12. Zone 23 °C COPRAS normalized matrix.

Material	Tm (°C) Score	ΔHfus (kJ/kg)	Vol. Latent Heat (MJ/m^3)	k (W/m·K)	Cp (kJ/kg·K)	ρ (kg/m^3)
n-Heptadecane (C17)	0.1111	0.0807	0.0586	0.0426	0.0705	0.0465
n-Octadecane (C18)	0.0707	0.0875	0.0640	0.0426	0.0705	0.0468
n-Nonadecane (C19)	0.0303	0.0897	0.0661	0.0426	0.0705	0.0471
Rubitherm RT27	0.0808	0.0430	0.0312	0.0426	0.0671	0.0465
Rubitherm RT31	0.0404	0.0412	0.0302	0.0426	0.0671	0.0469
Capric acid (C10)	0.0343	0.0538	0.0478	0.0384	0.0705	0.0568
Capric–Lauric eutectic	0.1111	0.0574	0.0488	0.0384	0.0705	0.0544
Fatty acid eutectic (C10–C12–C14)	0.0000	0.0538	0.0447	0.0384	0.0671	0.0532
1-Dodecanol	0.1111	0.0574	0.0441	0.0362	0.0671	0.0491
PEG-600	0.0909	0.0430	0.0447	0.0533	0.0638	0.0664
Glauber's salt ($Na_2SO_4 \cdot 10H_2O$)	0.0263	0.0911	0.1291	0.1279	0.0470	0.0906
$CaCl_2 \cdot 6H_2O$	0.0606	0.0681	0.1095	0.1279	0.0436	0.1027
$Na_2HPO_4 \cdot 12H_2O$	0.0000	0.0717	0.1084	0.1066	0.0470	0.0966
$Na_2CO_3 \cdot 10H_2O$	0.0303	0.0681	0.0935	0.1279	0.0470	0.0876
Form-stable paraffin/HDPE	0.1010	0.0430	0.0366	0.0533	0.0638	0.0544
Organic ester mixture	0.1010	0.0502	0.0427	0.0384	0.0671	0.0544

Table A13. Zone 23 °C COPRAS weighted matrix.

Material	Tm (°C) Score	ΔHfus (kJ/kg)	Vol. Latent Heat (MJ/m^3)	k (W/m·K)	Cp (kJ/kg·K)	ρ (kg/m^3)
n-Heptadecane (C17)	0.0528	0.0207	0.0079	0.0029	0.0023	0.0015
n-Octadecane (C18)	0.0336	0.0225	0.0086	0.0029	0.0023	0.0015
n-Nonadecane (C19)	0.0144	0.0230	0.0089	0.0029	0.0023	0.0016
Rubitherm RT27	0.0384	0.0111	0.0042	0.0029	0.0022	0.0015
Rubitherm RT31	0.0192	0.0106	0.0041	0.0029	0.0022	0.0015
Capric acid (C10)	0.0163	0.0138	0.0065	0.0026	0.0023	0.0019
Capric–Lauric eutectic	0.0528	0.0147	0.0066	0.0026	0.0023	0.0018
Fatty acid eutectic (C10–C12–C14)	0.0000	0.0138	0.0060	0.0026	0.0022	0.0018
1-Dodecanol	0.0528	0.0147	0.0059	0.0025	0.0022	0.0016
PEG-600	0.0432	0.0111	0.0060	0.0036	0.0021	0.0022
Glauber's salt ($Na_2SO_4 \cdot 10H_2O$)	0.0125	0.0234	0.0174	0.0087	0.0016	0.0030
$CaCl_2 \cdot 6H_2O$	0.0288	0.0175	0.0148	0.0087	0.0014	0.0034
$Na_2HPO_4 \cdot 12H_2O$	0.0000	0.0184	0.0146	0.0072	0.0016	0.0032
$Na_2CO_3 \cdot 10H_2O$	0.0144	0.0175	0.0126	0.0087	0.0016	0.0029
Form-stable paraffin/HDPE	0.0480	0.0111	0.0049	0.0036	0.0021	0.0018
Organic ester mixture	0.0480	0.0129	0.0058	0.0026	0.0022	0.0018

Table A14. Zone 23 °C VIKOR normalized matrix.

Material	Tm (°C) Score	ΔHfus (kJ/kg)	Vol. Latent Heat (MJ/m^3)	k (W/m·K)	Cp (kJ/kg·K)	ρ (kg/m^3)
n-Heptadecane (C17)	1.00000	0.79137	0.28767	0.06977	1.00000	0.00000
n-Octadecane (C18)	0.63636	0.92806	0.34247	0.06977	1.00000	0.00538
n-Nonadecane (C19)	0.27273	0.97122	0.36301	0.06977	1.00000	0.01075
Rubitherm RT27	0.72727	0.03597	0.01027	0.06977	0.87500	0.00000
Rubitherm RT31	0.36364	0.00000	0.00000	0.06977	0.87500	0.00753
Capric acid (C10)	0.30909	0.25180	0.17808	0.02326	1.00000	0.18280
Capric–Lauric eutectic	1.00000	0.32374	0.18836	0.02326	1.00000	0.13978
Fatty acid eutectic (C10–C12–C14)	0.00000	0.25180	0.14726	0.02326	0.87500	0.11828
1-Dodecanol	1.00000	0.32374	0.14041	0.00000	0.87500	0.04624
PEG-600	0.81818	0.03597	0.14726	0.18605	0.75000	0.35484
Glauber's salt ($Na_2SO_4 \cdot 10H_2O$)	0.23636	1.00000	1.00000	1.00000	0.12500	0.78495
$CaCl_2 \cdot 6H_2O$	0.54545	0.53957	0.80137	1.00000	0.00000	1.00000
$Na_2HPO_4 \cdot 12H_2O$	0.00000	0.61151	0.79110	0.76744	0.12500	0.89247
$Na_2CO_3 \cdot 10H_2O$	0.27273	0.53957	0.64041	1.00000	0.12500	0.73118
Form-stable paraffin/HDPE	0.90909	0.03597	0.06507	0.18605	0.75000	0.13978
Organic ester mixture	0.90909	0.17986	0.12671	0.02326	0.87500	0.13978

Table A15. Zone 23 °C VIKOR weighted matrix.

Material	Tm (°C) Score	ΔHfus (kJ/kg)	Vol. Latent Heat (MJ/m^3)	k (W/m·K)	Cp (kJ/kg·K)	ρ (kg/m^3)
n-Heptadecane (C17)	0.47500	0.20338	0.03884	0.00474	0.03300	0.00000
n-Octadecane (C18)	0.30227	0.23851	0.04623	0.00474	0.03300	0.00018
n-Nonadecane (C19)	0.12955	0.24960	0.04901	0.00474	0.03300	0.00035
Rubitherm RT27	0.34545	0.00924	0.00139	0.00474	0.02888	0.00000
Rubitherm RT31	0.17273	0.00000	0.00000	0.00474	0.02888	0.00025
Capric acid (C10)	0.14682	0.06471	0.02404	0.00158	0.03300	0.00603
Capric–Lauric eutectic	0.47500	0.08320	0.02543	0.00158	0.03300	0.00461
Fatty acid eutectic (C10–C12–C14)	0.00000	0.06471	0.01988	0.00158	0.02888	0.00390
1-Dodecanol	0.47500	0.08320	0.01896	0.00000	0.02888	0.00153
PEG-600	0.38864	0.00924	0.01988	0.01265	0.02475	0.01171
Glauber's salt ($Na_2SO_4 \cdot 10H_2O$)	0.11227	0.25700	0.13500	0.06800	0.00412	0.02590
$CaCl_2 \cdot 6H_2O$	0.25909	0.13867	0.10818	0.06800	0.00000	0.03300
$Na_2HPO_4 \cdot 12H_2O$	0.00000	0.15716	0.10680	0.05219	0.00412	0.02945
$Na_2CO_3 \cdot 10H_2O$	0.12955	0.13867	0.08646	0.06800	0.00412	0.02413
Form-stable paraffin/HDPE	0.43182	0.00924	0.00878	0.01265	0.02475	0.00461
Organic ester mixture	0.43182	0.04622	0.01711	0.00158	0.02888	0.00461

Table A16. Zone 23 °C TOPSIS normalized matrix.

Material	Tm (°C) Score	ΔHfus (kJ/kg)	Vol. Latent Heat (MJ/m^3)	k (W/m·K)	Cp (kJ/kg·K)	ρ (kg/m^3)
n-Heptadecane (C17)	0.3783	0.3115	0.2117	0.1485	0.2785	0.1778
n-Octadecane (C18)	0.2408	0.3378	0.2312	0.1485	0.2785	0.1789
n-Nonadecane (C19)	0.1032	0.3462	0.2386	0.1485	0.2785	0.1801
Rubitherm RT27	0.2752	0.1662	0.1126	0.1485	0.2653	0.1778
Rubitherm RT31	0.1376	0.1592	0.1089	0.1485	0.2653	0.1794
Capric acid (C10)	0.1169	0.2077	0.1725	0.1337	0.2785	0.2170
Capric–Lauric eutectic	0.3783	0.2215	0.1762	0.1337	0.2785	0.2078
Fatty acid eutectic (C10–C12–C14)	0.0000	0.2077	0.1615	0.1337	0.2653	0.2032
1-Dodecanol	0.3783	0.2215	0.1591	0.1262	0.2653	0.1877
PEG-600	0.3096	0.1662	0.1615	0.1856	0.2520	0.2540
Glauber's salt ($Na_2SO_4 \cdot 10H_2O$)	0.0894	0.3517	0.4662	0.4455	0.1857	0.3463
$CaCl_2 \cdot 6H_2O$	0.2064	0.2631	0.3952	0.4455	0.1724	0.3925
$Na_2HPO_4 \cdot 12H_2O$	0.0000	0.2769	0.3915	0.3713	0.1857	0.3694
$Na_2CO_3 \cdot 10H_2O$	0.1032	0.2631	0.3377	0.4455	0.1857	0.3348
Form-stable paraffin/HDPE	0.3439	0.1662	0.1321	0.1856	0.2520	0.2078
Organic ester mixture	0.3439	0.1938	0.1542	0.1337	0.2653	0.2078

Table A17. Zone 23 °C TOPSIS weighted matrix.

Material	Tm (°C) Score	ΔHfus (kJ/kg)	Vol. Latent Heat (MJ/m^3)	k (W/m·K)	Cp (kJ/kg·K)	ρ (kg/m^3)
n-Heptadecane (C17)	0.17971	0.08007	0.02857	0.01010	0.00919	0.00587
n-Octadecane (C18)	0.11436	0.08683	0.03122	0.01010	0.00919	0.00590
n-Nonadecane (C19)	0.04901	0.08896	0.03221	0.01010	0.00919	0.00594
Rubitherm RT27	0.13070	0.04270	0.01520	0.01010	0.00875	0.00587
Rubitherm RT31	0.06535	0.04092	0.01470	0.01010	0.00875	0.00592
Capric acid (C10)	0.05555	0.05338	0.02329	0.00909	0.00919	0.00716
Capric–Lauric eutectic	0.17971	0.05694	0.02378	0.00909	0.00919	0.00686
Fatty acid eutectic (C10–C12–C14)	0.00000	0.05338	0.02180	0.00909	0.00875	0.00670
1-Dodecanol	0.17971	0.05694	0.02147	0.00858	0.00875	0.00619
PEG-600	0.14704	0.04270	0.02180	0.01262	0.00832	0.00838
Glauber's salt ($Na_2SO_4·10H_2O$)	0.04248	0.09039	0.06293	0.03030	0.00613	0.01143
$CaCl_2·6H_2O$	0.09802	0.06761	0.05335	0.03030	0.00569	0.01295
$Na_2HPO_4·12H_2O$	0.00000	0.07117	0.05285	0.02525	0.00613	0.01219
$Na_2CO_3·10H_2O$	0.04901	0.06761	0.04559	0.03030	0.00613	0.01105
Form-stable paraffin/HDPE	0.16337	0.04270	0.01784	0.01262	0.00832	0.00686
Organic ester mixture	0.16337	0.04982	0.02081	0.00909	0.00875	0.00686

Table A18. Zone 23 °C MOORA normalized matrix.

Material	Tm (°C) Score	ΔHfus (kJ/kg)	Vol. Latent Heat (MJ/m^3)	k (W/m·K)	Cp (kJ/kg·K)	ρ (kg/m^3)
n-Heptadecane (C17)	0.37834	0.31154	0.21166	0.14852	0.27854	0.17778
n-Octadecane (C18)	0.24076	0.33785	0.23124	0.14852	0.27854	0.17893
n-Nonadecane (C19)	0.10318	0.34615	0.23858	0.14852	0.27854	0.18009
Rubitherm RT27	0.27516	0.16615	0.11256	0.14852	0.26528	0.17778
Rubitherm RT31	0.13758	0.15923	0.10889	0.14852	0.26528	0.17939
Capric acid (C10)	0.11694	0.20769	0.17251	0.13366	0.27854	0.21703
Capric–Lauric eutectic	0.37834	0.22154	0.17618	0.13366	0.27854	0.20779
Fatty acid eutectic (C10–C12–C14)	0.00000	0.20769	0.16150	0.13366	0.26528	0.20317
1-Dodecanol	0.37834	0.22154	0.15905	0.12624	0.26528	0.18770
PEG-600	0.30955	0.16615	0.16150	0.18564	0.25202	0.25397
Glauber's salt ($Na_2SO_4·10H_2O$)	0.08943	0.35169	0.46615	0.44555	0.18570	0.34632
$CaCl_2·6H_2O$	0.20637	0.26308	0.39519	0.44555	0.17243	0.39249
$Na_2HPO_4·12H_2O$	0.00000	0.27692	0.39152	0.37129	0.18570	0.36941
$Na_2CO_3·10H_2O$	0.10318	0.26308	0.33768	0.44555	0.18570	0.33477
Form-stable paraffin/HDPE	0.34395	0.16615	0.13214	0.18564	0.25202	0.20779
Organic ester mixture	0.34395	0.19385	0.15416	0.13366	0.26528	0.20779

Table A19. Zone 23 °C MOORA weighted matrix.

Material	Tm (°C) Score	ΔHfus (kJ/kg)	Vol. Latent Heat (MJ/m^3)	k (W/m·K)	Cp (kJ/kg·K)	ρ (kg/m^3)
n-Heptadecane (C17)	0.1797	0.0801	0.0286	0.0101	0.0092	0.0059
n-Octadecane (C18)	0.1144	0.0868	0.0312	0.0101	0.0092	0.0059
n-Nonadecane (C19)	0.0490	0.0890	0.0322	0.0101	0.0092	0.0059
Rubitherm RT27	0.1307	0.0427	0.0152	0.0101	0.0088	0.0059
Rubitherm RT31	0.0653	0.0409	0.0147	0.0101	0.0088	0.0059
Capric acid (C10)	0.0555	0.0534	0.0233	0.0091	0.0092	0.0072
Capric–Lauric eutectic	0.1797	0.0569	0.0238	0.0091	0.0092	0.0069
Fatty acid eutectic (C10–C12–C14)	0.0000	0.0534	0.0218	0.0091	0.0088	0.0067
1-Dodecanol	0.1797	0.0569	0.0215	0.0086	0.0088	0.0062
PEG-600	0.1470	0.0427	0.0218	0.0126	0.0083	0.0084
Glauber's salt ($Na_2SO_4·10H_2O$)	0.0425	0.0904	0.0629	0.0303	0.0061	0.0114
$CaCl_2·6H_2O$	0.0980	0.0676	0.0534	0.0303	0.0057	0.0130
$Na_2HPO_4·12H_2O$	0.0000	0.0712	0.0529	0.0252	0.0061	0.0122
$Na_2CO_3·10H_2O$	0.0490	0.0676	0.0456	0.0303	0.0061	0.0110
Form-stable paraffin/HDPE	0.1634	0.0427	0.0178	0.0126	0.0083	0.0069
Organic ester mixture	0.1634	0.0498	0.0208	0.0091	0.0088	0.0069

Table A20. Zone 23 °C PROMETHEE pairwise matrix.

Material	n-Heptadecane (C17)	n-Octadecane (C18)	n-Nonadecane (C19)	Rubitherm RT27	Rubitherm RT31	Capric Acid (C10)	Capric–Lauric Eutectic	Fatty Acid Eutectic (C10–C12–C14)	1-Dodecanol	PEG-600	Glauber's Salt ($Na_2SO_4 \cdot 10H_2O$)	$CaCl_2 \cdot 6H_2O$	$Na_2HPO_4 \cdot 12H_2O$	$Na_2CO_3 \cdot 10H_2O$	Form-Stable Paraffin/HDPE	Organic Ester Mixture
n-Heptadecane (C17)	0.000	0.173	0.345	0.365	0.549	0.485	0.137	0.640	0.149	0.308	0.392	0.314	0.550	0.439	0.276	0.229
n-Octadecane (C18)	0.043	0.000	0.173	0.278	0.418	0.355	0.179	0.510	0.191	0.264	0.219	0.176	0.413	0.301	0.275	0.229
n-Nonadecane (C19)	0.057	0.014	0.000	0.292	0.303	0.213	0.193	0.351	0.205	0.278	0.046	0.144	0.251	0.140	0.289	0.243
Rubitherm RT27	0.000	0.043	0.216	0.000	0.183	0.202	0.003	0.349	0.005	0.004	0.258	0.115	0.370	0.241	0.004	0.003
Rubitherm RT31	0.000	0.000	0.043	0.000	0.000	0.029	0.003	0.176	0.005	0.004	0.085	0.029	0.197	0.068	0.004	0.003
Capric acid (C10)	0.006	0.006	0.023	0.088	0.099	0.000	0.001	0.157	0.015	0.068	0.063	0.033	0.176	0.046	0.080	0.031
Capric–Lauric eutectic	0.005	0.177	0.350	0.236	0.419	0.348	0.000	0.504	0.015	0.174	0.392	0.249	0.504	0.374	0.142	0.093
Fatty acid eutectic (C10–C12–C14)	0.004	0.004	0.004	0.078	0.088	0.000	0.000	0.000	0.005	0.060	0.025	0.029	0.025	0.025	0.071	0.021
1-Dodecanol	0.002	0.174	0.347	0.223	0.406	0.347	0.000	0.493	0.000	0.164	0.387	0.245	0.500	0.370	0.131	0.082
PEG-600	0.020	0.106	0.278	0.081	0.264	0.259	0.018	0.408	0.024	0.000	0.297	0.154	0.409	0.280	0.018	0.021
Glauber's salt ($Na_2SO_4 \cdot 10H_2O$)	0.239	0.196	0.182	0.471	0.481	0.390	0.371	0.508	0.382	0.432	0.000	0.149	0.256	0.169	0.451	0.416
$CaCl_2 \cdot 6H_2O$	0.166	0.158	0.285	0.332	0.429	0.364	0.233	0.517	0.244	0.294	0.154	0.000	0.280	0.160	0.313	0.278
$Na_2HPO_4 \cdot 12H_2O$	0.145	0.137	0.134	0.330	0.341	0.249	0.231	0.256	0.242	0.292	0.004	0.023	0.000	0.044	0.310	0.276
$Na_2CO_3 \cdot 10H_2O$	0.135	0.127	0.124	0.302	0.312	0.221	0.202	0.357	0.214	0.264	0.017	0.004	0.145	0.000	0.282	0.248
Form-stable paraffin/HDPE	0.013	0.142	0.314	0.106	0.289	0.296	0.011	0.444	0.016	0.043	0.340	0.197	0.452	0.323	0.000	0.011
Organic ester mixture	0.005	0.134	0.307	0.144	0.327	0.285	0.000	0.433	0.005	0.084	0.344	0.202	0.457	0.327	0.049	0.000

Table A21. Zone 28 °C decision matrix.

Material	Tm (°C) Score	ΔHfus (kJ/kg)	Vol. Latent Heat (MJ/m³)	k (W/m·K)	Cp (kJ/kg·K)	ρ (kg/m³)
n-Heptadecane (C17)	0.25	225	173	0.2	2.1	770
n-Octadecane (C18)	1	244	189	0.2	2.1	775
n-Nonadecane (C19)	0.5	250	195	0.2	2.1	780
Rubitherm RT27	0.875	120	92	0.2	2	770
Rubitherm RT31	0.625	115	89	0.2	2	777
Capric acid (C10)	0.55	150	141	0.18	2.1	940
Capric–Lauric eutectic	0.5	160	144	0.18	2.1	900
Fatty acid eutectic (C10–C12–C14)	0.125	150	132	0.18	2	880
1-Dodecanol	0.5	160	130	0.17	2	813
PEG-600	0	120	132	0.25	1.9	1100
Glauber's salt ($Na_2SO_4 \cdot 10H_2O$)	0.45	254	381	0.6	1.4	1500
$CaCl_2 \cdot 6H_2O$	0.875	190	323	0.6	1.3	1700
$Na_2HPO_4 \cdot 12H_2O$	0.125	200	320	0.5	1.4	1600
$Na_2CO_3 \cdot 10H_2O$	0.5	190	276	0.6	1.4	1450
Form-stable paraffin/HDPE	0.625	120	108	0.25	1.9	900
Organic ester mixture	0.625	140	126	0.18	2	900

Table A22. Zone 28 °C COPRAS normalized matrix.

Material	Tm (°C) Score	ΔHfus (kJ/kg)	Vol. Latent Heat (MJ/m³)	k (W/m·K)	Cp (kJ/kg·K)	ρ (kg/m³)
n-Heptadecane (C17)	0.0308	0.0807	0.0586	0.0426	0.0705	0.0465
n-Octadecane (C18)	0.1231	0.0875	0.0640	0.0426	0.0705	0.0468
n-Nonadecane (C19)	0.0615	0.0897	0.0661	0.0426	0.0705	0.0471
Rubitherm RT27	0.1077	0.0430	0.0312	0.0426	0.0671	0.0465
Rubitherm RT31	0.0769	0.0412	0.0302	0.0426	0.0671	0.0469
Capric acid (C10)	0.0677	0.0538	0.0478	0.0384	0.0705	0.0568
Capric–Lauric eutectic	0.0615	0.0574	0.0488	0.0384	0.0705	0.0544
Fatty acid eutectic (C10–C12–C14)	0.0154	0.0538	0.0447	0.0384	0.0671	0.0532
1-Dodecanol	0.0615	0.0574	0.0441	0.0362	0.0671	0.0491
PEG-600	0.0000	0.0430	0.0447	0.0533	0.0638	0.0664
Glauber's salt ($Na_2SO_4 \cdot 10H_2O$)	0.0554	0.0911	0.1291	0.1279	0.0470	0.0906
$CaCl_2 \cdot 6H_2O$	0.1077	0.0681	0.1095	0.1279	0.0436	0.1027
$Na_2HPO_4 \cdot 12H_2O$	0.0154	0.0717	0.1084	0.1066	0.0470	0.0966
$Na_2CO_3 \cdot 10H_2O$	0.0615	0.0681	0.0935	0.1279	0.0470	0.0876
Form-stable paraffin/HDPE	0.0769	0.0430	0.0366	0.0533	0.0638	0.0544
Organic ester mixture	0.0769	0.0502	0.0427	0.0384	0.0671	0.0544

Table A23. Zone 28 °C COPRAS weighted matrix.

Material	Tm (°C) Score	ΔHfus (kJ/kg)	Vol. Latent Heat (MJ/m³)	k (W/m·K)	Cp (kJ/kg·K)	ρ (kg/m³)
n-Heptadecane (C17)	0.0146	0.0207	0.0079	0.0029	0.0023	0.0015
n-Octadecane (C18)	0.0585	0.0225	0.0086	0.0029	0.0023	0.0015
n-Nonadecane (C19)	0.0292	0.0230	0.0089	0.0029	0.0023	0.0016
Rubitherm RT27	0.0512	0.0111	0.0042	0.0029	0.0022	0.0015
Rubitherm RT31	0.0365	0.0106	0.0041	0.0029	0.0022	0.0015
Capric acid (C10)	0.0322	0.0138	0.0065	0.0026	0.0023	0.0019
Capric–Lauric eutectic	0.0292	0.0147	0.0066	0.0026	0.0023	0.0018
Fatty acid eutectic (C10–C12–C14)	0.0073	0.0138	0.0060	0.0026	0.0022	0.0018
1-Dodecanol	0.0292	0.0147	0.0059	0.0025	0.0022	0.0016
PEG-600	0.0000	0.0111	0.0060	0.0036	0.0021	0.0022
Glauber's salt ($Na_2SO_4 \cdot 10H_2O$)	0.0263	0.0234	0.0174	0.0087	0.0016	0.0030
$CaCl_2 \cdot 6H_2O$	0.0512	0.0175	0.0148	0.0087	0.0014	0.0034
$Na_2HPO_4 \cdot 12H_2O$	0.0073	0.0184	0.0146	0.0072	0.0016	0.0032
$Na_2CO_3 \cdot 10H_2O$	0.0292	0.0175	0.0126	0.0087	0.0016	0.0029
Form-stable paraffin/HDPE	0.0365	0.0111	0.0049	0.0036	0.0021	0.0018
Organic ester mixture	0.0365	0.0129	0.0058	0.0026	0.0022	0.0018

Table A24. Zone 18 °C VIKOR normalized matrix.

Material	Tm (°C) Score	ΔHfus (kJ/kg)	Vol. Latent Heat (MJ/m³)	k (W/m·K)	Cp (kJ/kg·K)	ρ (kg/m³)
n-Heptadecane (C17)	0.2500	0.7914	0.2877	0.0698	1.0000	0.0000
n-Octadecane (C18)	1.0000	0.9281	0.3425	0.0698	1.0000	0.0054
n-Nonadecane (C19)	0.5000	0.9712	0.3630	0.0698	1.0000	0.0108
Rubitherm RT27	0.8750	0.0360	0.0103	0.0698	0.8750	0.0000
Rubitherm RT31	0.6250	0.0000	0.0000	0.0698	0.8750	0.0075
Capric acid (C10)	0.5500	0.2518	0.1781	0.0233	1.0000	0.1828
Capric–Lauric eutectic	0.5000	0.3237	0.1884	0.0233	1.0000	0.1398
Fatty acid eutectic (C10–C12–C14)	0.1250	0.2518	0.1473	0.0233	0.8750	0.1183
1-Dodecanol	0.5000	0.3237	0.1404	0.0000	0.8750	0.0462
PEG-600	0.0000	0.0360	0.1473	0.1860	0.7500	0.3548
Glauber's salt ($Na_2SO_4 \cdot 10H_2O$)	0.4500	1.0000	1.0000	1.0000	0.1250	0.7849
$CaCl_2 \cdot 6H_2O$	0.8750	0.5396	0.8014	1.0000	0.0000	1.0000
$Na_2HPO_4 \cdot 12H_2O$	0.1250	0.6115	0.7911	0.7674	0.1250	0.8925
$Na_2CO_3 \cdot 10H_2O$	0.5000	0.5396	0.6404	1.0000	0.1250	0.7312
Form-stable paraffin/HDPE	0.6250	0.0360	0.0651	0.1860	0.7500	0.1398
Organic ester mixture	0.6250	0.1799	0.1267	0.0233	0.8750	0.1398

Table A25. Zone 28 °C VIKOR weighted matrix.

Material	Tm (°C) Score	ΔHfus (kJ/kg)	Vol. Latent Heat (MJ/m³)	k (W/m·K)	Cp (kJ/kg·K)	ρ (kg/m³)
n-Heptadecane (C17)	0.1188	0.2034	0.0388	0.0047	0.0330	0.0000
n-Octadecane (C18)	0.4750	0.2385	0.0462	0.0047	0.0330	0.0002
n-Nonadecane (C19)	0.2375	0.2496	0.0490	0.0047	0.0330	0.0004
Rubitherm RT27	0.4156	0.0092	0.0014	0.0047	0.0289	0.0000
Rubitherm RT31	0.2969	0.0000	0.0000	0.0047	0.0289	0.0002
Capric acid (C10)	0.2613	0.0647	0.0240	0.0016	0.0330	0.0060
Capric–Lauric eutectic	0.2375	0.0832	0.0254	0.0016	0.0330	0.0046
Fatty acid eutectic (C10–C12–C14)	0.0594	0.0647	0.0199	0.0016	0.0289	0.0039
1-Dodecanol	0.2375	0.0832	0.0190	0.0000	0.0289	0.0015
PEG-600	0.0000	0.0092	0.0199	0.0127	0.0248	0.0117
Glauber's salt ($Na_2SO_4 \cdot 10H_2O$)	0.2138	0.2570	0.1350	0.0680	0.0041	0.0259
$CaCl_2 \cdot 6H_2O$	0.4156	0.1387	0.1082	0.0680	0.0000	0.0330
$Na_2HPO_4 \cdot 12H_2O$	0.0594	0.1572	0.1068	0.0522	0.0041	0.0295
$Na_2CO_3 \cdot 10H_2O$	0.2375	0.1387	0.0865	0.0680	0.0041	0.0241
Form-stable paraffin/HDPE	0.2969	0.0092	0.0088	0.0127	0.0248	0.0046
Organic ester mixture	0.2969	0.0462	0.0171	0.0016	0.0289	0.0046

Table A26. Zone 28 °C TOPSIS normalized matrix.

Material	Tm (°C) Score	ΔHfus (kJ/kg)	Vol. Latent Heat (MJ/m³)	k (W/m·K)	Cp (kJ/kg·K)	ρ (kg/m³)
n-Heptadecane (C17)	0.1086	0.3115	0.2117	0.1485	0.2785	0.1778
n-Octadecane (C18)	0.4343	0.3378	0.2312	0.1485	0.2785	0.1789
n-Nonadecane (C19)	0.2171	0.3462	0.2386	0.1485	0.2785	0.1801
Rubitherm RT27	0.3800	0.1662	0.1126	0.1485	0.2653	0.1778
Rubitherm RT31	0.2714	0.1592	0.1089	0.1485	0.2653	0.1794
Capric acid (C10)	0.2389	0.2077	0.1725	0.1337	0.2785	0.2170
Capric–Lauric eutectic	0.2171	0.2215	0.1762	0.1337	0.2785	0.2078
Fatty acid eutectic (C10–C12–C14)	0.0543	0.2077	0.1615	0.1337	0.2653	0.2032
1-Dodecanol	0.2171	0.2215	0.1591	0.1262	0.2653	0.1877
PEG-600	0.0000	0.1662	0.1615	0.1856	0.2520	0.2540
Glauber's salt ($Na_2SO_4 \cdot 10H_2O$)	0.1954	0.3517	0.4662	0.4455	0.1857	0.3463
$CaCl_2 \cdot 6H_2O$	0.3800	0.2631	0.3952	0.4455	0.1724	0.3925
$Na_2HPO_4 \cdot 12H_2O$	0.0543	0.2769	0.3915	0.3713	0.1857	0.3694
$Na_2CO_3 \cdot 10H_2O$	0.2171	0.2631	0.3377	0.4455	0.1857	0.3348
Form-stable paraffin/HDPE	0.2714	0.1662	0.1321	0.1856	0.2520	0.2078
Organic ester mixture	0.2714	0.1938	0.1542	0.1337	0.2653	0.2078

Table A27. Zone 28 °C TOPSIS weighted matrix.

Material	Tm (°C) Score	ΔHfus (kJ/kg)	Vol. Latent Heat (MJ/m^3)	k (W/m·K)	Cp (kJ/kg·K)	ρ (kg/m^3)
n-Heptadecane (C17)	0.05157	0.08007	0.02857	0.01010	0.00919	0.00587
n-Octadecane (C18)	0.20629	0.08683	0.03122	0.01010	0.00919	0.00590
n-Nonadecane (C19)	0.10315	0.08896	0.03221	0.01010	0.00919	0.00594
Rubitherm RT27	0.18050	0.04270	0.01520	0.01010	0.00875	0.00587
Rubitherm RT31	0.12893	0.04092	0.01470	0.01010	0.00875	0.00592
Capric acid (C10)	0.11346	0.05338	0.02329	0.00909	0.00919	0.00716
Capric–Lauric eutectic	0.10315	0.05694	0.02378	0.00909	0.00919	0.00686
Fatty acid eutectic (C10–C12–C14)	0.02579	0.05338	0.02180	0.00909	0.00875	0.00670
1-Dodecanol	0.10315	0.05694	0.02147	0.00858	0.00875	0.00619
PEG-600	0.00000	0.04270	0.02180	0.01262	0.00832	0.00838
Glauber's salt ($Na_2SO_4 \cdot 10H_2O$)	0.09283	0.09039	0.06293	0.03030	0.00613	0.01143
$CaCl_2 \cdot 6H_2O$	0.18050	0.06761	0.05335	0.03030	0.00569	0.01295
$Na_2HPO_4 \cdot 12H_2O$	0.02579	0.07117	0.05285	0.02525	0.00613	0.01219
$Na_2CO_3 \cdot 10H_2O$	0.10315	0.06761	0.04559	0.03030	0.00613	0.01105
Form-stable paraffin/HDPE	0.12893	0.04270	0.01784	0.01262	0.00832	0.00686
Organic ester mixture	0.12893	0.04982	0.02081	0.00909	0.00875	0.00686

Table A28. Zone 28 °C MOORA normalized matrix.

Material	Tm (°C) Score	ΔHfus (kJ/kg)	Vol. Latent Heat (MJ/m^3)	k (W/m·K)	Cp (kJ/kg·K)	ρ (kg/m^3)
n-Heptadecane (C17)	0.10857	0.31154	0.21166	0.14852	0.27854	0.17778
n-Octadecane (C18)	0.43430	0.33785	0.23124	0.14852	0.27854	0.17893
n-Nonadecane (C19)	0.21715	0.34615	0.23858	0.14852	0.27854	0.18009
Rubitherm RT27	0.38001	0.16615	0.11256	0.14852	0.26528	0.17778
Rubitherm RT31	0.27143	0.15923	0.10889	0.14852	0.26528	0.17939
Capric acid (C10)	0.23886	0.20769	0.17251	0.13366	0.27854	0.21703
Capric–Lauric eutectic	0.21715	0.22154	0.17618	0.13366	0.27854	0.20779
Fatty acid eutectic (C10–C12–C14)	0.05429	0.20769	0.16150	0.13366	0.26528	0.20317
1-Dodecanol	0.21715	0.22154	0.15905	0.12624	0.26528	0.18770
PEG-600	0.00000	0.16615	0.16150	0.18564	0.25202	0.25397
Glauber's salt ($Na_2SO_4 \cdot 10H_2O$)	0.19543	0.35169	0.46615	0.44555	0.18570	0.34632
$CaCl_2 \cdot 6H_2O$	0.38001	0.26308	0.39519	0.44555	0.17243	0.39249
$Na_2HPO_4 \cdot 12H_2O$	0.05429	0.27692	0.39152	0.37129	0.18570	0.36941
$Na_2CO_3 \cdot 10H_2O$	0.21715	0.26308	0.33768	0.44555	0.18570	0.33477
Form-stable paraffin/HDPE	0.27143	0.16615	0.13214	0.18564	0.25202	0.20779
Organic ester mixture	0.27143	0.19385	0.15416	0.13366	0.26528	0.20779

Table A29. Zone 28 °C MOORA weighted matrix.

Material	Tm (°C) Score	ΔHfus (kJ/kg)	Vol. Latent Heat (MJ/m^3)	k (W/m·K)	Cp (kJ/kg·K)	ρ (kg/m^3)
n-Heptadecane (C17)	0.05157	0.08007	0.02857	0.01010	0.00919	0.00587
n-Octadecane (C18)	0.20629	0.08683	0.03122	0.01010	0.00919	0.00590
n-Nonadecane (C19)	0.10315	0.08896	0.03221	0.01010	0.00919	0.00594
Rubitherm RT27	0.18050	0.04270	0.01520	0.01010	0.00875	0.00587
Rubitherm RT31	0.12893	0.04092	0.01470	0.01010	0.00875	0.00592
Capric acid (C10)	0.11346	0.05338	0.02329	0.00909	0.00919	0.00716
Capric–Lauric eutectic	0.10315	0.05694	0.02378	0.00909	0.00919	0.00686
Fatty acid eutectic (C10–C12–C14)	0.02579	0.05338	0.02180	0.00909	0.00875	0.00670
1-Dodecanol	0.10315	0.05694	0.02147	0.00858	0.00875	0.00619
PEG-600	0.00000	0.04270	0.02180	0.01262	0.00832	0.00838
Glauber's salt ($Na_2SO_4 \cdot 10H_2O$)	0.09283	0.09039	0.06293	0.03030	0.00613	0.01143
$CaCl_2 \cdot 6H_2O$	0.18050	0.06761	0.05335	0.03030	0.00569	0.01295
$Na_2HPO_4 \cdot 12H_2O$	0.02579	0.07117	0.05285	0.02525	0.00613	0.01219
$Na_2CO_3 \cdot 10H_2O$	0.10315	0.06761	0.04559	0.03030	0.00613	0.01105
Form-stable paraffin/HDPE	0.12893	0.04270	0.01784	0.01262	0.00832	0.00686
Organic ester mixture	0.12893	0.04982	0.02081	0.00909	0.00875	0.00686

Table A30. Zone 28 °C PROMETHEE pairwise matrix.

Material	n-Heptadecane (C17)	n-Octadecane (C18)	n-Nonadecane (C19)	Rubitherm RT27	Rubitherm RT31	Capric Acid (C10)	Capric–Lauric Eutectic	Fatty Acid Eutectic (C10–C12–C14)	1-Dodecanol	PEG-600	Glauber's Salt ($Na_2SO_4 \cdot 10H_2O$)	$CaCl_2 \cdot 6H_2O$	$Na_2HPO_4 \cdot 12H_2O$	$Na_2CO_3 \cdot 10H_2O$	Form-Stable Paraffin/HDPE	Organic Ester Mixture
n-Heptadecane (C17)	0.000	0.000	0.000	0.236	0.246	0.157	0.137	0.224	0.149	0.340	0.029	0.098	0.134	0.094	0.232	0.186
n-Octadecane (C18)	0.399	0.000	0.238	0.338	0.467	0.413	0.417	0.623	0.429	0.739	0.290	0.192	0.526	0.366	0.453	0.407
n-Nonadecane (C19)	0.175	0.014	0.000	0.292	0.303	0.213	0.193	0.399	0.205	0.515	0.053	0.144	0.299	0.140	0.289	0.243
Rubitherm RT27	0.297	0.000	0.178	0.000	0.129	0.158	0.181	0.359	0.183	0.420	0.227	0.029	0.381	0.203	0.123	0.122
Rubitherm RT31	0.178	0.000	0.059	0.000	0.000	0.039	0.063	0.241	0.064	0.301	0.108	0.029	0.262	0.084	0.004	0.003
Capric acid (C10)	0.149	0.006	0.029	0.088	0.099	0.000	0.025	0.212	0.039	0.329	0.076	0.033	0.231	0.053	0.080	0.031
Capric–Lauric eutectic	0.123	0.004	0.004	0.107	0.117	0.020	0.000	0.207	0.015	0.325	0.053	0.033	0.207	0.029	0.099	0.049
Fatty acid eutectic (C10–C12–C14)	0.004	0.004	0.004	0.078	0.088	0.000	0.000	0.000	0.005	0.119	0.025	0.029	0.025	0.025	0.071	0.021
1-Dodecanol	0.120	0.001	0.001	0.093	0.103	0.018	0.000	0.197	0.000	0.316	0.048	0.029	0.203	0.025	0.088	0.039
PEG-600	0.020	0.019	0.019	0.038	0.048	0.017	0.018	0.019	0.024	0.000	0.021	0.025	0.021	0.021	0.018	0.021
Glauber's salt ($Na_2SO_4 \cdot 10H_2O$)	0.334	0.196	0.182	0.471	0.481	0.390	0.371	0.550	0.382	0.646	0.000	0.149	0.298	0.169	0.451	0.416
$CaCl_2 \cdot 6H_2O$	0.462	0.158	0.333	0.332	0.462	0.406	0.411	0.614	0.422	0.710	0.209	0.000	0.377	0.209	0.431	0.397
$Na_2HPO_4 \cdot 12H_2O$	0.145	0.137	0.134	0.330	0.341	0.249	0.231	0.256	0.242	0.351	0.004	0.023	0.000	0.044	0.310	0.276
$Na_2CO_3 \cdot 10H_2O$	0.254	0.127	0.124	0.302	0.312	0.221	0.202	0.405	0.214	0.501	0.024	0.004	0.194	0.000	0.282	0.248
Form-stable paraffin/HDPE	0.191	0.012	0.072	0.020	0.030	0.047	0.070	0.249	0.075	0.297	0.104	0.025	0.258	0.080	0.000	0.011
Organic ester mixture	0.183	0.004	0.064	0.057	0.068	0.036	0.059	0.238	0.064	0.338	0.108	0.029	0.262	0.084	0.049	0.000

References

1. Balaras, C.A.; Dascalaki, E.G.; Patsioti, M.; Droutsa, K.G.; Kontoyiannidis, S.; Cholewa, T. Carbon and Greenhouse Gas Emissions from Electricity Consumption in European Union Buildings. *Buildings* **2024**, *14*, 71. [CrossRef]
2. Wang, G.; Luo, T.; Luo, H.; Liu, R.; Liu, Y.; Liu, Z. A comprehensive review of building lifecycle carbon emissions and reduction approaches. *City Built Environ.* **2024**, *2*, 12. [CrossRef]
3. U.S. Department of Energy. Building Technologies Office Peer Review Program 2024. 2024. Available online: https://www.energy.gov/sites/default/files/2024-11/BTOPeerReview2024_Program.pdf (accessed on 10 November 2025).
4. Ma, Z.; Yan, Z.; He, M.; Zhao, H.; Song, J. A review of the influencing factors of building energy consumption and the prediction and optimization of energy consumption. *AIMS Energy* **2025**, *13*, 35–85. [CrossRef]
5. Jha, S.K.; Sankar, A.; Zhou, Y.; Ghosh, A. Incorporation of Phase Change Materials in Buildings. *Constr. Mater.* **2024**, *4*, 676–703. [CrossRef]
6. Raj, C.R.; Suresh, S.; Bhavsar, R.R.; Singh, V.K. Recent developments in thermo-physical property enhancement and applications of solid solid phase change materials: A review. *J. Therm. Anal. Calorim.* **2020**, *139*, 3023–3049. [CrossRef]
7. Pereira, J.; Souza, R.; Oliveira, J.; Moita, A. Phase change materials in residential buildings: Challenges, opportunities, and performance. *Materials* **2025**, *18*, 2063. [CrossRef]
8. Shah, D.A. Phase Change Materials (PCMs) for passive cooling: Performance in urban buildings. *Int. J. Sci. Res. Arch.* **2024**, *13*, 3529–3548. [CrossRef]
9. Nandi, A.; Biswas, N.; Datta, A.; Manna, N.K.; Mandal, D.K.; Biswas, S. A comprehensive review on enhanced phase change materials (PCMs) for high-performance thermal energy storage: Progress, challenges, and future perspectives. *J. Therm. Anal. Calorim.* **2025**, *150*, 8933–8976. [CrossRef]
10. Sharma, S.; Singh, R. Improving thermal comfort and reducing peak cooling loads in high-rise buildings using phase change materials. *Int. J. Sustain. Build. Technol.* **2025**, *7*, 29–37.
11. Rastogi, M.; Chauhan, A.; Vaish, R.; Kishan, A. Selection and performance assessment of Phase Change Materials for heating, ventilation and air-conditioning applications. *Energy Convers. Manag.* **2015**, *89*, 260–269. [CrossRef]
12. Turnpenny, J.R.; Etheridge, D.W.; Reay, D.A. Novel ventilation cooling system for reducing air conditioning in buildings: Part I: Testing and theoretical modelling. *Appl. Therm. Eng.* **2000**, *20*, 1019–1037. [CrossRef]
13. Parameshwaran, R.; Harikrishnan, S.; Kalaiselvam, S. Energy efficient PCM-based variable air volume air conditioning system for modern buildings. *Energy Build.* **2010**, *42*, 1353–1360. [CrossRef]
14. Sun, X.; Zhang, Q.; Medina, M.A.; Lee, K.O. Energy and economic analysis of a building enclosure outfitted with a phase change material board (PCMB). *Energy Convers. Manag.* **2014**, *83*, 73–78. [CrossRef]
15. Whiffen, T.R.; Riffat, S.B. A review of PCM technology for thermal energy storage in the built environment: Part II. *Int. J. Low-Carbon Technol.* **2013**, *8*, 159–164. [CrossRef]
16. Kang, Y.; Kim, S. Climate-Responsive Optimization of Phase Change Materials for Energy-Efficient Buildings. *J. Therm. Anal. Calorim.* **2025**, *46*, 122. Available online: https://link.springer.com/content/pdf/10.1007/s10765-025-03596-4.pdf (accessed on 10 November 2025).
17. Cui, H.; Xu, Z.; Liang, Y.; Yang, H. Optimisation of Design for PCM-Enhanced Buildings Through Parametric Modelling and Multi-objective Methodology. In *Key Themes in Energy Management: A Compilation of Current Practices, Research Advances, and Future Opportunities*; Springer Nature: Cham, Switzerland, 2024; pp. 297–314.
18. Jiao, K.; Lu, L.; Zhao, L.; Wang, G. Towards Passive Building Thermal Regulation: A State-of-the-Art Review on Recent Progress of PCM-Integrated Building Envelopes. *Sustainability* **2024**, *16*, 6482. [CrossRef]
19. Felix, P.G.; Rajagopal, V.; Kumaresan, K. Applicability of MCDM Algorithms for the Selection of Phase Change Materials for Thermal Energy Storage Heat Exchangers. *J. Mech. Eng.* **2021**, *67*, 611–621. [CrossRef]
20. Bajwa, A.U.R.; Siriwardana, C.; Shahzad, W.; Naeem, M.A. Material selection in the construction industry: A systematic literature review on multi-criteria decision making. *Environ. Syst. Decis.* **2025**, *45*, 8. [CrossRef]
21. Al-Sulaiman, F.A. A methodological concept for phase change material selection based on multi-criteria decision making (MCDM): A case study. *Energy* **2022**, *165*, 1085–1096.
22. Nadeem, A.; Rakhman, K.; Hossain, M.A. Phase Change Materials Ranking by Using the Analytic Hierarchy Process. In Proceedings of the 3rd European and Mediterranean Structural Engineering and Construction Conference (EURO-MED-SEC-03), Limassol, Cyprus, 3–8 August 2020; ISEC Press: Fargo, ND, USA, 2020.
23. Kumar, A.; Singh, R. A Hybrid Decision-Making Method for the Selection of a Phase Change Material Based on Experimental Data. *ASME J. Therm. Sci. Eng. Appl.* **2022**, *12*, 41020.
24. Mehling, H. Review and analysis of existing approaches to investigate property degradation of phase change materials and development of a new systematic approach. *Appl. Sci.* **2023**, *13*, 8682. [CrossRef]
25. Vallejo-Coral, E.C.; Garzón, R.; Ortega López, M.D.; Martínez-Gómez, J.; Moya, M. Determine the Profiles of Power Consumption in Commercial Buildings in a Very Hot Humid Climate Using a Temporary Series. *Sustainability* **2024**, *16*, 9770. [CrossRef]

26. Martínez-Gómez, J.; Nicolalde, J.F. Brake Disc Material Selection Based on MCDM and Simulation. *Processes* **2025**, *13*, 1287. [CrossRef]
27. Villacreses, G.; Jijón, D.; Nicolalde, J.F.; Martínez-Gómez, J.; Betancourt, F. Multicriteria Decision Analysis of Suitable Location for Wind and Photovoltaic Power Plants on the Galápagos Islands. *Energies* **2023**, *16*, 29. [CrossRef]
28. Saaty, T.L. Decision making—The analytic hierarchy and network processes (AHP/ANP). *J. Syst. Sci. Syst. Eng.* **2004**, *13*, 1–35. [CrossRef]
29. Goswami, S.S.; Behera, D.K.; Afzal, A.; Razak Kaladgi, A.; Khan, S.A.; Rajendran, P.; Subbiah, R.; Asif, M. Analysis of a robot selection problem using two newly developed hybrid MCDM models of TOPSIS-ARAS and COPRAS-ARAS. *Symmetry* **2021**, *13*, 1331. [CrossRef]
30. Radovanovic, M.; Ranđelović, A.; Jokić, Ž. Application of hybrid model fuzzy AHP-VIKOR in selection of the most efficient procedure for rectification of the optical sight of the long-range rifle. *Decis. Mak. Appl. Manag. Eng.* **2020**, *3*, 131–148. [CrossRef]
31. Taherdoost, H.; Madanchian, M. Multi-criteria decision making (MCDM) methods and concepts. *Encyclopedia* **2023**, *3*, 77–87. [CrossRef]
32. Raju, S.S.; Murali, G.B.; Patnaik, P.K. Ranking of Al-CSA composite by MCDM approach using AHP–TOPSIS and MOORA methods. *J. Reinf. Plast. Compos.* **2020**, *39*, 721–732. [CrossRef]
33. Agrawal, N. Multi-criteria decision-making toward supplier selection: Exploration of PROMETHEE II method. *Benchmarking Int. J.* **2022**, *29*, 2122–2146. [CrossRef]

energies

MDPI

Article

Optimising Behavioural Control Based on Actual HVAC Use in Naturally Ventilated Buildings

Jiajing Wu [1,2], Rongxin Qiu [3], Xiaoyu Ying [1,2,*], Shuqin Chen [3] and Xueyuan Zhao [4]

1 Department of Architecture, School of Spatial Planning and Design, Hangzhou City University, Hangzhou 310015, China; wujj@hzcu.edu.cn
2 Zhejiang Engineering Research Center of Building's Digital Carbon Neutral Technology, Hangzhou 310015, China
3 Department of Architecture, College of Civil Engineering and Architecture, Zhejiang University, Hangzhou 310058, China
4 Department of Architecture, The College of Urban Construction, Zhejiang Shuren University, Hangzhou 310015, China
* Correspondence: yingxiaoyu@hzcu.edu.cn

Abstract

Multi-objective optimisation is essential for balancing building energy efficiency and thermal comfort. Existing research primarily focuses on passive optimisation strategies that assume fixed behavioural patterns of a 'rational occupant'. However, these studies often overlook the impact of stochastic occupant behaviour on building energy efficiency and thermal comfort. Furthermore, they fail to consider the inherent randomness, variability, dynamic nature, and feedback mechanisms of individual actions. As a result, this oversight can lead to suboptimal energy efficiency, insufficient thermal comfort, and a poor user experience. This study examines a naturally ventilated research building equipped with split-type air conditioning in China's hot summer and cold winter climate zone. The research develops a rapid prediction model for air conditioning (AC) energy consumption and thermal comfort based on actual HVAC behaviours, incorporating the AC and natural ventilation (NV) operation schedules. The model utilises Artificial Neural Networks (ANNs), importance analysis, and batch simulation. Furthermore, a multi-objective optimisation decision-making model is developed to balance building AC energy consumption and indoor environmental thermal comfort, using the NSGA-II algorithm. The results indicate that when building design parameters comply with the current energy-saving design standards, behavioural optimisation can lead to a 31.4% reduction in energy use for building AC systems while enhancing thermal comfort by 37.5%. Furthermore, by implementing integrated optimisation strategies, comfort can be improved by as much as 92.6% without raising energy consumption.

Keywords: HVAC behaviour; behavioural control; multi-objective optimisation; energy-saving; thermal comfort

1. Introduction

Public buildings are major contributors to urban architectural energy consumption, significantly impacting city-wide energy efficiency and total carbon emissions. Research findings from the Building Energy Conservation Research Center of Tsinghua University indicate that in 2020, China's total energy consumption for public buildings (excluding heating in northern regions) amounted to 346 million tons of standard coal equivalent, accounting for 33% of total building energy consumption, while associated electricity consumption was 102.21 billion kWh [1]. In 2021, the total energy consumption throughout the entire life cycle of China's building sector (encompassing material production, construction, operation, and demolition) increased by 5.0% compared to 2020, while carbon emissions rose by 4.8%. These statistics reveal a striking growth rate in energy consumption and carbon emissions [2].

Research efforts to reduce building energy consumption have gained considerable attention due to increasing energy consumption and emissions. An expanding group of scholars focused on practical approaches to decreasing building energy consumption from various angles [3,4]. For instance, Ding et al. [5] developed an urban-level energy consumption database based on residential and office building prototypes. Their research commenced with creating typical building energy consumption models using EnergyPlus v8.8.0, followed by integrating stochastic simulation and data-driven regression analysis to generate the energy consumption database. Accordingly, a web platform was designed to provide energy-saving solutions for urban planning and existing building retrofits.

Meanwhile, as people's living standards improve, research has increasingly shifted toward enhancing building thermal comfort [6–8]. Consequently, scholars have investigated approaches integrating multi-objective optimisation with building simulation software, including energy conservation, thermal comfort, building envelope performance, and carbon emissions reduction [9,10]. Specific methodologies involve applying tools such as DOE-2 simulation [11], machine learning [12–14], and optimisation algorithms [15,16] to achieve multi-objective optimisation of buildings. For example, Wang et al. [17] proposed a multi-objective optimisation framework based on SVR-NSGA-II: it combines support vector regression (SVR) modelling with the NSGA-II algorithm to optimise building performance after generating energy consumption data through OpenStudio. The results show that the optimised building significantly improves energy efficiency and thermal comfort. Liu et al. [18] conducted multi-objective optimisation design research on a university activity centre in China, using Grasshopper modelling and the Ladybug and Honeybee plugins to simulate energy consumption and thermal comfort, and applying the Octopus algorithm for optimisation. As a result, a 58.8% reduction in annual energy consumption and a 53.0% increase in thermal comfort duration were achieved.

Despite progress in studying traditional multi-objective optimisation problems (MOPs), many shortcomings remain. Traditional research is based on the 'rational person' assumption, simplifying individual behaviour patterns into fixed and regular schedules for building simulations [19,20]. However, this approach neglects critical factors such as the randomness, variability, dynamic nature, and feedback mechanisms of human behaviours. As a result, it becomes challenging to reflect the complexity of human behaviour in real-world scenarios accurately. Consequently, significant discrepancies arise between the simulated values for building energy consumption and indoor thermal environments—derived from fixed schedules [21,22]—and the actual measured values. This gap ultimately leads to suboptimal outcomes when implementing multi-objective optimisation strategies.

To address the aforementioned issues, scholars have begun to focus on behavioural model research, exploring methods for more precise energy consumption optimisation [23]. Various models have been proposed to describe the HVAC behaviours, including deterministic models [24], statistical models, stochastic models, and occupant behaviour action models [25]. Among these, the stochastic model has been shown to offer a superior simulation of human energy consumption patterns [26,27], thereby significantly enhancing the accuracy of energy consumption and thermal comfort predictions [28–30].

Furthermore, in traditional multi-objective optimisation research, the predominant focus has been on passive strategies, often overlooking the effective management of individual behaviours [31]. This limited viewpoint does not adequately address issues such as energy wastage, excessive comfort, and low energy efficiency. Jeong et al. [32] and Azar et al. [33] have indicated that relying solely on passive strategies and equipment optimisation is insufficient for achieving optimal energy savings, as user behaviour significantly impacts building energy consumption. Optimising user behaviour patterns has been shown to reduce building energy consumption while enhancing thermal comfort. For example, Hamed et al. [34] achieved significant energy savings by optimising household energy consumption behaviours, such as adjusting clothing choices, increasing natural ventilation, promptly turning off lights, and effectively managing air conditioning usage times. These efforts resulted in an overall energy consumption reduction of 19.39% while improving thermal comfort. However, many of these studies are based on fixed behavioural patterns. To align research outcomes more closely with actual energy consumption, Haidar et al. [35] explored multi-objective optimisation strategies that consider real HVAC behavioural characteristics. Xiang et al. [36] developed an HVAC behaviour model to describe the behavioural characteristics of a real-world HVAC, including air conditioning and window operation. Using this model, they implemented behaviour optimisation strategies through simulation. The results showed that these optimised solutions could reduce energy demand by 50.2% to 60.2% while decreasing the duration of indoor thermal discomfort by 3.52% to 11.09% compared to the benchmark case.

In general, passive multi-objective optimisation strategies based on the "rational agent" (which uses fixed schedules) have produced significant results, but these strategies remain limited. Most studies on behavioural optimisation rely on the rational-agent assumption, treating user behaviour as a predictable process. However, there is a notable oversight regarding users' stochastic actions and dynamic adaptations in real-world scenarios. Furthermore, research on multi-objective optimisation problems (MOPs) for behavioural optimisation using real-world behavioural data is even more scarce.

Based on the stochastic sequence of HVAC behaviour for a research building, which reflects actual behavioural characteristics (proposed in the preliminary research [37]), this study identifies the important influencing factors of building AC energy consumption and indoor thermal comfort as decision variables. These factors were extracted from three aspects: building design scheme, internal disturbance elements, and adaptive behaviours. A rapid multi-objective prediction model for building AC energy consumption and indoor thermal comfort is constructed using EnergyPlus batch simulation and an artificial neural network approach. Combining the NSGA-II genetic algorithm with the control parameter method, a multi-objective optimisation decision-making model is established for generating optimisation strategies. This study proposes a multi-objective optimisation method that comprehensively considers actual HVAC behavioural characteristics and user demands, aiming to significantly improve indoor thermal comfort while reducing air conditioning energy consumption.

In terms of content, this study innovatively adopts the coupling of air conditioning and natural ventilation (specifically, window operation) as a pivotal indicator for behavioural optimisation. This approach accurately captures the effects of the occupants' combined HVAC behaviours on building energy consumption and the indoor thermal environment. A comprehensive technical framework is established for the methodology, encompassing data acquisition, model training, and optimisation decision-making. A rapid prediction model is developed using an artificial neural network (ANN) based on EnergyPlus (E+) batch simulation results. This provides a scientifically rigorous and computationally efficient basis for formulating multi-objective optimisation strategies. This integrated approach enhances the practicality and forward-looking value of the research.

2. Methodology

The research technical route of the paper is shown in Figure 1. This study proposes a multi-objective optimisation method based on actual HVAC behaviours to optimise energy consumption and thermal comfort for air conditioning (AC) in naturally ventilated research buildings in China's hot summer and cold winter climates. The research quantitatively represents real HVAC behaviours in EnergyPlus simulations [37] by developing a model for predicting HVAC stochastic behaviours. A rapid prediction model for building AC energy consumption and indoor thermal comfort was developed using a BP neural network based on batch simulations. Using the NSGA-II algorithm, a multi-objective optimisation decision-making model was established for energy conservation and enhanced thermal comfort. This model effectively generates traditional passive optimisation strategies while providing users with behavioural optimisation strategies and integrated recommendations. This approach offers comprehensive and scientific support for making energy-saving retrofits and operational management decisions, delivering significant practical value.

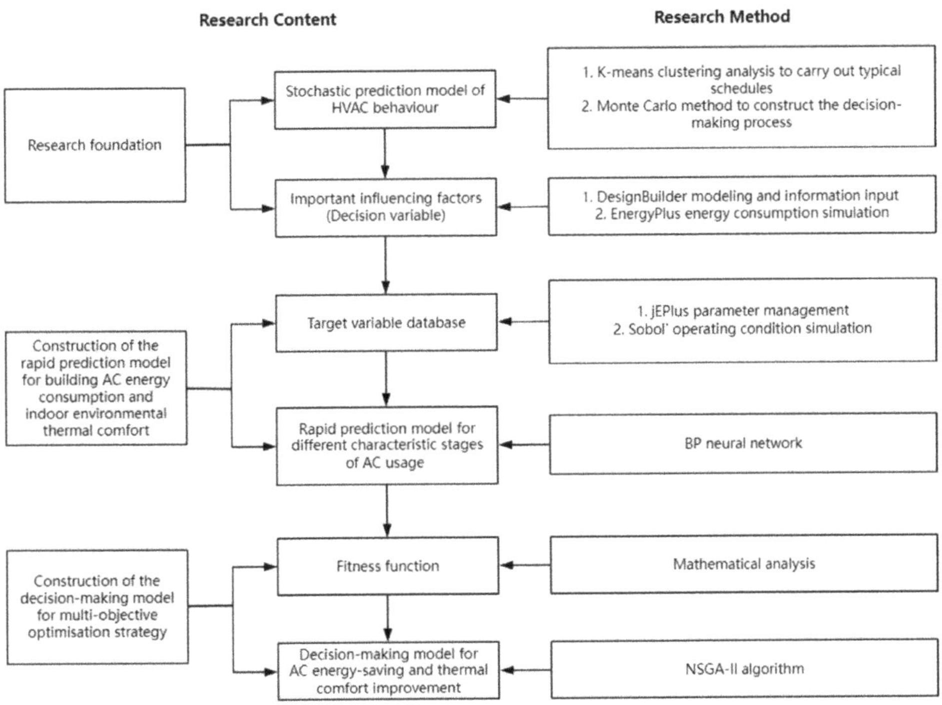

Figure 1. The research technical route.

2.1. Construction of a Multi-Objective Database

The multi-objective database, constructed under the various operating scenarios, is the foundation for developing a rapid prediction model for AC energy consumption and indoor environmental thermal comfort. This study begins by quantifying actual HVAC behaviours in building simulations using measured values. This approach helps to avoid the prediction errors arising from traditional fixed-schedule methods and enhances simulation accuracy. Subsequently, an importance analysis is performed to identify the key factors influencing building AC energy consumption and thermal comfort. Finally, batch simulations are executed using jEPlus v2.1.0 to create a 'decision variable—target variable' mapping database across different operating scenarios.

2.1.1. Methods of Quantifying and Characterising Actual HVAC Behaviours in Building Simulation

This paper is based on preliminary research conducted at a naturally ventilated research building in Hangzhou, China, which created a stochastic prediction model to accurately represent the HVAC system's actual behaviours. This model was developed by analysing indoor thermal environment parameters, AC energy consumption data, and window opening/closing behaviour metrics. These measurements were collected from January 2017 to January 2018.

This study divides the year into seven characteristic stages of AC usage to effectively capture the seasonal variations in HVAC behaviours induced by annual climate changes, as presented in Table 1 [37]. The k-means clustering analysis generates typical operation schedules and their probabilities of occurrence for different functional rooms (e.g., collective offices and individual offices) within each characteristic stage of AC usage. Subsequently, annual stochastic HVAC operation sequences are generated via Monte Carlo simulation. Stochastic behaviour sequences are transformed into EnergyPlus-compatible stochastic behavioural schedules using Python 3.7. This establishes a method for quantifying real HVAC behaviours in building simulations. The study demonstrates a significant improvement in prediction accuracy by simulating the annual AC energy consumption of the case building under stochastic behavioural and fixed schedules, then comparing the results with measured values. This finding underscores the efficacy and necessity of the proposed methodology.

Table 1. Characteristic stage division of AC usage [37].

Characteristic Stages of AC Usage		Duration Date
Cooling season	Mid-summer	21 June~10 September
	Early and late summer	11 May~20 June 11 September~30 September
	Late spring and early autumn	21 April~10 May 1 October~20 October
Transitional period		11 April~20 April 21 October~31 October
Heating season	Early spring and late autumn	1 November~20 November 21 March~10 April
	Early and late winter	11 February~20 March 21 November~10 December
	Mid-winter	11 December~10 February

2.1.2. Proposal of Decision Variables Based on Importance Analysis

An excessive number of input items can lead to neural network instability and reduced prediction accuracy. Therefore, it is crucial to carefully select decision variables when developing rapid prediction models for AC energy consumption and indoor thermal comfort. The preliminary study [37] identified 28 initial influencing factors across three dimensions: building design scheme, internal building disturbance elements, and adaptive behaviours. These factors were used to identify the key influencing factors of the target variables—AC energy consumption and indoor thermal comfort—which will serve as the decision variables for the rapid prediction model.

Measurements obtained from the case building revealed interactions and mutual influences between the air conditioning (AC) and the natural ventilation (NV) operations in practical operation, and these coupling patterns significantly impact the target variables. Consequently, the preliminary study incorporated the coupling pattern between AC and NV operation as an initial influencing factor within the adaptive behaviour category. Five typical coupling modes were proposed, as shown in Table 2.

Table 2. The description of coupling patterns between NV and AC operation.

Coupling Pattern	Feature Description
M0	A default pattern, reflecting the actual operational characteristics of AC and NV operation.
M1	Fixed ventilation in the morning: From 8:00 to 9:00, regardless of whether the air conditioner is on or off, the exterior windows are opened for a fixed period
M2	Fixed ventilation overnight: From 17:00 to 8:00 the following day, regardless of whether the air conditioner is on or off, the exterior windows are open for a fixed period.
M3	Opposition all day: The opening status of exterior windows and air conditioning is always in opposition between 0:00 and 24:00.
M4	Opposition overnight: From 17:00 to 8:00 the following day, the opening status of exterior windows and air conditioning is always in opposition; from 8:00 to 17:00, regardless of whether the air conditioner is on or off, the exterior windows remain closed.

The 28 initial influencing factors were considered independent variables, and batch simulations were performed using EnergyPlus with the assistance of jEPlus for parameter management. Specifically, the AC cooling and heating loads, as well as the uncomfortable hours of each air-conditioning zone, were simulated. The uncomfortable hours in this study refer to the duration during which the Predicted Mean Vote (PMV) index within the corresponding air-conditioning zone falls outside the comfort range while occupants are present. Based on the thermal sensation vote results from 1471 valid questionnaires (distributed and collected in the research building over 34 seasonal typical days from January 2017 to January 2018) and real-time thermal environment parameters, the actual Mean Thermal Sensation Vote (TSV) and the PMV were calculated and compared. Result analysis indicated that the PMV index specified in ASHRAE Standard 55-2017 [38] is suitable for assessing the subjective thermal sensation of occupants in research buildings in Hangzhou. Accordingly, this study adopted $[-0.5, 0.5]$ as the PMV comfort range for the simulation.

The total AC load of the entire building can be directly summed and output from the simulation results. However, the uncomfortable hours of individual zones cannot be directly used to characterise the thermal comfort of the entire building. To quantitatively evaluate the indoor thermal comfort of the entire building, this study proposed the "comprehensive uncomfortable hours" as a discomfort index, which is calculated using Equation (1). A higher value of this index indicates poorer thermal comfort. The target variables, namely the annual AC load per unit area and annual comprehensive uncomfortable hours (UCH) of the case building, were then derived through simulation and calculations, quantifying the building's AC energy consumption and indoor thermal comfort, respectively.

$$T_u = \frac{\sum_{i=1}^{n} S_i \cdot t_i}{\sum_{i=1}^{n} S_i}, \quad (1)$$

where

T_u is the comprehensive uncomfortable hours (UCH) for the entire building;
n is the number of air-conditioned rooms in the building;
S_i is the area of the ith room;
t_i is the simulated uncomfortable hours in the ith room.

Through batch simulations and calculations under 2000 multiple operating conditions, the target variables (i.e., building AC load per unit area and comprehensive uncomfortable hours) of the case building were obtained. Utilising random forest regression and residual mean methods, the importance scores of each independent variable (initial influencing factor) concerning each target variable were determined, and relative importance was calculated via normalisation. As shown in Table 3, 14 important influencing factors were selected as decision variables for the target variables based on a numerical analysis of their relative importance, with each factor demonstrating a relative importance greater than 0.5%. The remaining 14 factors were classified as non-important influencing factors; their values were set as fixed optimal values based on correlation analysis results and were not treated as variables.

Table 3. Decision variables and their relative importance [37].

Code	Decision Variable	Unit	Relative Importance of AC Energy Consumption	Relative Importance of Thermal Comfort
1	Cooling setpoint temperature	°C	100.0%	100.0%
2	Air permeability performance	ach	67.0%	4.7%
3	Heating setpoint temperature	°C	46.6%	33.7%
4	People density in collective offices	m^2/person	17.3%	0.3%
5	Coupling pattern between NV and AC operation	-	7.8%	0.9%
6	SHGC of the external window	-	4.5%	2.8%
7	Power density of equipment	W/person	2.9%	1.0%
8	Heat transfer coefficient of the external wall	$W/(m^2 \cdot K)$	2.5%	1.4%
9	Heat transfer coefficient of the external window	$W/(m^2 \cdot K)$	1.8%	0.8%
10	Heat transfer coefficient of the roof	$W/(m^2 \cdot K)$	1.5%	1.3%
11	Window-to-wall ratio of the south elevation	-	1.3%	0.7%
12	Window-to-wall ratio of the north elevation	-	1.2%	0.5%
13	Heat transfer coefficient of the internal wall	$W/(m^2 \cdot K)$	0.9%	0.9%
14	Building orientation	°	0.8%	0.7%

A multi-objective database was constructed under the various operating scenarios based on the aforementioned decision variables. Artificial neural networks utilised the database to develop and validate a rapid prediction model for target variables.

2.1.3. Construction of a "Decision Variable-Target Variable" Mapping Database Based on Batch Simulation

The study uses the decision variables shown in Table 3 as input variables, setting each variable as a dynamic parameter with the value ranges shown in Table 2, as defined by jEPlus's parameter management. The Sobol' sequence method is then employed to generate 5000 random samples, with each sample recording the values of the input variables.

The target variable serves as the output vector and, together with the corresponding input vector, forms a complete data sample. Ultimately, 4000 samples were randomly selected for the train set and the remaining 1000 for the test set. As the input variables have different units and their numerical values vary significantly, the original data must undergo normalisation preprocessing to avoid issues such as increased error and reduced convergence during neural network learning. This involves uniformly quantising the values into the [0, 1] range using Equation (2).

$$x'_i = \frac{x_i - min(x_i)}{max(x_i) - min(x_i)}, \quad (2)$$

where

x'_i is the value of the *i*th input variable after preprocessing;

x_i is the original value of the *i*th input variable before processing;

$max(x_i)$ and $min(x_i)$ are the maximum and minimum values of the *i*th input variable in all data samples.

2.2. Construction of a Multi-Objective Rapid Prediction Model Based on ANN

Due to the limitations of the parameter setting mechanism in EnergyPlus software, specific parameters that significantly influence building AC energy consumption and thermal comfort cannot be varied continuously. Therefore, when these dynamic parameters are combined within their specified ranges to generate multiple operating conditions, these non-continuous variables inevitably result in 'interval values' that cannot be fully covered. Exploring optimisation strategies for building AC energy-saving and indoor thermal comfort is time-consuming and labour-intensive compared to traditional simulation methods. Therefore, it is necessary to construct a rapid prediction model with sufficient accuracy, based on the 'decision variable—target variable' mapping database established in Section 2.1.3. This approach reduces CPU runtime and human resource costs while addressing EnergyPlus' limitations; EnergyPlus has difficulty simulating results directly when dynamic parameters are set to 'interval values'.

According to the characteristic stage division of AC usage defined in Section 2.1.1, the following six stages are identified: mid-summer, early and late summer, late spring and early autumn, early spring and late autumn, early and late winter, and mid-winter, and rapid prediction models for each characteristic stage are developed with the corresponding dates representing the prediction cycle.

The prediction model is developed based on the BP neural network and consists of an input, hidden, and output layer. The input layer comprises 13 nodes, and the output layer comprises two nodes, corresponding to the 14 decision variables and two target variables shown in Table 3. Figure 2 shows the structural diagram of the neural network model.

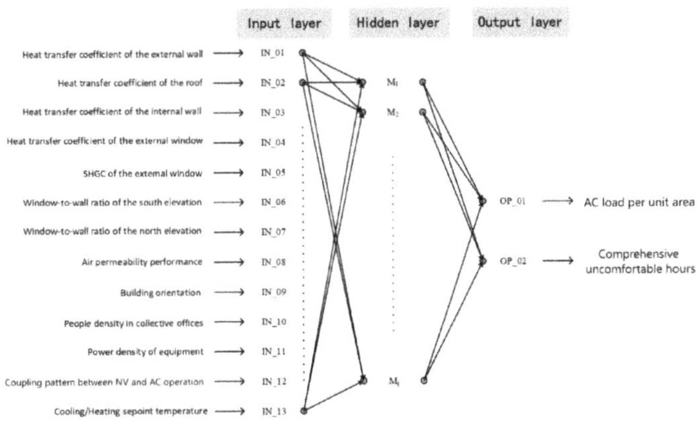

Figure 2. BP neural network model structure.

According to Equation (3), the boundary values of the reference number of hidden layer nodes n1 are calculated. Trial calculations are performed starting from the minimum number of nodes, $min(n_1)$, and are gradually increased to the maximum number of nodes, $max(n_1)$. The relative error results of the test set after each training session are compared to select the optimal number of hidden layer nodes.

$$n_1 = \sqrt{n+m} + a, \tag{3}$$

where

n is the number of input nodes;

m is the number of output nodes;

a is a constant with a value between [1, 10].

Rapid prediction models are constructed based on the BP neural network for different characteristic stages of AC usage. The input and output variables and their functional mapping relationships are shown in Table 4. The influencing factors corresponding to input items 1 to 11 are derived from the building design scheme and internal disturbance elements. These values are consistent in the rapid prediction model for different characteristic stages, i.e., $\forall i \in [1, 11]$, $x_{i,H1} = x_{i,H2} = x_{i,H3} = x_{i,C1} = x_{i,C2} = x_{i,C3}$. However, the influence factors corresponding to input items 12 and 13 are derived from the adaptive behaviours. They take independent values within their respective ranges in the prediction models for different characteristic stages.

Table 4. Inputs, outputs, and function mapping relationships of the rapid prediction model at each characteristic stage of AC usage.

Characteristic Stages of AC Usage	Code	Input	AC Load per Unit Area		Comprehensive Uncomfortable Hours	
			Output	Function Mapping Relationship	Output	Function Mapping Relationship
Late spring and early autumn	C1	$x_{i,C1}$, $i \in [1, 13]$	$y_{1,C1}$	$y_{1,C1} = F_{C1}(x_{1,C1}, x_{2,C1}, \ldots, x_{13,C1})$	$y_{2,C1}$	$y_{2,C1} = P_{C1}(x_{1,C1}, x_{2,C1}, \ldots, x_{13,C1})$
Early and late summer	C2	$x_{i,C2}$, $i \in [1, 13]$	$y_{1,C2}$	$y_{1,C2} = F_{C2}(x_{1,C2}, x_{2,C2}, \ldots, x_{13,C2})$	$y_{2,C2}$	$y_{2,C2} = P_{C2}(x_{1,C2}, x_{2,C2}, \ldots, x_{13,C2})$
Mid-summer	C3	$x_{i,C3}$, $i \in [1, 13]$	$y_{1,C3}$	$y_{1,C3} = F_{C3}(x_{1,C3}, x_{2,C3}, \ldots, x_{13,C3})$	$y_{2,C3}$	$y_{2,C3} = P_{C3}(x_{1,C3}, x_{2,C3}, \ldots, x_{13,C3})$
Early spring and late autumn	H1	$x_{i,H1}$, $i \in [1, 13]$	$y_{1,H1}$	$y_{1,H1} = F_{H1}(x_{1,H1}, x_{2,H1}, \ldots, x_{13,H1})$	$y_{2,H1}$	$y_{2,H1} = P_{H1}(x_{1,H1}, x_{2,H1}, \ldots, x_{13,H1})$
Early and late winter	H2	$x_{i,H2}$, $i \in [1, 13]$	$y_{1,H2}$	$y_{1,H2} = F_{H2}(x_{1,H2}, x_{2,H2}, \ldots, x_{13,H2})$	$y_{2,H2}$	$y_{2,H2} = P_{H2}(x_{1,H2}, x_{2,H2}, \ldots, x_{13,H2})$
Mid-winter	H3	$x_{i,H3}$, $i \in [1, 13]$	$y_{1,H3}$	$y_{1,H3} = F_{H3}(x_{1,H3}, x_{2,H3}, \ldots, x_{13,H3})$	$y_{2,H3}$	$y_{2,H3} = P_{H3}(x_{1,H3}, x_{2,H3}, \ldots, x_{13,H3})$

The study uses a BP neural network to develop a prediction model. JAVA SE 9 is applied as the programming language for the development platform. The BP neural network's error tolerance is set to 1×10^{-5}, and the Adam optimiser [39,40] is used for training. The maximum number of iterations is set to 2000.

2.3. Multi-Objective Optimisation Strategy Decision-Making Model Based on NSGA-II

Reducing AC energy consumption while improving indoor thermal comfort is a multi-objective minimisation problem, with building AC load per unit area and comprehensive uncomfortable hours as the optimisation targets.

This paper takes a naturally ventilated research building in a hot summer and cold winter climate zone in China as its case study. It explores AC energy-saving and thermal comfort optimisation strategies from three perspectives: passive optimisation strategies during the building design phase, behavioural optimisation strategies during the building operation phase, and integrated optimisation strategies considering both phases. It provides a specific technical chain for the multi-objective optimisation of similar buildings.

2.3.1. Proposal of the Fitness Function

(1) Construction of the Fitness Function

This paper proposes a decision-making model to conserve AC energy while enhancing indoor thermal comfort. The model uses a fitness function to quantify energy consumption and thermal comfort performance in different operating conditions throughout the year. This provides a basis for accurately evaluating different optimisation strategies. The optimisation strategy for case building comprises the optimised values of various decision variables. Therefore, the fitness function is constructed based on the rapid prediction model developed in Section 2.2. The decision variables act as the independent variables, while the target variables (annual AC load per unit area and the comprehensive uncomfortable hours) act as the dependent variables.

While decision variables related to building design scheme and internal disturbance elements typically remain unchanged, occupants' adaptive behaviour changes with climate variations across different characteristic stages of AC usage. Therefore, the values of the decision variables related to occupants' adaptive behaviours vary within their respective ranges at each stage.

Equation (4) outlines the fitness function used to evaluate the target variables. The equation is based on the rapid prediction model's fundamental information and functional mapping relations, as described in Table 4, and it employs the NSGA-II algorithm.

$$\begin{cases} y_1 = y_{1,H1} + y_{1,H2} + y_{1,H3} + y_{1,C1} + y_{1,C2} + y_{1,C3} \\ y_2 = y_{2,H1} + y_{2,H2} + y_{2,H3} + y_{2,C1} + y_{2,C2} + y_{2,C3} \end{cases}, \tag{4}$$

where

y_1 is the total annual AC load per unit area of the case building;

y_2 is the annual comprehensive uncomfortable hours of the case building.

(2) Range of Values for Decision Variables

In multi-objective optimisation problems, the range of values for the decision variables defines the space where optimisation strategies can be adjusted. This paper studies multi-objective optimisation strategies using the energy-saving renovation of a case building. As a result, certain variables must be fixed based on the actual characteristics of the case building, and the remaining decision variables must be searched for the optimal combination.

The main façade of the case building is oriented north–south, with a window-to-wall ratio of 0.4 for south and north elevations. The aforementioned default parameters remain unchanged. Referring to the typical values for internal disturbance elements provided in the appendix of the current *Design standard for energy efficiency of public buildings (GB 50189-2015)* [41], the people density in collective offices is set to 10 m^2/person and the power density of equipment to 150 W/person.

The upper limits of the decision variables related to the building design scheme are based on the actual characteristics of the case building and a reference public building typical of a hot summer and cold winter climate zone in the 1980s. This consideration applies except for the building orientation and window-to-wall ratio of the north and south elevations. The lower limits are set under the thermal performance requirements for the building envelope specified in the Technical standard for nearly zero energy buildings (GB/T 51350-2019) [42]. The range is based on technical measures currently used for parameters not explicitly addressed by existing energy-saving regulations and standards.

The decision variables and their corresponding ranges required to optimise AC energy savings and indoor thermal comfort in the case building are presented in Table 5.

Table 5. Description of decision variables.

Type	ID	Decision Variable	Unit	Range of Values
Building design scheme	IN_01	Heat transfer coefficient of the external wall	W/(m²·K)	[0.15, 2.50]
	IN_02	Heat transfer coefficient of the roof	W/(m²·K)	[0.15, 3.00]
	IN_03	Heat transfer coefficient of the internal wall	W/(m²·K)	[1.00, 5.00]
	IN_04	Heat transfer coefficient of the external window	W/(m²·K)	[2.20, 6.40]
	IN_05	SHGC of the external window	-	[0.15, 1.00]
	IN_06	Window-to-wall ratio of the south elevation	-	{0.4}
	IN_07	Window-to-wall ratio of the north elevation	-	{0.4}
	IN_08	Air permeability performance	ach	[0.1, 0.8]
	IN_09	Building orientation	°	{0}
Internal disturbance elements	IN_10	People density in collective offices	m²/person	{10}
	IN_11	Power density of equipment	W/person	{150}
Occupants' adaptive behaviours	IN_12	Coupling pattern between NV and AC operation	-	{M0, M1, M2, M3, M4}
	IN_13	Cooling/heating setpoint temperature	°C	[18, 30]

(3) Acquisition of Reference Values for Optimisation Objectives

To systematically evaluate the effectiveness of different optimisation strategies in improving energy conservation and indoor thermal comfort in the case building, this study conducts a comparative analysis based on the following standards and methods:

- The thermal performance limits for building envelopes specified in the *Design standard for energy efficiency of public buildings (GB 50189-2015)* [41] and the *Technical standard for nearly zero energy buildings (GB/T 51350-2019)* [42] are used as the baseline;
- The reference standards of occupants' behaviour are based on the typical fixed schedule of 08:00–17:00 and the stochastic behavioural schedules generated with the method described in Section 2.1.1. The method was established in the preliminary research [37];
- Based on the two types of building envelope structure and HVAC behavioural patterns, the AC load per unit area and the comprehensive uncomfortable hours of the case building are worked out by EnergyPlus simulation and calculation. Equation (5) transforms the total uncomfortable hours into a proportion of those hours to enable comparative analysis. As shown in Table 6, the optimisation target for comparative analysis is set using four typical reference points: B1, B2, B3, and B4.

$$P_u = \frac{\sum_{i=1}^{n} S_i \cdot t_i}{\sum_{i=1}^{n} S_i \cdot k_i}, \tag{5}$$

where

P_u is the proportion of uncomfortable hours;

n is the number of air-conditioned rooms in the building;

S_i is the area of the ith room;

t_i is the simulated uncomfortable hours in the ith room;

t_i is the number of hours occupied by occupants in the ith room.

Table 6. Basic information about the reference points of optimisation objectives.

Reference Point	Reference Standard for Thermal Performance of Building Envelopes	HVAC Behavioural Pattern	AC Load per Unit Area (kWh/m^2)	Proportion of Uncomfortable Hours
B1	Design standard for energy efficiency of public buildings (GB 50189-2015)	Stochastic behavioural schedule	59.326	84.82%
B2		Fixed schedule	75.536	80.65%
B3	Technical standard for nearly zero energy buildings (GB/T 51350-2019)	Stochastic behavioural schedule	44.366	70.93%
B4		Fixed schedule	55.942	63.00%

2.3.2. Algorithm Implementation of the Decision-Making Model Based on NSGA-II

The paper utilises the JAVA platform to develop a decision-making model for multi-objective optimisation based on NSGA-II to improve energy conservation and indoor thermal comfort. The model features an interactive interface designed to meet user needs and includes the following functionality:

(1) Evaluation of target variables;
(2) Control of the range of values of decision variables;
(3) Control of the range of values of target variables;
(4) Configuration of NSGA-II parameters.

Applying the NSGA-II algorithm with a population size of N and only one iteration is equivalent to randomly sampling N solutions from the function space. As the number of iterations increases, the population continuously approaches the actual Pareto front, and the final Pareto solution set is typically considered the global optimum.

The paper employs the NSGA-II algorithm, repeatedly adjusting the population size and the number of iterations to obtain different distributions of the objective variable across the entire function space. Examining how the objective variable is distributed across various iterations helps determine the optimisation potential within the decision variable value range.

3. Results

3.1. Construction and Calibration of the Multi-Objective Rapid Prediction Model

This study employs a BP neural network to create a rapid prediction model for building AC energy consumption and indoor thermal comfort. Subsequently, it is used to propose a fitness function. This fitness function is closely related to various characteristic stages of AC usage. Notably, the input variables (particularly those relating to adaptive behaviours) vary independently across different stages. Consequently, it is essential to create independent rapid prediction models for each stage to accurately reflect the corresponding AC energy consumption and thermal comfort characteristics.

The 'decision variable—target variable' mapping database constructed in Section 2.1.3 trains six rapid prediction models using the BP neural network for different characteristic stages of AC usage, as described in Section 2.2. Table 7 displays the findings of the linear fitting of the predicted and simulated results. The mean relative error is calculated using Equation (6) for verification.

$$y = \frac{1}{N}\sum_{i=1}^{N} \frac{|x_{m'} - x_m|}{x_m} \times 100\%, \tag{6}$$

where

N is the total number of samples in the test set;
y is the mean relative error (MRE);

$x_{m'}$ is the predicted AC load per unit area, or the predicted comprehensive uncomfortable hours, worked out by the rapid prediction model based on the input vector in the test set;

x_m is the corresponding AC load per unit area, or comprehensive uncomfortable hours, of the input vector in the test set.

Table 7. Linear fitting of the predicted and simulated results at different characteristic stages of AC usage.

Code of the Prediction Model	Characteristic Stages of AC Usage	Hidden Layer Nodes	Number of Iterations	Linear Fitting of the Predicted and the Simulated Results	
				AC load per unit area	Comprehensive uncomfortable hours
C1	Late spring and early autumn	13	1900	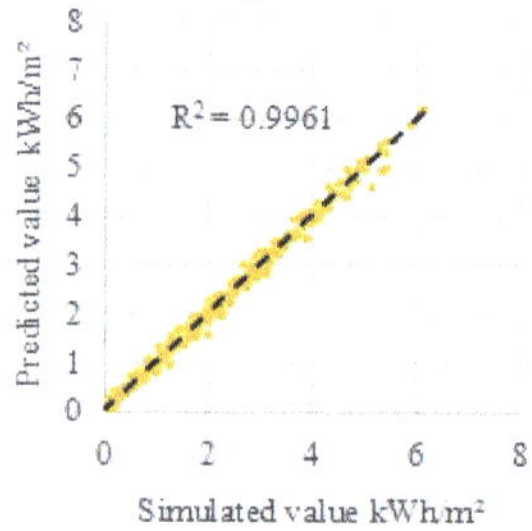	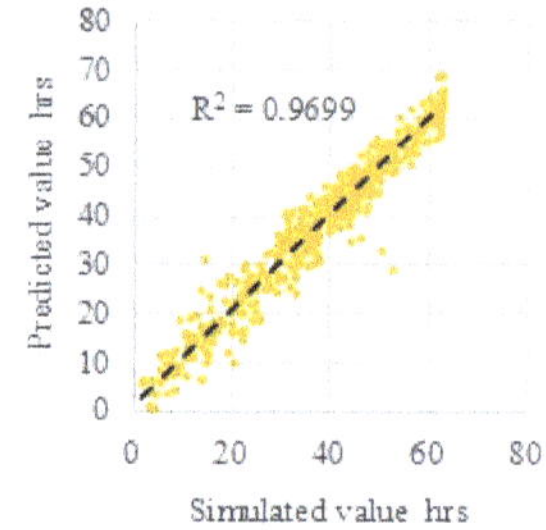
		MRE		4.1%	6.9%
C2	Early and late summer	9	1000	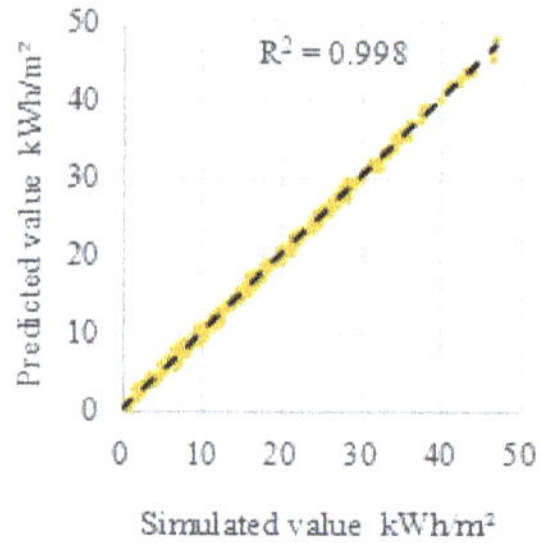	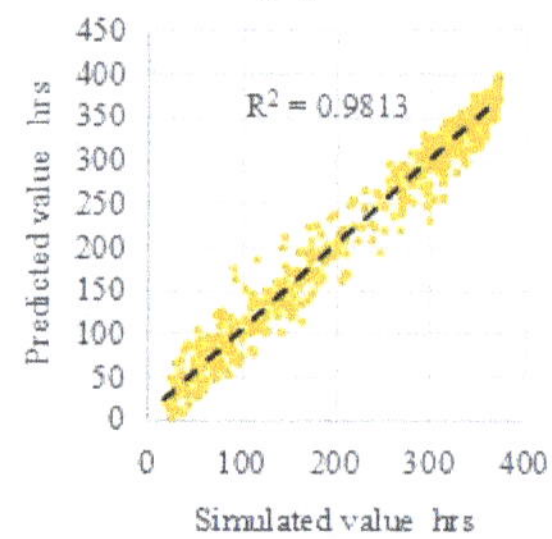
		MRE		3.1%	8.1%
C3	Mid-summer	11	2000	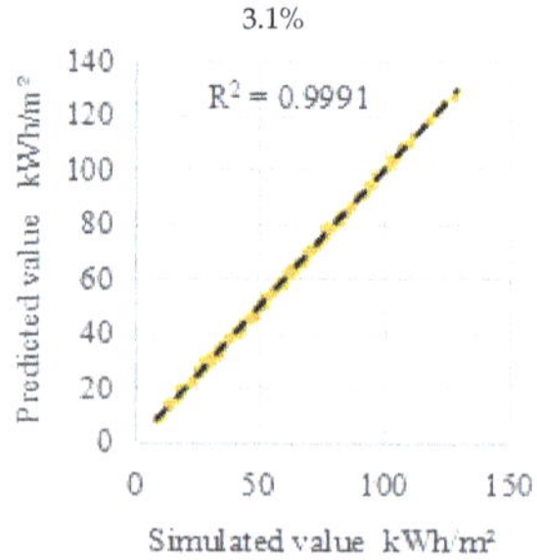	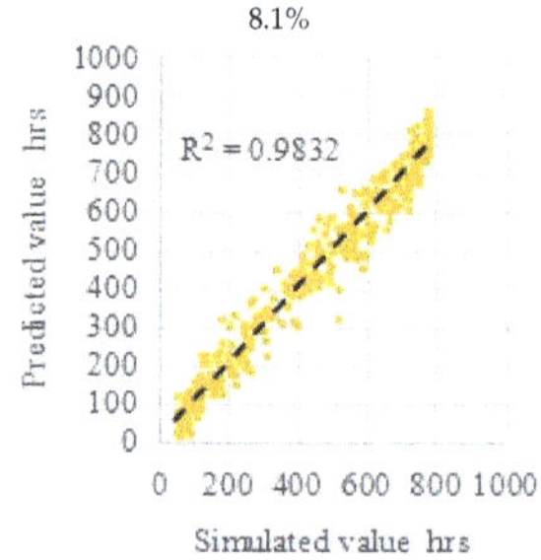
		MRE		1.1%	7.9%
H1	Early spring and late autumn	10	1610	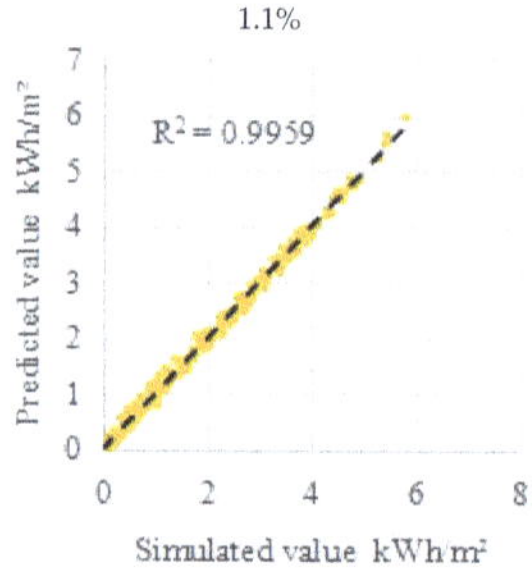	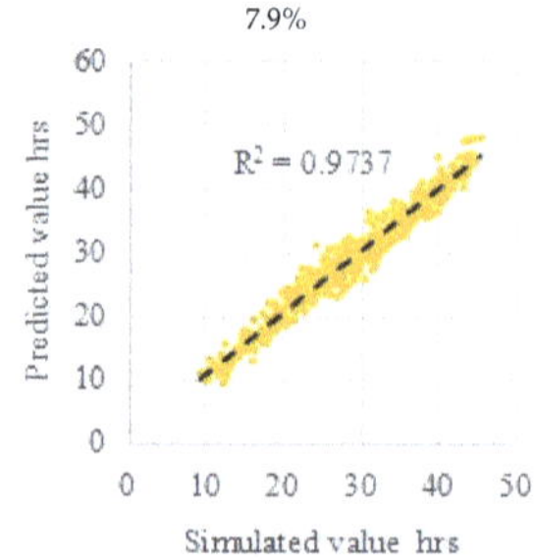
		MRE		4.4%	4.0%

Table 7. *Cont.*

Code of the Prediction Model	Characteristic Stages of AC Usage	Hidden Layer Nodes	Number of Iterations	Linear Fitting of the Predicted and the Simulated Results	
H2	Early and late winter	11	1000	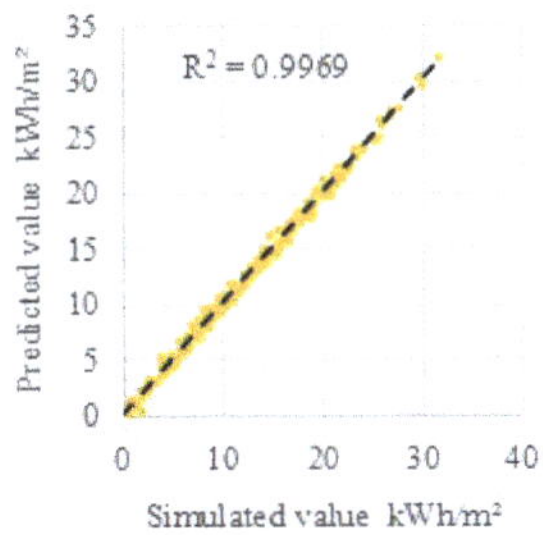	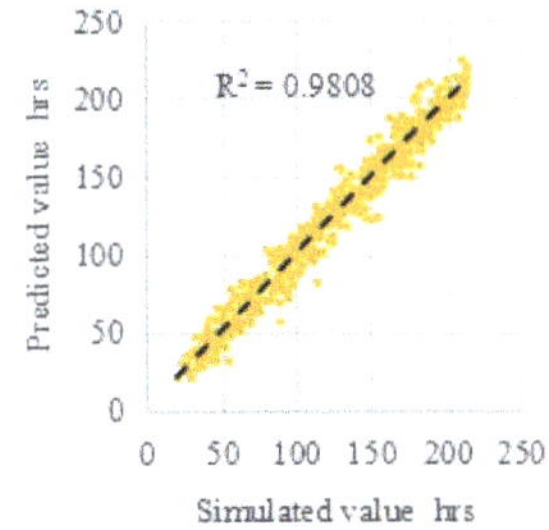
		MRE		4.3%	6.1%
H3	Mid-winter	10	1850	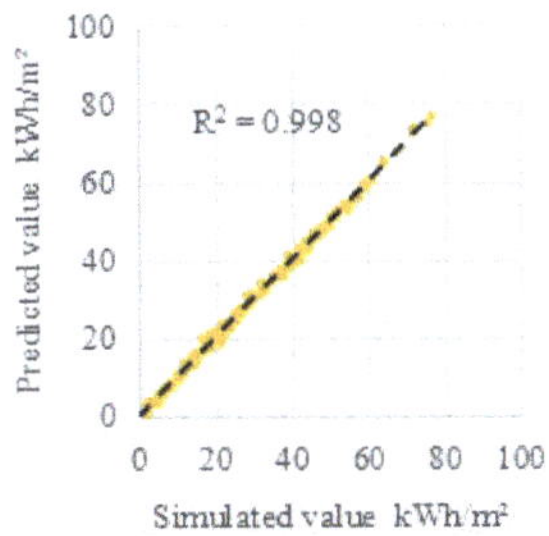	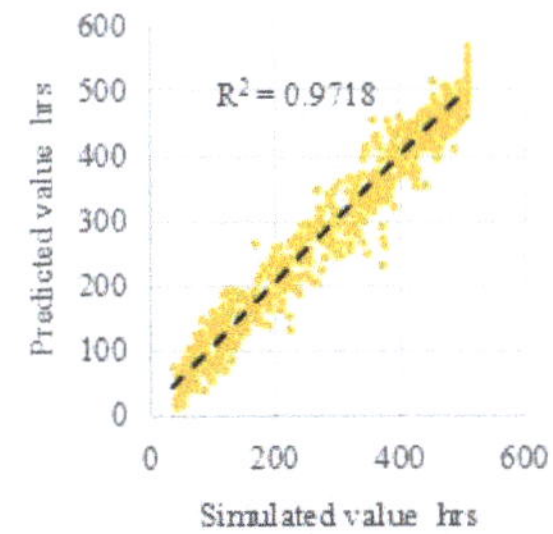
		MRE		4.0%	9.7%

As shown in Table 7, the verification results of the multi-objective rapid prediction model indicate that at different characteristic stages of AC usage, the mean relative error for AC load per unit area is below 5%, while the mean relative error for comprehensive uncomfortable hours is under 10%. Consequently, the rapid prediction model developed in this study, based on the BP neural network, can accurately predict the case building's AC energy consumption and indoor thermal comfort levels.

3.2. Construction of the Multi-Objective Optimisation Decision-Making Model

As outlined in Section 2.3, the study employs the JAVA platform to construct a multi-objective optimisation decision-making model for AC energy conservation and indoor thermal comfort, utilising the NSGA-II method. This model is the basis for analysing the case building, which is presented as an example. The resulting optimisation strategies are classified into three categories: passive, behavioural, and integrated optimisation strategies. Passive optimisation strategies are primarily applied during the building design phase. They optimise decision variables derived from the building design scheme, such as heat transfer coefficients of the exterior walls and the roofs, and air permeability performance. The goal is to reduce AC energy consumption and enhance thermal comfort. At the same time, decision variables related to the adaptive behaviours (such as cooling/heating set-point temperatures and coupling pattern between NV and AC operation) remain fixed. Behavioural optimisation strategies focus on the building operation phase. These strategies aim to achieve energy-saving goals by optimising users' HVAC behaviours while keeping decision variables related to the building design scheme unchanged. Integrated optimisation strategies combine passive and behavioural approaches, allowing for the simultaneous optimisation of decision variables associated with building design and adaptive behaviours. These strategies have proven effective for retrofitting existing buildings and designing new ones comprehensively, facilitating optimal energy savings and improved thermal comfort throughout a building's entire lifecycle.

In the context of multi-objective optimisation strategies during the operation phase of buildings, the decision variables derived from adaptive behaviours are designated as independent variables, while those originating from the building design scheme are assigned as fixed values. While investigating integrated optimisation strategies, decision variables derived from the building design scheme and adaptive behaviours are designated as independent variables. The range of values for the decision variables is defined in Table 8. When applying the aforementioned decision-making model to actual scheme decision-making, upper and lower limits can be assigned to the values of the decision variables. These limits can be adjusted based on each item's optimisation potential and economic costs.

Table 8. Range of values for the decision variables.

Type	Decision Variable	Code	Unit	Range of Values For Behavioural Optimisation	For Integrated Optimisation
Building design scheme	Heat transfer coefficient of the external wall	IN_01	$W/(m^2{\cdot}K)$	Fixed values under different operating conditions	[0.15, 2.50]
	Heat transfer coefficient of the roof	IN_02	$W/(m^2{\cdot}K)$		[0.15, 3.00]
	Heat transfer coefficient of the internal wall	IN_03	$W/(m^2{\cdot}K)$		[1.00, 5.00]
	Heat transfer coefficient of the external window	IN_04	$W/(m^2{\cdot}K)$		[2.20, 6.40]
	SHCG of the external window	IN_05	-		[0.15, 1.00]
	Window-to-wall ratio of the south elevation	IN_06	-	{0.4}	{0.4}
	Window-to-wall ratio of the north elevation	IN_07	-	{0.4}	{0.4}
	Air permeability performance	IN_08	ach	Fixed values under different operating conditions	[0.1, 0.8]
	Building orientation (northwards deflection)	IN_09	°	{0}	{0}
Internal disturbance elements	People density in collective offices	IN_09	m^2/person	{10}	{10}
	Power density of equipment	IN_10	W/person	{150}	{150}
Adaptive behaviours	Coupling pattern between NV and AC operation	IN_12	-	{M0, M1, M2, M3, M4}	{M0, M1, M2, M3, M4}
	Cooling setpoint temperature	IN_13	°C	[18, 30]	[18, 30]
	Heating setpoint temperature		°C	[18, 30]	[18, 30]

4. Discussion

4.1. Generation of Behavioural Optimisation Strategies Based on the Decision-Making Model

This section explores multi-objective behavioural optimisation strategies. Consequently, the decision variables associated with this process are designated independent variables, while those derived from the building design scheme are assigned as fixed values.

This section references the building's actual conditions, baseline conditions in the 1980s, baseline conditions in the 2000s, standard conditions from the 2015 edition of the '*Design standard for energy efficiency of public buildings (GB 50189)*' [41], and high and low standard conditions from the '*Technical standard for nearly zero energy buildings (GB/T 51350-2019)*' [42]. The use of these references is essential in establishing six typical operating conditions. Subsequently, these conditions assess the potential for AC energy conservation and thermal comfort optimisation of the case building under behavioural strategies. The fixed values of the decision variables derived from the building design scheme corresponding to these six operating conditions are shown in Table 9.

In implementing the behavioural optimisation strategy, only two decision variables are utilised as independent variables: the coupling pattern between NV and AC operation and the cooling/heating setpoint temperature. The values of these two variables vary across the six characteristic stages of AC usage, resulting in the execution of distinct behavioural strategies for each stage, as opposed to their consistent application within the computational cycle. The multi-objective optimisation decision-making model using the NSGA-II algorithm is executed multiple times with various population sizes and iteration counts. This approach aims to assess the optimisation potential of the target variables while implementing a behavioural optimisation strategy across six typical operating conditions, as shown in Figure 3.

Table 9. Fixed values of decision variables derived from the building design scheme under typical operating conditions.

Building Design Parameters		IN_01	IN_02	IN_03	IN_04	IN_05	IN_08
		Heat Transfer Coefficient [W/(m²·K)]				SHGC of the External Window	Air Permeability (ach)
Code	Operating Condition	External Wall	Roof	Internal Wall	External Window		
P01	Actual condition of the case building	2.42	2.98	3.93	5.78	0.819	0.7
P02	Baseline condition of typical buildings in the 1980s	2.0	1.5	2.0	6.4	0.8	0.8
P03	Baseline condition of typical buildings in the 2000s	1.0	0.7	2.0	3.0	0.5	0.5
P04	Standard condition from GB 50189-2015	0.8	0.5	2.0	2.6	0.4	0.3
P05	Low standard condition from GB/T 51350-2019	0.4	0.35	2.0	2.2	0.15	0.1
P06	High standard condition from GB/T 51350-2019	0.15	0.15	2.0	2.2	0.15	0.1

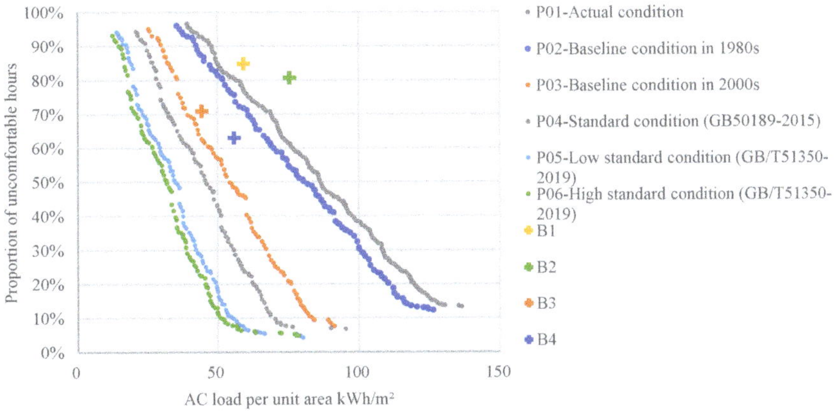

Figure 3. Optimisation potential of target variables under behavioural optimisation strategies controlled for typical operating conditions.

Table 5 illustrates how the optimisation potential of behavioural strategies varies under six typical operating conditions and across four reference points. It has been demonstrated that in the actual condition P01 and the baseline condition P02, the behavioural strategies can be employed to achieve the optimisation targets of reference points B1 and B2. However, they cannot exert dominance over reference points B3 and B4. In the standard conditions P03, P04, P05, and P06, the behavioural strategies can achieve the optimisation objectives of reference points B1, B2, B3, and B4. In each typical operating condition, the implementation effectiveness of the behavioural optimisation strategies is analysed with reference points as optimisation objectives. The quantitative characterisation of the optimisation potential of the target variables is shown in Table 10.

Among the six typical operating conditions, P04 indicates the thermal performance limits for building envelopes in current public buildings, as specified in the Design standard for energy efficiency of public buildings (GB 50189-2015) [41]. It is a widely utilised metric in the design of such buildings. This study thus employs the P04 condition as a case study, assuming that the primary optimisation objective is to ensure indoor thermal comfort. By analysing the Pareto frontier, optimal strategies can be obtained for different thermal comfort levels. Furthermore, the relationships between these optimal solutions and the four optimisation reference points can be identified. This study employs 10% intervals to explore optimal strategies for maximising AC energy savings through behavioural modification at different thermal comfort levels, specifically when the case building is in P04 conditions, as shown in Table 11.

Table 10. Quantitative characterisation of the multi-objective optimisation potential in six typical operating conditions.

Parameter	Reference Point	Reference Value	Optimisation Potential Compared with the Reference Point					
			P01	P02	P03	P04	P05	P06
AC load per unit area	B1	59.326 kWh/m^2	52.86%	40.30%	56.75%	34.06%	76.22%	78.85%
	B2	75.536 kWh/m^2	62.97%	53.11%	66.03%	48.21%	81.32%	83.39%
	B3	44.366 kWh/m^2	36.96%	20.17%	42.17%	11.83%	68.20%	71.72%
	B4	55.942 kWh/m^2	50.00%	36.68%	54.13%	30.07%	74.78%	77.57%
Proportion of uncomfortable hours	B1	84.82%	92.01%	85.58%	90.98%	84.27%	95.01%	94.29%
	B2	80.65%	91.59%	84.84%	90.51%	83.46%	94.76%	94.00%
	B3	70.93%	90.44%	82.76%	89.21%	81.19%	94.04%	93.18%
	B4	63.00%	89.24%	80.59%	87.86%	78.83%	93.29%	92.32%

Table 11. Optimal behavioural optimisation strategies for thermal comfort levels in the P04 operating condition.

Optimisation Objective		Optimal Behavioural Strategy: Cooling/Heating Setpoint Temperature (°C) ǀ Coupling Pattern Between NV and AC Operation						Reference Point Dominance ("√" for Dominated, "×" for Non-Dominated)			
Proportion of Uncomfortable Hours	AC Load per Unit Area (kWh/m^2)	H1	H2	H3	C1	C2	C3	B1	B2	B3	B4
10%	69.72	24 ǀ M4	25 ǀ M3	25 ǀ M4	26 ǀ M3	25 ǀ M4	24 ǀ M3	×	×	×	√
20%	63.28	19 ǀ M1	18 ǀ M4	25 ǀ M4	26 ǀ M3	25 ǀ M4	24 ǀ M4	×	×	×	√
30%	56.76	23 ǀ M1	23 ǀ M4	24 ǀ M4	26 ǀ M3	30 ǀ M2	24 ǀ M3	√	√	×	×
40%	51.45	19 ǀ M4	18 ǀ M4	25 ǀ M4	25 ǀ M3	30 ǀ M2	25 ǀ M4	√	√	√	√
50%	45.99	19 ǀ M4	24 ǀ M4	25 ǀ M4	27 ǀ M4	25 ǀ M4	30 ǀ M2	√	√	×	√
60%	38.62	24 ǀ M4	25 ǀ M3	25 ǀ M4	25 ǀ M0	30 ǀ M2	30 ǀ M2	√	√	√	√
70%	31.84	19 ǀ M3	18 ǀ M0	25 ǀ M4	26 ǀ M3	30 ǀ M2	30 ǀ M2	√	√	√	×
80%	28.52	19 ǀ M4	18 ǀ M3	23 ǀ M4	26 ǀ M3	30 ǀ M2	30 ǀ M2	√	√	×	×
90%	24.38	24 ǀ M4	25 ǀ M3	25 ǀ M4	26 ǀ M3	25 ǀ M4	24 ǀ M3	×	×	×	×

As illustrated in Table 11, the dominance relationship between the optimal strategy and the four reference points varies depending on the optimisation objective. In the P04 operating condition, the appropriate behavioural strategy can dominate all four reference points. This means the target variables should ensure that the AC load per unit area does not exceed 44.37 kWh/m^2, and the proportion of uncomfortable hours remains below 63.00%. The area of the function space where these target variables meet the requirements is highlighted in red in Figure 4. Executing the multi-objective decision-making model based on the NSGA-II algorithm is essential to clarify the solution set within this function space. The optimisation objectives are defined by controlling the range of values for the target variables. We modify the population size and iteration number to conduct multiple calculations, and Table 12 presents the portion of the behavioural strategies that satisfy these objectives.

Decision-making models and multiple rounds of iteration allowed the identification of 481 non-overlapping solutions within the optimisation objective range (as shown by the red area in Figure 4). Each solution represents a distinct behavioural optimisation strategy. The numerical distribution of these target variable solutions is summarised and analysed, providing insights into the characteristics of the optimised value for two decision variables derived from adaptive behaviours at different characteristic stages of AC usage. These characteristics are demonstrated in Tables 13 and 14.

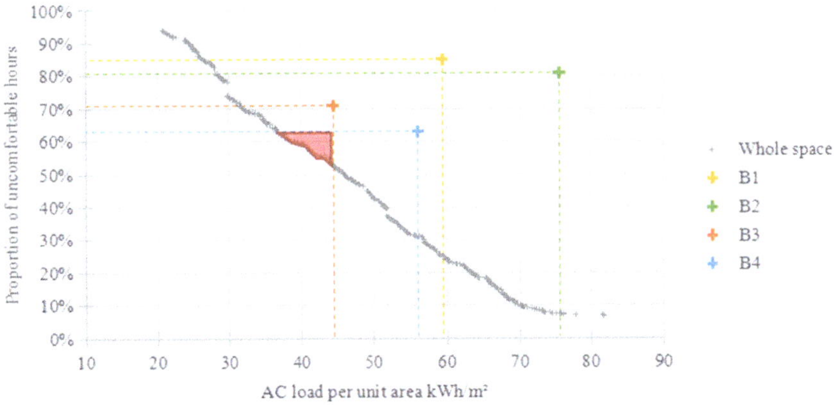

Figure 4. The function space of target variables dominating the four reference points in the P04 operation condition.

Table 12. Samples of the behavioural strategies meeting the optimisation objectives in the P04 operating condition.

Optimisation Objective		Behavioural Optimisation Strategy: Cooling/Heating Setpoint Temperature (°C) I Coupling Pattern Between NV and AC Operation						Reference Point Dominance ("√" for Dominated)			
Proportion of Uncomfortable Hours	**AC Load per Unit Area (kWh/m²)**	**H1**	**H2**	**H3**	**C1**	**C2**	**C3**	**B1**	**B2**	**B3**	**B4**
53%	44.32	19 I M1	24 I M3	24 I M4	27 I M0	25 I M4	30 I M2	√	√	√	√
55%	44.23	20 I M1	22 I M3	25 I M4	26 I M0	25 I M3	30 I M2	√	√	√	√
57%	43.48	21 I M2	18 I M0	18 I M0	27 I M1	30 I M2	24 I M3	√	√	√	√
60%	44.05	20 I M0	18 I M0	18 I M3	29 I M1	30 I M4	24 I M2	√	√	√	√
62%	40.7	22 I M1	20 I M3	25 I M4	29 I M2	26 I M4	30 I M2	√	√	√	√

Table 13. Numerical distribution of the cooling/heating setpoint temperatures in different characteristic stages of AC usage when the optimisation objective is satisfied.

Operating Condition	Numerical Distribution of the Cooling/Heating Setpoint Temperature		
Heating	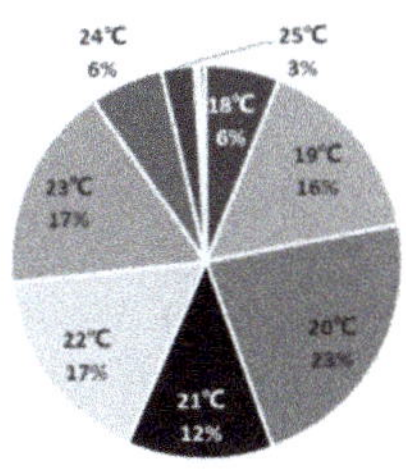 H1: Early spring and late autumn	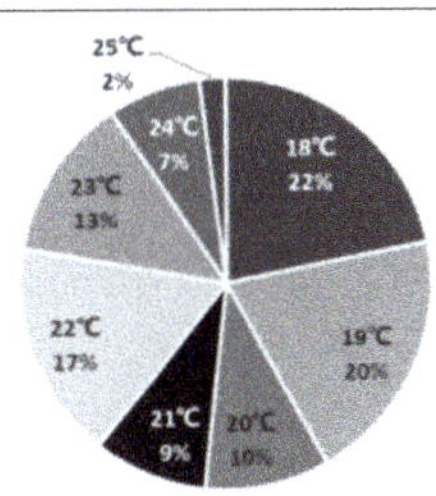 H2: Early and late winter	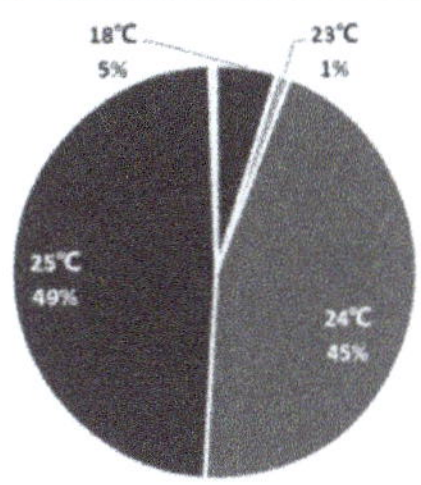 H3: Mid-winter
Cooling	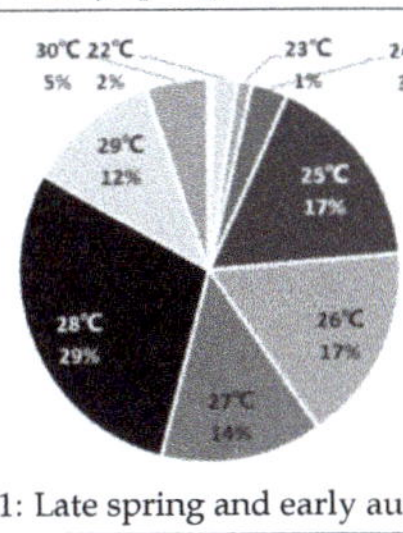 C1: Late spring and early autumn	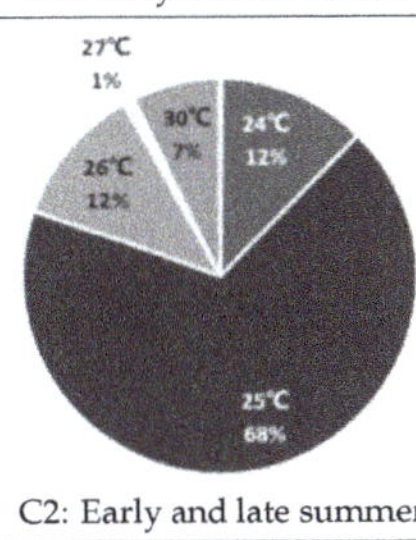 C2: Early and late summer	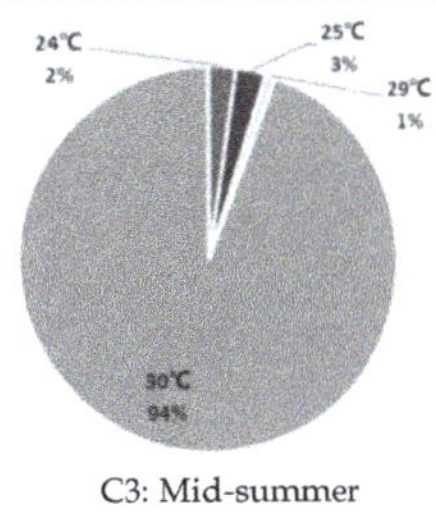 C3: Mid-summer

Table 14. Numerical distribution of the coupling pattern between NV and AC operation in different characteristic stages of AC usage when the optimisation objective is satisfied.

<table>
<tr><th>Operating Condition</th><th colspan="3">Numerical Distribution of the Coupling Pattern Between NV and AC Operation</th></tr>
<tr><td>Heating</td><td>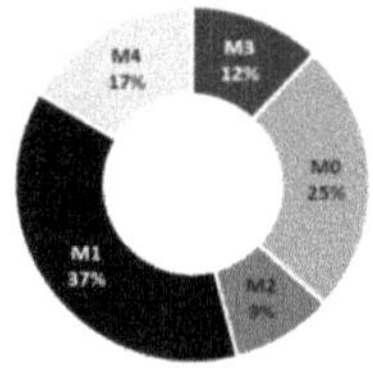

H1: Early spring and late autumn</td><td>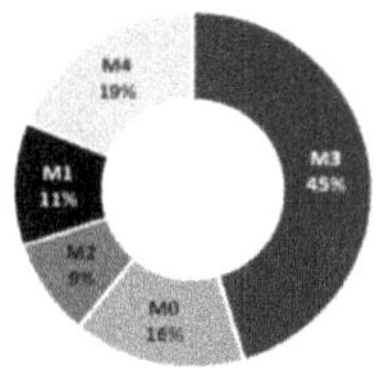

H2: Early and late winter</td><td>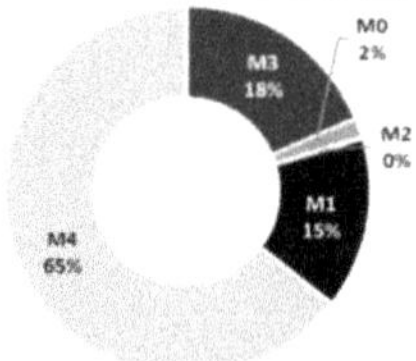

H3: Mid-winter</td></tr>
<tr><td>Cooling</td><td>

C1: Late spring and early autumn</td><td>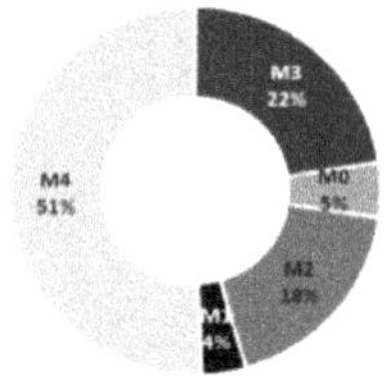

C2: Early and late summer</td><td>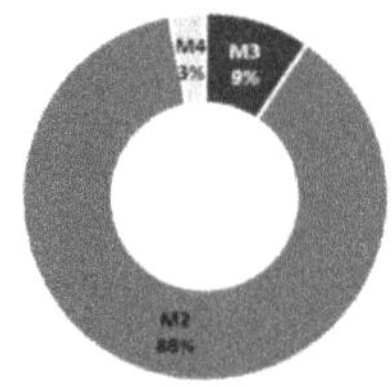

C3: Mid-summer</td></tr>
</table>

The AC system's set temperature is limited in different characteristic stages of AC usage to achieve the optimisation objective of simultaneously dominating four reference points. In heating conditions, air conditioning should be set to the following temperatures: 19–23 °C during the late autumn and early spring, 18–24 °C during the early and late winter, and 24–25 °C during the mid-winter stage. In cooling conditions, air conditioning should be adjusted to these temperatures: 25–29 °C during late spring and early autumn, 24–26 °C during early and late summer, and 30 °C during mid-summer.

In order to meet the optimisation objectives of simultaneously dominating four reference points, the suitability of different coupling patterns between NV and AC operation varies at different characteristic stages of AC usage. In heating conditions, the M1 pattern (fixed ventilation in the morning) is recommended during the late autumn and early spring, the M3 pattern (opposition all-day) is recommended during the early and late winter, and the M4 pattern (opposition overnight) is recommended during the mid-winter stage. In cooling conditions, the M3 or M2 pattern (fixed ventilation overnight) is recommended during late spring and early autumn, the M4 pattern is recommended during early summer and late summer, and the M2 pattern is recommended during mid-summer.

4.2. Generation of Integrated Optimisation Strategies Based on the Decision-Making Model

Behavioural optimisation strategies involve exploring the optimal combinations of stochastic HVAC behaviours while maintaining fixed thermal performance conditions of the building envelope. The focus is on behavioural management during the case building's operational phase. When implementing integrated optimisation strategies, the objective is to identify the optimal solutions for the thermal performance of the building envelope during the design phase and to encourage behavioural adjustments from occupants during the operational phase. Consequently, all decision variables derived from the building design scheme and the adaptive behaviours are considered independent variables. The parameters under consideration are as follows: the heat transfer coefficient of the external wall, the roof, the internal wall, and the external window; the solar heat gain coefficient (SHGC) of the external window; air permeability performance; and the coupling pattern between NV and AC operation. The ranges of values for decision variables are defined in Table 8.

To explore integrated optimisation strategies, it is necessary first to establish specific optimisation objectives. For instance, the coordinates of a reference point (C) can be defined as (x_C, y_C) with the optimisation objective tailored according to this reference point. Configuring different population sizes and iteration numbers is possible based on different control ranges within the target variable's value domain. As demonstrated, integrated optimisation strategies that meet these objectives can be obtained by repeatedly utilising the multi-objective optimisation decision-making model based on the NSGA-II algorithm.

This study adopts the annual AC load per unit area corresponding to reference point C as the AC energy-saving target and investigates optimisation strategies for indoor thermal comfort at the corresponding AC energy consumption level. Therefore, the range of the target variable, the annual AC load per unit area (OP_01), is controlled, while the comprehensive uncomfortable hours (OP_02) range remains unrestricted. Specifically, the upper limit of the AC load per unit area, *max_OP_01*, is set to $(x_C + k)$, and the lower limit, *min_OP_01*, is set to $(x_C - k)$. It is crucial to note that the upper and lower limits must be distinct; hence, they are set to similar but different values, with k defined as necessary for the solution.

Taking reference point B1 in the optimisation objective as an example, a value is designated to reference point C. The annual AC load per unit area corresponding to B1 is 59.33 kWh/m^2; therefore, *OP_01* is assigned the value of 59.33 kWh/m^2. With k_1 defined as 0.5, the upper and lower bounds of the target variable domain are adjusted as follows: *max_OP_01* = −59.83, *min_OP_01* = 58.83, *max_OP_02* = 2000, and *min_OP_02* = 0. The optimisation potential of the integrated optimisation strategy that can dominate reference point B1 is demonstrated in Figure 5 under this optimisation objective.

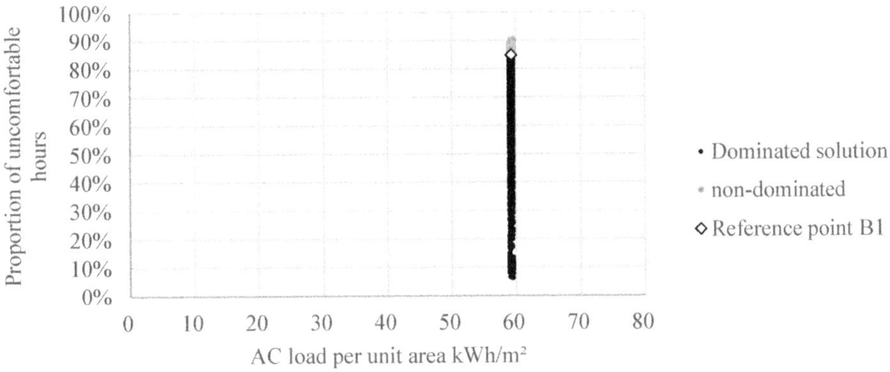

Figure 5. Optimisation potential of the integrated optimisation based on the AC energy consumption of the reference point B1.

Under the optimisation objective, implementing integrated strategies will achieve the optimal indoor thermal comfort level, characterised by a proportion of uncomfortable hours of 6.23%. As shown in Table 15, delineating disparate thermal comfort optimisation objectives at 10% intervals achieves the integrated optimisation strategies for distinct thermal comfort levels.

Table 15. Samples of the integrated optimisation strategies for different thermal comfort levels (based on the AC energy consumption of reference point B1).

Proportion of Uncomfortable Hours	Integrated Optimisation Strategy						Behavioural Optimisation Strategy					
	Heat Transfer Coefficient [W/(m²·K)]				SHGC of the External Window	Air Permeability (ach)	Cooling/Heating Setpoint Temperature (°C) | Coupling Pattern Between NV and AC Operation					
	External Wall	Roof	Internal Wall	External Window			H1	H2	H3	C1	C2	C3
6.23%	0.15	0.15	1.06	2.2	0.33	0.1	24 I M0	24 I M4	24 I M2	25 I M0	24 I M1	24 I M1
10.00%	0.15	0.31	2.81	2.85	0.29	0.12	26 I M3	25 I M1	24 I M0	26 I M3	25 I M4	24 I M1
20.00%	0.36	0.31	3.37	2.77	0.29	0.11	27 I M2	22 I M2	24 I M0	27 I M0	25 I M2	24 I M1
30.00%	0.41	0.51	4.24	2.4	0.22	0.18	27 I M3	21 I M1	24 I M4	20 I M3	25 I M0	24 I M1
40.00%	0.72	0.48	3.88	3.94	0.53	0.1	29 I M0	22 I M4	19 I M0	27 I M4	24 I M0	24 I M2
50.00%	1.6	0.49	1.06	3.34	0.59	0.29	29 I M1	25 I M4	18 I M0	26 I M1	29 I M4	24 I M4
60.00%	1.69	0.67	3.27	3.11	0.93	0.26	20 I M0	23 I M1	18 I M0	27 I M0	29 I M2	24 I M2
70.00%	0.42	1.11	1.2	3.68	0.25	0.59	28 I M0	19 I M2	22 I M1	23 I M3	29 I M2	25 I M2
80.00%	0.91	0.69	1.28	3.78	0.62	0.53	24 I M4	18 I M1	21 I M3	25 I M4	26 I M3	26 I M3
84.82%	1.38	1.61	4.45	3.1	0.47	0.15	28 I M1	20 I M2	29 I M4	25 I M4	28 I M4	30 I M1

Establishing clear optimisation objectives, implementing a multi-objective optimisation decision-making model based on the NSGA-II algorithm for building AC energy consumption and indoor thermal comfort, and deriving integrated strategies that meet these optimisation objectives are all possible. This model provides a framework for designing the thermal performance of building envelopes during the building design phase and managing occupants' HVAC behaviour during the building operation phase.

5. Conclusions

The paper focuses on constructing and applying a multi-objective optimisation decision-making model for AC energy consumption and indoor thermal comfort. A stochastic prediction model of HVAC behaviour has been developed, and the influencing factors have been identified through preliminary studies. Based on these findings, the paper proposes a multi-objective optimisation decision-making model. The model aims to provide scientific foundations and practical tools for the AC energy conservation and thermal comfort optimisation strategies for naturally ventilated research buildings in China's hot summer and cold winter climate zones.

(1) Take the case building as an example. A rapid prediction model for building air conditioning (AC) energy consumption and indoor thermal comfort was constructed based on a BP neural network. Using important influencing factors of the target variables as decision variables, the model efficiently and accurately predicts the AC load per unit area and the comprehensive uncomfortable hours for the case building under the various operating scenarios, providing data support for multi-objective optimisation decision-making.
(2) A multi-objective optimisation decision-making model regarding AC energy consumption and indoor thermal comfort was constructed. Implementing the model generates passive optimisation strategies for the design phase, behavioural optimisation strategies for the operational phase, and integrated optimisation strategies for both phases of the case building. Furthermore, it can output corresponding strategy combinations based on predefined optimisation objectives, providing a scientific basis for building energy retrofitting and operational management.
(3) Research was conducted on the application of multi-objective optimisation decision-making models, examining behavioural and integrated optimisation strategies under different operating conditions:
    - The P04 operating condition, as outlined in the current *Design standard for energy efficiency of public buildings*, serves as a case study. Substantial AC energy savings and improvements in thermal comfort can be realised by adjusting the stage-

based cooling/heating setpoint temperature and the coupling pattern between NV and AC operations. This has led to meeting the optimisation objectives for B1, B2, B3, and B4. Additionally, it has been demonstrated that reducing AC energy consumption to 40.7 kWh/m^2 is possible while reducing the proportion of uncomfortable hours to 53%.

- Utilising B1 as the reference point, the multi-objective optimisation decision-making model regulates AC energy consumption limits. It combines passive and behavioural strategies to create integrated optimisation strategies for varying indoor thermal comfort levels. Consequently, the proportion of uncomfortable hours has been reduced to 6.23%.

However, the study does have limitations. These primarily involve the lack of general applicability of the multi-objective optimisation decision-making model, the incompleteness of the behaviour prediction model, the overlooking of the impact of individual temperature regimes on indoor setpoint temperature, and the absence of a practical value assessment of the optimisation strategy. First, the model mainly focuses on small and medium-sized research buildings in specific regions, necessitating improvements to its general applicability. Second, the behaviour prediction model does not adequately consider the various occupant behaviour patterns, such as spatial movement and clothing adjustment, which impact accuracy. Third, the setpoint temperatures of all AC zones are set as a single uniform variable, resulting in consistent target temperatures across all rooms and disregarding the differentiated thermal environment needs of different individuals. Finally, the optimisation strategies lacked a cost–benefit analysis, limiting their feasibility in practical applications.

Future work will take three approaches to address these limitations. First, the model's applicability will be expanded to cover more regions and building types, and in-depth research will be conducted on the behavioural characteristics of different users. Second, improving the accuracy of behavioural prediction models requires capturing the diversity of user behaviour through a broader range of actual measurement data. Third, for different AC zones, separate variables for cooling/heating setpoint temperatures will be established, and variables for PMV thermal comfort range limits will be added, aiming to achieve differentiated control of indoor thermal requirements. Finally, cost analysis will be incorporated into evaluating optimisation strategies to ensure their economic feasibility and practical value. This will provide more comprehensive guidance for practical applications.

Author Contributions: Conceptualisation, J.W.; methodology, S.C.; software, J.W.; validation, R.Q. and J.W.; formal analysis, X.Y.; investigation, R.Q.; resources, X.Y.; data curation, J.W. and S.C.; writing—original draft, J.W.; writing—review and editing, X.Y. and R.Q.; visualisation, J.W. and X.Z.; supervision, X.Y.; project administration, S.C.; funding acquisition, J.W. and X.Z. All authors have read and agreed to the published version of the manuscript.

Funding: This research was supported by the Hangzhou Joint Fund of the Zhejiang Provincial Natural Science Foundation of China under Grant No. LHZY24A010003, the University Cultivation Fund of the Hangzhou City University under Grant No. J-202311, the 2023 Zhejiang Provincial Department of Education General Research Project under Grant No. Y202351545, and the Key Research Project in the Fields of High-end Equipment, New Materials, and Green Energy in Hangzhou under Grant No. 2023SZD0070.

Data Availability Statement: The data presented in this study are available on request from the corresponding author.

Conflicts of Interest: The authors declare no conflicts of interest.

References

1. Tsinghua University, Building Energy Research Center. *Annual Research Report on Building Energy Efficiency in China: Public Buildings*; China Architecture & Building Press: Beijing, China, 2022.
2. Tsinghua University, Building Energy Research Center. *Annual Research Report on Building Energy Efficiency in China: Urban Residential Buildings*; China Architecture & Building Press: Beijing, China, 2023.
3. Dang, Y.C.; Yao, J. Multi-Objective Optimization Study of Residential Energy Consumption Based on Edh and the Stochastic Operation of Air Conditioning. *J. Green Build.* **2024**, *19*, 195–220. [CrossRef]
4. Zhang, L.; Zhong, B. Partition-based building energy analysis and optimization of energy conservation. In Proceedings of the 1st International Conference on Computational Intelligence, Theory, Systems and Applications (CCITSA), Shanghai, China, 24–27 August 2015; pp. 171–176. [CrossRef]
5. Ding, C.; Zhou, N. Using residential and office building archetypes for energy-efficiency solutions in an Urban Scale: A China Case Study. *Energies* **2020**, *13*, 3210. [CrossRef]
6. Carlucci, S.; Pagliano, L. A review of indices for the long-term evaluation of general thermal comfort conditions in buildings. *Energy Build.* **2012**, *53*, 194–205. [CrossRef]
7. Balbis-Morejón, M.; Rey-Hernández, J.M.; Amaris-Castilla, C.; Velasco-Gómez, E.; San José-Alonso, J.F.; Rey-Martínez, F.J. Experimental study and analysis of thermal comfort in a university campus building in tropical climate. *Sustainability* **2020**, *12*, 8886. [CrossRef]
8. Tian, X.-Y.; Liu, W.-W.; Liu, J.-W.; Yu, B.; Zhang, J. Effects of personal heating on thermal comfort: A review. *J. Cent. South Univ.* **2022**, *29*, 2279–2300. [CrossRef]
9. Shakeel, H.M.; Iram, S.; Hill, R.; Farid, H.M.A.; Sheikh-Akbari, A.; Saleem, F. A machine learning-based intelligent framework for predicting energy efficiency in next-generation residential buildings. *Buildings* **2025**, *15*, 1275. [CrossRef]
10. Al-Tamimi, N. Passive design strategies for energy-efficient buildings in the Arabian desert. *Front. Built Environ.* **2022**, *7*, 805603. [CrossRef]
11. Duan, Y.; Zhang, T.; Yang, Y.; Li, P.; Mo, W.; Jiao, Z.; Gao, W. A multi-objective approach to optimizing the geometry and envelope of rural dwellings for energy demand, thermal comfort, and daylight in cold regions of China: A case study of Shandong province. *Energy Convers. Manag.* **2024**, *322*, 119128. [CrossRef]
12. Yaman, G.Ö. Determination of optimum passive design parameters for industrial buildings in different climate zones using an ANN–PSO energy-performance optimization model. *Sustainability* **2025**, *17*, 2357. [CrossRef]
13. Xu, S.; Chen, Y.Z.; Liu, J.L.; Kang, J.; Gao, J.F.; Qin, Y.C.; Tan, W.J.; Li, G.M. Comprehensive improvement of energy efficiency and indoor environmental quality for university library atrium—A multi-objective fast optimization framework. *Front. Archit. Res.* **2025**, *14*, 449–470. [CrossRef]
14. Hosamo, H.H.; Tingstveit, M.S.; Nielsen, H.K.; Svennevig, P.R.; Svidt, K. Multi-objective optimization of building energy consumption and thermal comfort based on integrated BIM framework with machine learning-NSGA II. *Energy Build.* **2022**, *277*, 112479. [CrossRef]
15. Shao, T.; Zheng, W.X.; Zhang, K. Research on the establishment of rural house optimal design pattern in northwest cold region under multi-objective conditions. *J. Therm. Sci. Technol.* **2024**, *23*, 300–310. [CrossRef]
16. Song, L.W. Performance optimization of high-rise residential buildings in cold regions considering energy consumption. *Discov. Appl. Sci.* **2025**, *7*, 105. [CrossRef]
17. Wang, A.; Xiao, Y.; Liu, C.; Zhao, Y.; Sun, S. Multi-objective optimization of building energy consumption and thermal comfort based on SVR-NSGA-II. *Case Stud. Therm. Eng.* **2024**, *50*, 105368. [CrossRef]
18. Liu, M.; Que, Y.F.; Yang, N.X.; Yan, C.Y.; Liu, Q.B. Research on Multi-Objective Optimization Design of University Student Center in China Based on Low Energy Consumption and Thermal Comfort. *Energies* **2024**, *17*, 2082. [CrossRef]
19. Liu, X.B.; Wu, Y.Y.; Wu, H. Enhancing HVAC energy management through multi-zone occupant-centric approach: A multi-agent deep reinforcement learning solution. *Energy Build.* **2024**, *303*, 113770. [CrossRef]
20. Li, W.; Koo, C.; Hong, T.; Oh, J.; Cha, S.H.; Wang, S.W. A novel operation approach for the energy efficiency improvement of the HVAC system in office spaces through real-time big data analytics. *Renew. Sustain. Energy Rev.* **2020**, *127*, 109885. [CrossRef]
21. Ao, J.Y.; Du, C.Q.; Jing, M.Y.; Li, B.Z.; Chen, Z.Y. A method of integrating air-conditioning usage models into building simulations for predicting residential cooling energy consumption. *Buildings* **2024**, *14*, 2026. [CrossRef]
22. Yao, J. Modelling and simulating occupant behaviour on air conditioning in residential buildings. *Energy Build.* **2018**, *175*, 1–10. [CrossRef]
23. Virote, J.; Neves-Silva, R. Stochastic models for building energy prediction based on occupant behavior assessment. *Energy Build.* **2012**, *53*, 183–193. [CrossRef]
24. Uddin, M.N.; Wei, H.-H.; Chi, H.L.; Ni, M. Influence of occupant behavior for building energy conservation: A systematic review study of diverse modeling and simulation approach. *Buildings* **2021**, *11*, 41. [CrossRef]

25. Diao, L.; Sun, Y.; Chen, Z.; Chen, J. Modeling energy consumption in residential buildings: A bottom-up analysis based on occupant behavior pattern clustering and stochastic simulation. *Energy Build.* **2017**, *147*, 47–66. [CrossRef]
26. Fabi, V.; Andersen, R.K.; Corgnati, S. Verification of stochastic behavioural models of occupants' interactions with windows in residential buildings. *Build. Environ.* **2015**, *94*, 394–404. [CrossRef]
27. Ren, X.; Yan, D.; Hong, T. Air-conditioning usage conditional probability model for residential buildings. *Energy Build.* **2020**, *206*, 109571. [CrossRef]
28. Luo, M.H.; Jiang, K.Y.; Wang, J.L.; Feng, W.; Ma, L.; Shi, X.D.; Zhou, X. Data-driven thermal preference prediction model with embodied air-conditioning sensors and historical usage behaviors. *Build. Environ.* **2022**, *220*, 109269. [CrossRef]
29. Chen, S.; Zhang, H.; Pan, Y.; Zhang, X.; Fang, F. Research on simulation methodology of stochastic usage behavior of air conditioners in residential buildings by agent-based modeling. *Build. Sci.* **2017**, *33*, 37–44.
30. Zhang, Z.C.; Yao, J.; Zheng, R.Y. Multi-objective optimization of building energy saving based on the randomness of energy-related occupant behavior. *Sustainability* **2024**, *16*, 1935. [CrossRef]
31. Li, R.; Shari, Z.; Ab Kadir, M.Z.A. A Review of Multi-objective Optimization Research on Building Performance from the Perspective of Bibliometric Analysis. *Energy Build.* **2024**, *11*, e42480. [CrossRef]
32. Jeong, B.; Kim, J.; Ma, Z.; Cooper, P.; de Dear, R. Identification of environmental and contextual driving factors of air-conditioning usage behaviour in Sydney residential buildings. *Buildings* **2021**, *11*, 122. [CrossRef]
33. Azar, E.; O'Brien, W.; Carlucci, S.; Hong, T.; Sonta, A.; Kim, J.; Andargie, M.S.; Abuimara, T.; El Asmar, M.; Jain, R.K.; et al. Simulation-aided occupant-centric building design: A critical review of tools, methods, and applications. *Energy Build.* **2020**, *224*, 110292. [CrossRef]
34. Hamed, M.M.; Alkhreasha, A.; Alshaer, A.; Olabi, A.G. Promoting sustainable development goals through energy-related behaviors of household occupants: Fostering sustainable energy solutions in developing countries. *Renew. Sustain. Energy Rev.* **2025**, *213*, 115511. [CrossRef]
35. Haidar, N.; Tamani, N.; Ghamri-Doudane, Y.; Bouju, A. Towards a new graph-based occupant behavior modeling in smart buildings. In Proceedings of the 15th International Wireless Communications & Mobile Computing Conference (IWCMC), Tangier, Morocco, 24–28 June 2019; pp. 1413–1418. [CrossRef]
36. Xiang, J.; Liu, H.; Li, X.; Jones, P.; Perisoglou, E. Multi-objective optimization of ultra-low energy housing in the hot-summer cold-winter climate zone of China based on a probabilistic behavioral model. *Buildings* **2023**, *13*, 1172. [CrossRef]
37. Wu, J.J.; Chen, S.Q.; Ying, X.Y.; Shu, J.B. Influencing Factors on Air Conditioning Energy Consumption of Naturally Ventilated Research Buildings Based on Actual HVAC Behaviours. *Buildings* **2023**, *13*, 2710. [CrossRef]
38. *ANSI/ASHRAE Standard 55-2017*; Thermal Environmental Conditions for Human Occupancy. American Society of Heating, Refrigerating and Air-Conditioning Engineers (ASHRAE): Atlanta, GA, USA, 2017.
39. Feng, Y.; Li, Y. An Overview of Deep Learning Optimization Methods and Learning Rate Attenuation Methods. *Hans J. Data Min.* **2018**, *8*, 186–200. [CrossRef]
40. Kingma, D.P.; Ba, J. Adam: A Method for Stochastic Optimization. In Proceedings of the 3rd International Conference on Learning Representations (ICLR), San Diego, CA, USA, 7–9 May 2015.
41. *GB 50189-2015*; Design Standard for Energy Efficiency of Public Buildings. China Architecture & Building Press: Beijing, China, 2015.
42. *GB/T 51350-2019*; Technical Standard for Nearly Zero Energy Buildings. China Standards Press: Beijing, China, 2019.

energies

MDPI

Article

From Data to Action: A Methodological Approach to Address Energy Poverty in Private Multi-Family Buildings

Alberto Lodovico Ghiberti [1], Giorgio Dutto [1], Maria Ferrara [1,2] and Enrico Fabrizio [1,2,*]

[1] TEBE Research Group, Department of Energy, Politecnico di Torino, 10129 Torino, Italy; alberto.ghiberti@polito.it (A.L.G.); maria.ferrara@polito.it (M.F.)
[2] FULL Future Urban Legacy Lab, Politecnico di Torino, 10129 Torino, Italy
* Correspondence: enrico.fabrizio@polito.it

Abstract

Achieving the decarbonization of the building stock by 2050 requires not only technological innovation but also strategies capable of addressing energy poverty, which threatens to exclude millions of households from a fair transition. Measuring this phenomenon remains challenging: at the European level, monitoring systems rely mainly on aggregated statistics, useful for territorial comparisons but often too approximate to describe the conditions of individual households and dwellings. This paper proposes a building-scale methodology that integrates socio-economic and technical data collected directly through surveys, interviews, and utility bills. The approach was applied to a private multi-family building built in the early twentieth century in Turin (Italy), involving 16 households. Results indicate that 31% of households exceed the 10% energy expenditure threshold, with heating emerging as the main cost driver. Correlation analyses suggest that single parameters such as income or dwelling size are not sufficient on their own to explain vulnerability, whereas the integration of socio-technical factors provides a more detailed picture of household conditions. Based on this evidence, four intervention strategies were developed, ranging from the insulation of the envelope to the installation of photovoltaics, conceived to be implemented progressively according to real technical and economic constraints. The novelty of this study lies in linking building-scale evidence with concrete design solutions, bridging the gap between measurement and action.

Keywords: energy poverty; climate neutrality; multi-family buildings; energy efficiency; survey; residential buildings; space heating; building renovation; interviews

1. Introduction

The decarbonization of the building stock by 2050 represents one of the most ambitious challenges set by the European Union within the framework of the Green Deal. Achieving this objective requires a profound transformation of energy systems, still largely dependent on fossil fuels, with wide-ranging economic and social implications. Within this transition, energy poverty (EP) emerges as a central barrier: households living in inefficient dwellings and struggling to afford energy costs risk being excluded from a fair and inclusive energy transition [1–3].

In 2022, more than 41 million European citizens reported being unable to keep their homes adequately warm [4], with direct consequences for well-being, health, and social inclusion. EP is commonly defined as "the inability to access adequate energy services" [5,6]

https://doi.org/10.3390/en18236194

and is recognized as a complex and multidimensional condition, resulting from the interaction of insufficient household income, high energy expenditures, and poor efficiency of buildings and technical systems. Additional factors such as housing precarity, unemployment, health problems, or the presence of dependent children and elderly people can further exacerbate household vulnerability [7,8].

1.1. Legislative Framework and Policy Initiatives

The issue of EP was first introduced into the European regulatory framework with Directive 2009/72/EC, which incorporated the concept into energy policies. Since then, the European Union has progressively reinforced its attention to this phenomenon, recognizing its social and structural implications. With the Energy Efficiency Directive 2012/27/EU [9] and the Energy Performance of Buildings Directive (EPBD) 2018/844/EU [10], the reduction in EP was explicitly linked to building renovation policies, promoting measures to improve the performance of the existing building stock [11].

A major step forward is represented by the recent recast of the EPBD (Directive 2023/2407) [12], adopted in March 2024, which requires Member States to implement specific instruments to address EP. These include transparent monitoring systems, targeted incentives for the worst-performing buildings, and financial support for vulnerable households. The focus is therefore not only on improving energy efficiency, but also on ensuring a just transition, acknowledging that many households lack the financial capacity to cover upfront renovation costs and cannot fully benefit from tax deduction mechanisms [13–15].

Despite this regulatory progress, EP remains widespread, with particularly high incidence in some countries: Bulgaria (27%), Lithuania (23%), and Greece (17%) [4]. In Italy, the National Observatory on EP estimated that by the end of 2021 the phenomenon affected 2.2 million households (8.5% of the total), a percentage that further increased in 2022 due to rising energy prices [16].

Similar socio-technical dynamics have also been observed in non-European contexts, where energy poverty and energy deprivation are increasingly framed within broader debates on energy justice and housing inequality [17,18].

Policy responses among Member States have been heterogeneous, partly due to the absence of a shared definition [19]. The most common instrument used to fight EP is direct subsidies to households, which is considered a temporary measure [20]. Alongside these, more structural incentives have been introduced, which allow households to recover part of the energy renovation costs over the years [21]. Nevertheless, these measures are often not fully accessible to low-income households, given the need to anticipate expenses.

Overall, while such policies provide immediate relief and stimulate long-term investments, they often do not allow a precise identification of the most vulnerable households, nor a full understanding of the mechanisms driving EP. This suggests the importance of developing robust measurement methods capable of analyzing EP at a finer scale, thereby supporting more targeted and effective policy interventions.

At EU level, several dedicated initiatives have been launched to support the identification and mitigation of energy poverty, including Horizon 2020 projects focusing on monitoring and policy design. A notable example is the ENPOR project ("Actions to Mitigate Energy Poverty in the Private Rented Sector") [22], which develops guidelines, support schemes and an Energy Poverty Dashboard to identify and address energy poverty in the private rented sector across seven Member States. However, these initiatives mainly operate at program and territorial level and do not provide a standardized, field-based diagnostic framework at the scale of individual buildings. This reinforces the need for operational micro-scale methodologies, such as the one proposed in this paper, which can generate building-level evidence to complement EU monitoring tools.

1.2. Literature Framework and Measurement Insights

The lack of uniformity among European countries, both in the definition of EP and in the strategies adopted to address it, reflects country-specific socio-economic conditions and policy priorities [23]. This heterogeneity is also evident in the scientific literature [24]. Early studies mainly focused on quantifying the diffusion of the phenomenon, with particular attention to the share of energy expenditure in household income or to the inability to meet basic energy needs due to supply constraints. A fundamental contribution in this regard is that of Boardman (1991), who defined EP as the condition in which a household spends more than 10% of its income on energy bills [25]. This indicator, still widely used today, represented a turning point in the debate by explicitly linking energy issues with social welfare.

Over the past three decades, the debate has broadened across multiple disciplines, generating a wide range of definitions that reflect economic, sociological, and technical perspectives [23,26]. Some studies have proposed methods and indicators to measure the incidence of EP [27], while others have focused on the socio-demographic characteristics of affected households, thereby contributing to a more articulated understanding of the issue [28].

To provide a clearer analytical structure, the existing literature on energy poverty can be broadly grouped into two main scales of analysis. Macro-scale approaches rely on national and European datasets (EU-SILC, Eurostat, ARERA), enabling territorial comparisons and long-term monitoring. Micro-scale approaches, by contrast, are based on direct field data as household surveys, interviews, utility bills and building inspections, and allow for a detailed socio-technical understanding at the dwelling level. The review presented below follows this macro–micro distinction.

Macro-scale approaches have been widely used to produce urban or regional maps of energy vulnerability [29–31]. A notable example is the study of Bologna [32], which combined EUROSTAT data, municipal EPC records, and information from the Italian regulatory authority (ARERA) to identify the most vulnerable areas.

More recently, Desvallées [33] and colleagues have demonstrated how geospatial mapping techniques, when based on household- and neighborhood-level data, can effectively support policy planning. However, when only aggregated data are available, it becomes difficult to identify the specific households experiencing EP.

Micro-scale approaches, by contrast, are based on the direct collection of data through questionnaires, interviews, and building surveys [6,33–35]. This perspective allows researchers to capture the interplay between technical, economic, and social factors, providing a more detailed representation of household conditions. Several contributions have demonstrated that the building scale is particularly suitable for integrating socio-economic conditions with the physical and energy characteristics of dwellings [36].

For instance, Grdenić et al. [37] analyzed 102 households in Croatia, showing how combining socio-economic and technical data enables the identification of targeted solutions; while Fabbri and Camboni [30,32] highlighted, in the case of social housing, how building inefficiencies directly affect vulnerability and how renovation strategies must be aligned with residents' needs.

Other authors have experimented with multidimensional approaches. At the European level, Menyhért showed that the use of different indicators (10% threshold, LIHC, subjective measures) [38] leads to the identification of different vulnerable groups, thereby underlining the challenges of comparability across contexts.

A recurring issue thus concerns not only the lack of consistency across methodologies, but also the heterogeneity of measurement criteria, which can lead to divergent and difficult-to-compare results [39]. In this context, datasets derived from national statistical surveys, while useful for territorial comparisons, often provide an overly approximate representation of EP. An alternative is offered by building-scale surveys, which, through questionnaires, interviews, and technical assessments, make it possible to collect more detailed information, closer to the actual living conditions of households.

Although micro data cannot fully replace large-scale statistical analyses, they provide an essential contribution by highlighting their limitations and suggesting measurements more closely aligned with everyday reality. From this perspective, direct surveys should not be considered as an isolated approach, but as a complementary and potentially more effective tool to support targeted policies and intervention strategies that are genuinely responsive to the needs of households.

Within this macro–micro landscape, the present study adopts a micro-scale perspective, relying on detailed household-level data that cannot be captured through aggregated statistical indicators.

1.3. The Role of Building Quality in EP

Several studies have shown that the physical characteristics of buildings, and, in particular, the performance of the building envelope, have a significant impact on the risk of EP. Dwellings with poor insulation or inadequate energy performance lead to higher energy consumption and costs, which are often unsustainable for households and exacerbate situations of social vulnerability.

A number of authors have underlined that the inefficiency of the existing building stock is one of the main structural drivers of EP in Europe [40]. In particular, buildings constructed before the introduction of thermal performance regulations are characterized by high levels of heat loss, making it difficult to ensure minimum comfort conditions without excessive expenditure.

Fabbri et al. (2023) [41], analyzing a social housing complex in Italy, demonstrated that progressively increasing the insulated surface of the building envelope results in a marked reduction in heating demand and, consequently, of EP risk, although with different effects between winter and summer seasons. Gradual renovation scenarios can represent a realistic strategy for the existing building stock, especially in social housing contexts where resources are limited and interventions must minimize disruption for residents.

Empirical evidence from various European countries further confirms that inadequate housing conditions, such as small dwellings, dampness, thermal dispersion, and poor construction quality, are strongly correlated with higher levels of EP. This indicates that tackling EP requires not only income support measures but also structural interventions to improve housing conditions.

From this perspective, the joint assessment of households' socio-economic characteristics and the technical quality of buildings, as proposed in this study, is essential to develop targeted intervention strategies that are both effective and socially equitable.

1.4. Objectives and Contributions of the Research

As discussed above, the building scale represents a meaningful level of analysis where the technical characteristics of dwellings and the socio-economic conditions of households converge. This research builds on this perspective by proposing a building-scale approach to measuring EP, based on the direct collection of field data. The methodology is explicitly conceived as a micro-scale approach, relying on surveys, interviews, utility bills and technical inspections carried out directly at household and dwelling level. This fine-grained evidence enables an analysis of socio-technical mechanisms that remain invisible in macro-scale indicators.

The innovative aspect lies in the systematic integration of technical, energy, and socio-economic information, which enables a close examination of how building features and household conditions interact in shaping vulnerability.

The case study is located in Turin (Piedmont, Italy) and concerns a representative early 20th-century private multi-family building. Within this context, socio-economic, technical, and energy information was systematically collected through surveys, interviews, and documentary research. This dataset enables the assessment of EP incidence and the identification of the main factors influencing vulnerability, while also providing a concrete basis for developing potential intervention scenarios.

The main contributions of the research can be summarized as follows:

- Methodological: presentation of a building-scale approach capable of integrating technical, energy, and socio-economic dimensions within a single interpretative framework.
- Empirical–applicative: analysis of a residential building representative of the historic housing stock found in many European cities, offering insights into the relationship between building quality and energy vulnerability.
- Operational: demonstration of how field data can be directly translated into actionable intervention strategies, linking building renovation options with the actual conditions and needs of residents.

From this perspective, the study contributes to bridging the gap between analysis and action, providing tools to support more targeted and effective policies for reducing EP.

2. Materials and Methods

2.1. Methodological Framework

The methodology adopted in this study relies on a direct field approach aimed at collecting reliable and multidimensional data on EP at the building scale. It integrates information on the architectural and physical features of dwellings, household energy consumption and technical systems, together with socio-economic characteristics such as family composition, income, and household size.

The methodological process was structured into six main phases:

1. Case study identification and selection;
2. Household survey and utility bills collection (combining socio-economic, behavioral, and technical information);
3. Reconstruction of household energy balance;
4. Definition and Calculation of EP Indicator (using the Boardman 10% threshold);
5. Indicators and calculation procedure;
6. Identification of potential mitigation strategies.

This stepwise organization ensures clarity and replicability, making the methodology adaptable to different contexts and building stocks.

The overall process is illustrated in Figure 1, which summarizes the input data, the assessment phases, the EP measurement, and the resulting output strategies.

Some limitations should be acknowledged. The method relies on the active cooperation of households and on the availability of socio-economic and energy data. Missing information (e.g., incomplete meter readings or unavailable criteria for heat cost allocation) and uneven participation may reduce completeness. Nevertheless, even under partial participation, the methodology demonstrated robustness and feasibility, generating datasets suitable for both targeted interventions and comparative research.

Overall, the methodology has a twofold value: it enables the identification of practical and targeted strategies by linking consumption profiles with socio-economic vulnerability, while at the same time producing micro-scale datasets that can complement and validate national or regional statistics, thus enhancing the transferability of results. It should also be emphasized that the methodology was primarily conceived for application in the private residential sector. In the case of public housing, the approach could be replicated; however, the interaction with public authorities may affect data collection processes. Specifically, certain information, such as original design documents, historical records, and consumption data, may be more readily available due to the existence of institutional archives, thereby potentially facilitating the implementation of the method.

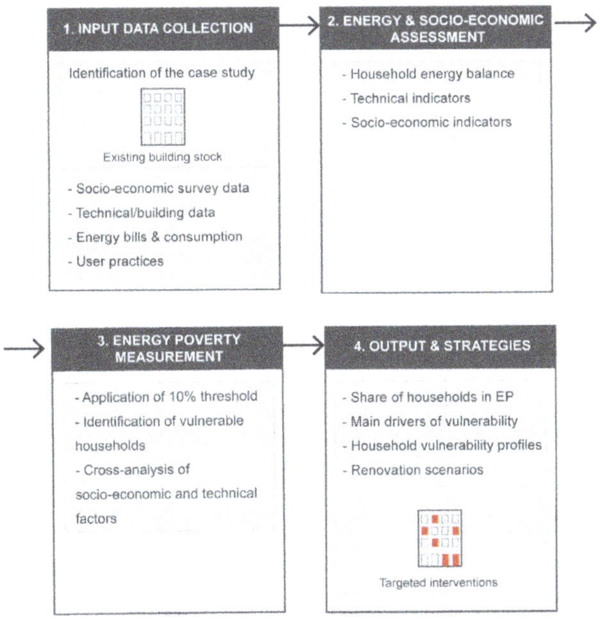

Figure 1. Flowchart of the proposed methodology, showing input data, assessment phases, EP measurement, and output strategies.

2.2. Case Study Identification and Selection

The identification of a suitable case study represents an essential step, as it directly influences both the methodological approach and the type of results obtained. Two main criteria were adopted for the selection process. The first criteria concerned the representativeness of the building stock. Since a large share of the European residential heritage was constructed before the introduction of energy efficiency regulations, it was considered essential to focus on a building that could reflect these characteristics. According to national statistics, more than 70% of the housing stock in Italy was built before 1980 and approximately 25% before 1945 [42]. Before any regulatory requirements on thermal performance were established. The choice of an early 20th-century multi-family building therefore ensured that the analysis would be relevant for a wide segment of the national context.

The second criterion related to the feasibility of data collection. The methodology relies on the active cooperation of residents, who are required to provide both socio-economic and energy-related information. Previous field-based studies on EP have shown that successful data collection is strongly conditioned by local trust and willingness to collaborate, sometimes supported by institutional actors [43]. For this reason, particular attention was given to selecting a case where such cooperation was achievable. While participation can remain uneven, leading to potential gaps or inconsistencies in the dataset, this approach allows for the acquisition of detailed information that would otherwise be unattainable through purely statistical or indirect methods.

2.3. Household Survey and Utilitybills Collection

Household data were collected through semi-structured questionnaires and interviews, administered directly to at least one member of each family. The survey design was informed by previous research on EP [27,37,43] and articulated into three thematic sections:

- Energy use: Past electricity and gas consumption, heating and cooling practices, appliance stock, and operating habits.
- Perceptions of the energy crisis: Household awareness of rising energy expenditure and behavioral strategies adopted to reduce consumption.
- Socio-economic data: Household composition, education, employment status, monthly income, and arrears in energy bill payments.

The questionnaire, developed at the FULL Laboratory of the Polytechnic of Turin, consisted of about 130 questions and combined closed and open-ended items, ensuring comparability across households while allowing for additional qualitative insights. Each session lasted 20–35 min and included the collection of utility bills. An extract of the questionnaire structure is provided in Figure 2, illustrating the comprehensiveness of the survey design.

Introduction
Geometric data

o How many square meters is the house?
....

o How is the house composed?
☐ Kitchen ☐ Living room ☐ Room 1 ☐ Storage room

Part 1
Electrical energy

o What is the electricity consumption of the last two/three years?

o What appliances are in the house?
☐ Washing machine ☐ Fridge ☐ Radio ☐ PC
☐ Microonde ☐ Dishwasher ☐ Oven ☐ Tv

o Do you use appliances at night and on Saturdays and Sundays?

Cooling

o At what setpoint does the cooling system maintain?

Heating

o Is the whole house heated or are there areas not heated?
☐ Fully heated ☐ Presence of unheated areas

o Is the heating independent or central?
☐ Autonomous ☐ Centralized

Part 2
Energy crisis: consequences and strategies

o In this period of energy crisis, has it adopted any particular strategies to save electricity consumption?

o Once the energy crisis is over, will you continue to adopt these strategies?

Part 3
Socio-demographic profile

o How many people live permanently in the house (including yourself)?

o What is the age of such people?
....

o What is your profession?
....

o Family monthly income
☐ < 1500 ☐ 1500-25 ☐ 2500-5000 ☐ > 5000

o Country of birth
....

o In the case of rented accommodation, what is the monthly rent?
....

o Qualification
....

o How many years have you lived in this house?
☐ less than 5 years ☐ 5-10 years ☐ more than 10 years

Figure 2. Extract of the questionnaire used during the interviews.

Utility bills served as the primary source of quantitative data on household energy use, providing annual values of consumption and associated costs. These were cross-checked and complemented with questionnaire responses to identify installed systems (heating, cooling, and domestic hot water) and their patterns of use. Special attention was devoted to appliances: respondents provided details on type, efficiency class, and operating hours, which allowed electricity demand to be estimated in a way that reflected actual household behavior rather than relying solely on standardized assumptions.

This integration of survey information and bills provided a coherent dataset for reconstructing household energy balances. This approach made it possible to distinguish the contributions of different energy carriers:

- gas for heating, cooking, and hot water;
- electricity for appliances and lighting.

In cases where electricity was used for cooking or hot water, these components were isolated to avoid distortions in the allocation of end uses. Figure 3 illustrates the framework for household energy balance, highlighting the flow from energy carriers to end uses.

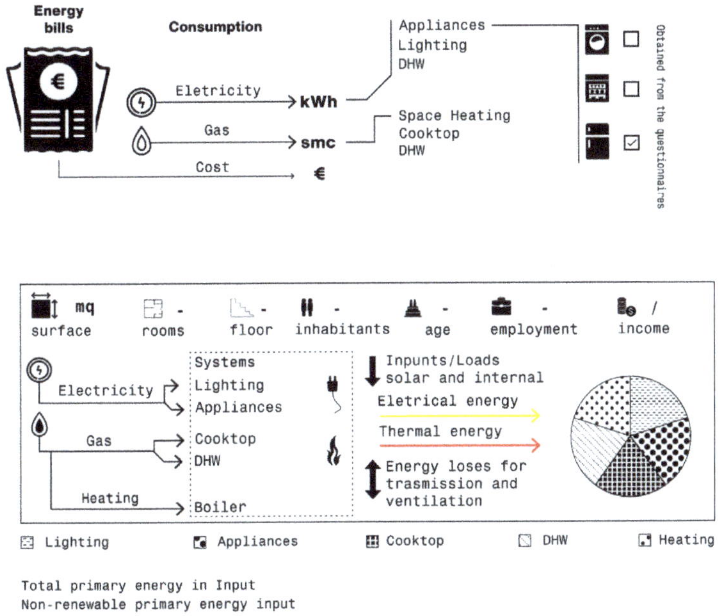

Figure 3. Framework for household energy balance.

2.4. Energy Balance Reconstruction

Household energy balances were reconstructed by combining bill-derived consumption values with the efficiencies of technical systems, thus estimating delivered energy for each end use.

Electricity demand associated with appliances was refined using the additional appliance data (type, efficiency, and daily use). In households with induction cooking or electric water heating, these loads were calculated separately to ensure correct allocation. This step provided a comprehensive overview of household consumption profiles, linking technical system performance with behavioral and socio-economic factors.

2.5. Definition and Calculation of EP Indicator

Once all data had been collected, the analysis phase was carried out to measure EP. This study applied the indicator proposed by Boardman [25], which defines a household as being in EP if it spends more than 10% of its income on energy services, including costs for space heating, domestic hot water, lighting, appliances, and cooking. To apply this threshold, both household income and total energy-related expenditures were required. This is a widely used indicator in international literature, offering a clear and comparable benchmark, although it does not capture all dimensions of the phenomenon.

It should be acknowledged that, while the Boardman threshold remains a robust and historically significant reference, it only reflects the share of energy expenditure relative to household income. As such, it does not account for other dimensions of EP, such as thermal comfort, arrears on energy bills, or non-monetary forms of deprivation.

Nevertheless, the comprehensive design of the questionnaire, which collected information on household perceptions, payment arrears, and comfort levels, provides the potential to construct additional multidimensional indicators of EP. Although these were not applied in the present study, they represent an important extension of the methodology and could be exploited in future research.

To further characterize household energy profiles, primary energy values were also calculated for each dwelling, using the national conversion factors established by DM 26-06-2015 Requisiti Minimi [44]. This step complemented the economic assessment with a technical indicator, providing a more comprehensive representation of household energy conditions.

2.6. Indicators and Calculation Procedure

The measurement follows the following steps.

The definition of a limit income, which represents the risk of ending up in EP, defined as

$$I_{EPR} = I \times t_{ep} \text{ [€/year]} \quad (1)$$

where

I_{EPR} is the limit income EP risk [€/year];

I is the annual family income [€/year];

t_{ep} is the EP risk threshold, set to 10% from the Boardman's definition.

Annual energy expenditure for electrical and thermal uses

$$C_{ENE} = C_{ENE,gas} + C_{ENE,ele} \text{ [€]} \quad (2)$$

where

C_{ENE} is the total Energy expenditure for electrical and thermal uses;

$C_{ENE,gas}$ is the obtained from the sum of consumption dedicated to heating, cooktop and DHW if presented as

$$C_{ENE,gas} = (Q_h + Q_{cook} + Q_{DHW}) \times C_{gas} \text{ [€]} \quad (3)$$

where

Q_h is the heating consumption [kWh_g];

Q_{cook} is the cooktop consumption [kWh_g];

Q_{DHW} is the domestic hot water consumption [kWh_g];

C_{gas} is the natural gas cost factor [€/kWh_g];

$C_{ENE,ele}$ is the obtained from the sum of consumption dedicated to appliances, lighting and DHW if presented as

$$C_{ENE,ele} = (E_{app} + E_{Light} + E_{DHW}) \times C_{ele} \text{ [€]} \quad (4)$$

where

E_{app} is the electricity consumption dedicated to appliances [kWh_e];

E_{Light} is the electricity consumption dedicated to lighting [kWh_e];

E_{DHW} is the electricity consumption dedicated to domestic hot water [kWh_e];
C_{ele} is the energy cost factor [€/kWh_e].
If $C_{ENE} > I_{EPR}$ possible risk of EP.

Additional indicators considered are the total primary energy ($E_{p,gl,tot}$) of the energy carriers, which incorporates both the non-renewable component ($E_{P,gl,nren}$) and the renewable component ($E_{P,gl,ren}$). To calculate the total primary and non-renewable energy input, the household's electricity and gas consumption are multiplied by the corresponding conversion factors indicated in DM 26-06-2015 Requisiti Minimi [44].

Primary energy is obtained by multiplying the input energy vector by its own conversion factor

$$E_{P,gl,nren} = E_{ENE,gas} \times f_{P,nren,g} + E_{ENE,ele} \times f_{P,nren,ele} \quad (5)$$

where
$E_{ENE,gas}$ is the natural gas input energy carrier [kWh_g/m^2a];
$f_{P,nren,gas}$ is the natural gas non-renewable conversion factor of 1,05 [-];
$f_{P,ren,gas}$ is the natural gas renewable conversion factor of 0 [-];
$f_{P,tot,gas}$ is the natural gas total conversion factor of 1.05 [-];
and
$E_{ENE,ele}$ is the electricity input energy carrier [kWh_e/m^2a];
$f_{P,nren,ele}$ is the electricity non-renewable conversion factor of 1.95;
$f_{P,ren,ele}$ is the electricity renewable conversion factor of 0.47;
$f_{P,tot,ele}$ is the electricity total conversion factor of 2.42.

2.7. Identification of Potential Mitigation Strategies

The final step of the methodology concerns the identification of potential strategies to mitigate EP. This step builds directly on the previous phases, as it relies on the integrated analysis of household profiles, socio-economic vulnerability, and building quality. In particular, the identification of households experiencing EP, together with the assessment of their dwelling characteristics and system performance, provides the basis for defining suitable intervention options.

The process does not aim to prescribe specific measures in advance, but to establish a structured framework that connects the evidence collected, including energy balances, building envelope quality, system efficiencies, and socio-economic indicators, to possible response strategies. These may range from technical measures on the building and its systems, to behavioral recommendations or financial support mechanisms adapted to household capacity. The concrete outcomes of this step are inherently case-specific and are presented in the Development of strategies section. Methodologically, however, its relevance lies in showing how field-based micro-scale evidence can be operationalized into a decision-support framework that moves beyond the measurement of EP and towards the formulation of actionable solutions.

3. Case Study

3.1. Building Characteristics

The case study concerns a six-story masonry private multi-family building located in Turin, originally built in 1928 by the Molini Dora company (Torino, Italy). The building comprises a total of 30 apartments (five per floor) with mixed residential and commercial use. Figure 4 illustrates the exterior of the building, which has two main entrances due to the slope of the ground. The overall condition is generally acceptable, although several deficiencies were observed, including deteriorated balconies, plaster detachment, and aluminum window frames with poor thermal performance.

Figure 4. Picture of the case study building.

The main physical and technical features of the building are summarized in Table 1. In addition to its morphological characteristics, the building is equipped with a centralized heating system. For the estimation of domestic hot water efficiency, typical coefficients were assumed: 0.82 for gas water heaters and 1.0 for electric water heaters, reflecting the heterogeneous technological solutions currently in place.

Table 1. Characteristics of the case study.

Type of building	Private multi-family building
Construction company	Molini Dora
Year of construction	1928
Number of floors	6
Number of apartments per floor	5
Height building	21 m
Building's cubage	11,572 m^3
Total free area of the building	1251 m^2
Area covered by the building	556 m^2
Free area of the yard	695 m^2
Type of structure	load-bearing masonry
Heating systems	centralized heating
Intended purpose	residential and commercial

3.2. *Architectural History and Transformations*

An extensive archival investigation was conducted to reconstruct the history and transformations of the building. Original project drawings preserved in the municipal archives, building permits (SCIA, DIA), and other official records were analyzed, complemented by on-site surveys and direct consultation with residents.

The main changes documented over time include, the following:

- Subdivision of internal spaces to create additional dwellings;
- Increase in the number of units per floor;
- Installation of an external elevator;
- Conversion of the ground floor from residential to commercial use.

These transformations are illustrated in Figure 5, which presents the timeline of modifications. Importantly, the historical analysis revealed that no energy retrofitting interventions have ever been carried out. This strongly suggests that the building still performs with very limited energy efficiency, in line with its original construction characteristics.

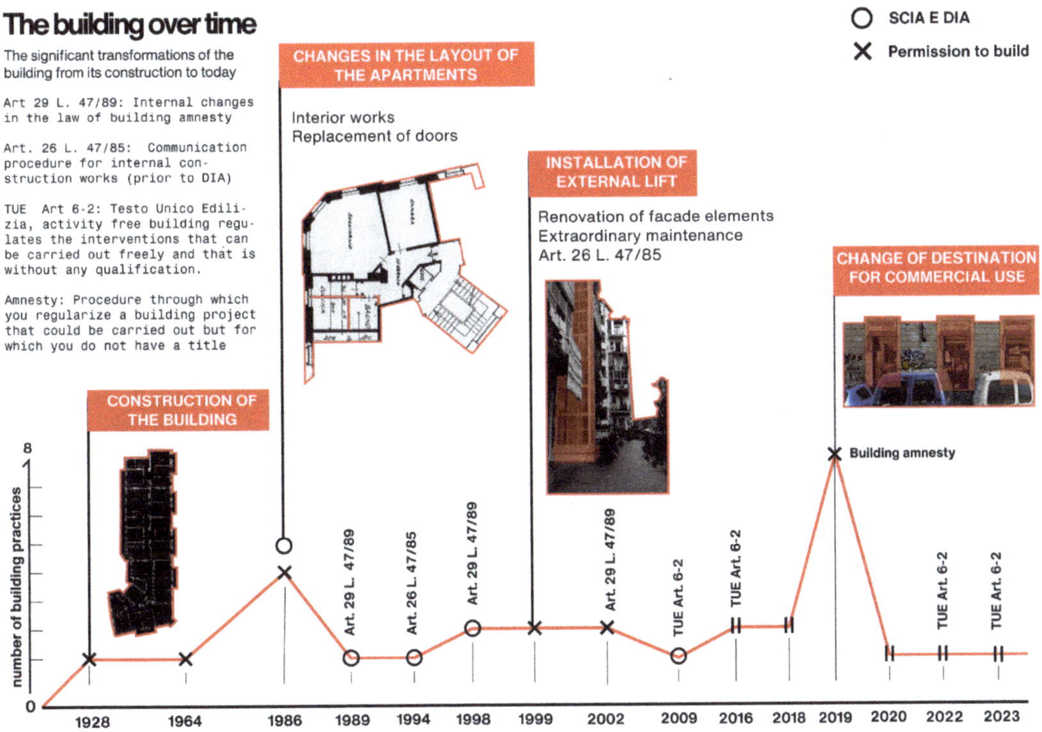

Figure 5. Timeline of transformations of the case study building.

4. Results

4.1. Survey Participation and Socio-Demographic Profile

The survey achieved a participation rate of 55%, corresponding to 16 households and 31 individuals. This level of response was considered sufficient to capture both the social and energy-related dimensions of the case study.

In terms of household structure, 30% were single-person households and 37% were couples, indicating a predominance of small household sizes. The age distribution revealed a concentration in the youngest and oldest groups, highlighting two segments of the population typically associated with higher vulnerability. Occupational status reflected a similar pattern, with 31% of respondents being retired individuals and 24% students.

Tenure status represented an additional dimension of potential vulnerability: 56% of households were tenants and 44% were homeowners. This distinction has direct implications for energy retrofitting, as tenants generally have limited capacity to influence investment decisions. The average annual household income was €21,150, a value that already points to economic fragility and may restrict access to renovation incentives.

4.2. Energy Balance at Household Scale

All apartments were connected to a centralized gas-fired heating system, while domestic hot water (DHW) production was more heterogeneous. Some households relied on individual gas boilers, others on electric water heaters, introducing variability that later proved significant in the analysis of consumption profiles.

To illustrate the analytical process adopted in this study, Household 5 was selected as a representative example, since its values were close to the sample average. The case provides a practical demonstration of how consumption data were collected, disaggregated, and integrated into a comprehensive household energy balance.

Electricity consumption, reconstructed from utility bills, is reported in Table 2. The household displayed a stable monthly demand of approximately 110 kWh, with a total cost of €176.25 over the three months considered. This corresponded to an annual specific demand of 29.3 kWh_e/m^2.

Table 2. Electricity consumption from household 5 bills.

Year	Month	Consumption [kWh_e]	n. Days	€/kWh_e	Euro
2021	November	108	30	0.58	62.1
2021	December	111	31	0.57	63.7
2022	January	111	31	0.45	50.45
Total		330	92		176.25
	Average monthly consumption [kWh_e]			110	
	Average annual consumption [kWh_e/m^2]			29.3	

Natural gas consumption for cooking and DHW (excluding space heating) is summarized in Table 3. Annual demand amounted to 39.14 kWh_g/m^2, with an associated cost of about €190. When space heating is included, total gas consumption increases substantially, confirming the dominant role of heating in the overall household energy balance.

Table 3. Natural gas consumption from household 5 bills (excluding space heating).

Year	Month	smc	Consumption [kWh_g]	n. Days	€/kWh_g	Euro
2022	April–August	54	577.26	122	0.12	70.92
2022	August–December	54	577.26	114	0.20	118.96
Total		108	1154.52	236		190
	Average monthly consumption [kWh_g]				Average annual consumption [kWh_g/m^2]	
	146.78				39.14	

Beyond aggregated electricity use, the methodology allowed a detailed disaggregation by end use, thanks to the integration of questionnaire information on appliances and operating habits. As shown in Table 4, appliances accounted for the majority of electricity demand (23.5 kWh_e/m^2a), while lighting represented a smaller but non-negligible component (5.7 kWh_e/m^2a).

The complete household energy balance is illustrated in Figure 6. The diagram shows the flow of energy carriers (input), technical systems, and final end uses. The accompanying pie chart highlights that heating represented by far the largest share of energy demand, overshadowing all other end uses combined.

Table 4. Analysis of the calculation of energy dedicated to household appliances.

Appliance	Consumption kWh_e	n. Device	Weekly Use	Monthly Consumption kWh_e	Annual Consumption kWh_e
Washing machine (cycle)	1.35	1	2	10.61	129.73
Fridge (hour)	0.82	1	168	23.01	300.00
Dishwasher	1.05	1	2	8.40	100.80
Oven (hour)	1.20	1	3	28.80	345.60
Toaster (cycle)	0.03	1	4	0.48	5.76
Television (hour)	0.3	1	21	6	72.00
Mobile phones	0.05	2	7	5.60	73.00
Printer	0.1	1	2	0.8	9.6
Playstation	0.14	1	8	4.48	58.24
Router wifi	0.288	1	168	8.06	96.77
				85.64	1061.77
Average monthly consumption [kWh]			Average annual consumption [kWh/m^{2}a]		
85.64			23.5		

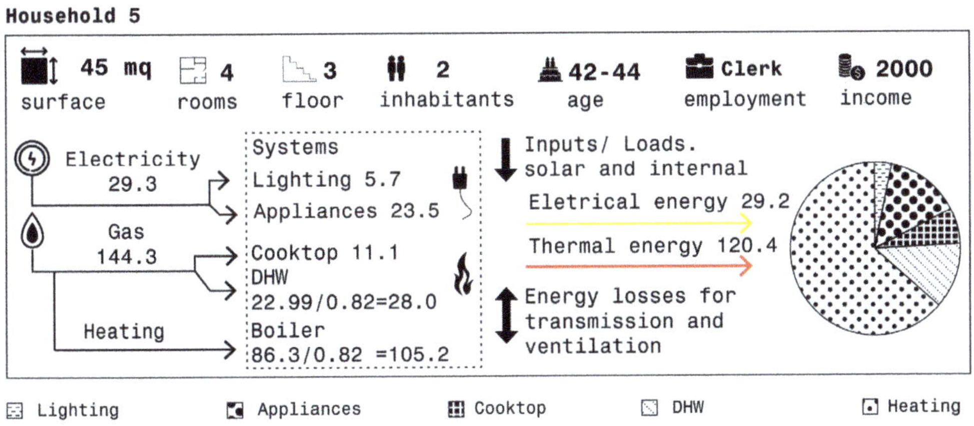

Total primary energy in Input =(29.24*2.42+144.34*1.05) = 222.32 kWh/m²a
Non-renewable primary energy input =(29.24*1.95+144.34*1.05)= 208.57 kWh/m²a

Figure 6. Example of energy balance for a household.

4.3. Energy Balance at Building Scale

Extending the analysis to all 16 households confirmed the trends observed at household level. Average electricity demand was 19.6 kWh_e/m^2a, while natural gas demand reached 326 kWh_g/m^2a (Table A1, Appendix A). Energy expenditure, reconstructed from utility bills and centralized heating records, is summarized in Table A2 (Appendix A). On average, heating represented by far the largest financial burden, with an annual cost of €1333 per household. Variations in other end uses were smaller and largely explained by appliance ownership and supplier tariffs.

Primary energy demand, calculated according to national conversion factors [44] ranged from 95.53 to 257.49 kWh/m^2a, with an average value of 172.68 kWh/m^2a (Table A3, Appendix A). This variability reflects both technical factors, such as system efficiency and envelope quality, and behavioral aspects, including occupancy patterns and user practices.

For completeness, additional representative examples of household energy balances are provided in Appendix A (Figure A1).

4.4. Measuring EP

The integration of socio-economic and technical data enabled a direct assessment of EP at the household level. Table 5 reports the key variables required for the calculation, including household size, income, energy consumption, and annual energy expenditure. Applying the Boardman 10% threshold [25], five households, corresponding to 31% of the sample, were identified as energy poor. Among these, the share of income spent on energy ranged from 10.2% to a maximum of 12.3%, confirming that for some household's energy costs represent a substantial economic burden. Households in energy poverty that exceed the 10% threshold are shown in bold in the table.

Table 5. Summary table of technical and economic indicators used to measure E.P.

Apartments			Income		Energy	Energy Costs	E. P.
Households	Inhabitants	Size [m^2]	Monthly €/m	Yearly €/m	Primary Energy [kWh/m^2]	Yearly €/a	>10%
1	3	58	2200	26,400	239.50	2109	8.0
2	2	72	1400	16,800	104.60	1398	8.3
3	2	63	2000	24,000	95.53	1640	6.8
4	1	59	1600	19,200	236.01	2130	**11.1**
5	2	45	2000	24,000	213.77	1611	6.7
6	1	58	1600	19,200	254.94	2231	**11.6**
7	2	54	1400	16,800	236.90	1639	9.8
8	2	59	1400	16,800	257.49	1960	**11.7**
9	3	72	2200	26,400 *	130.78	2201	8.3
10	2	118	2000	24,000	176.93	2202	9.2
11	2	75	2000	24,000	162.61	1714	7.1
12	1	76	1200	14,400	133.46	1772	**12.3**
13	1	75	1200	14,400	110.96	1142	7.9
14	1	80	1600	19,200	113.84	1674	8,7
15	3	76	2200	26,400	199.86	2398	9.1
16	3	75	2200	26,400 *	198.11	2683	**10.2**

Data marked with * indicate the absence of specific information for that household and the resulting estimated value.

A closer look at the socio-demographic characteristics of the affected households highlights important patterns of vulnerability. As illustrated in Figure 7, energy-poor households were concentrated among the youngest (<30 years) and oldest (>70 years) residents, suggesting a dual age-related fragility. Educational attainment also emerged as a relevant factor, with the majority of vulnerable households having at most a high school diploma. In terms of tenure status, 60% of energy-poor households were tenants, reflecting limited decision-making power regarding energy retrofits. Finally, income distribution shows that 40% of vulnerable households reported monthly earnings below €1500, underlining the overlap between EP and broader economic fragility.

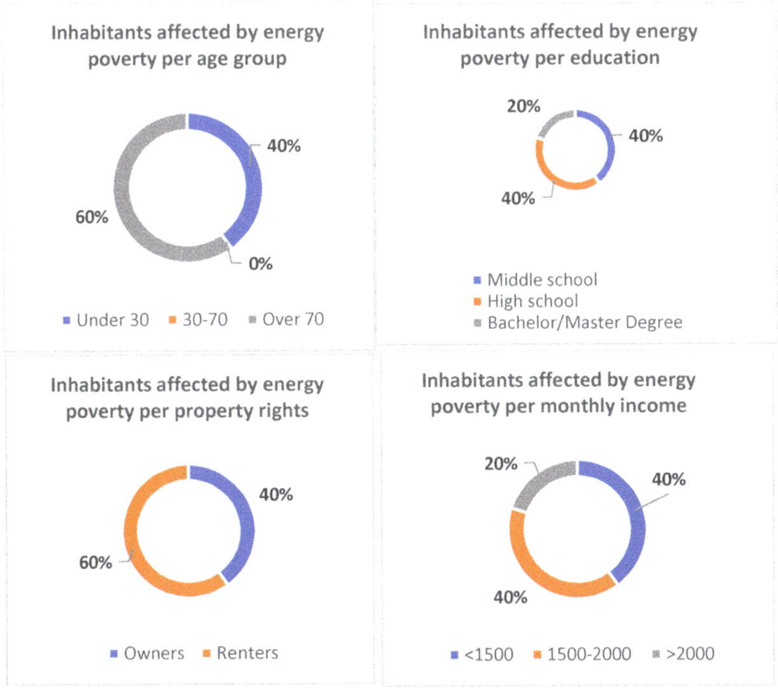

Figure 7. Characteristics of households in EP.

4.5. Correlations

To further investigate the relationship between socio-economic and technical variables, a set of correlation analyses was carried out. The objective was to assess whether simple parameters, such as income or dwelling size, could sufficiently explain household vulnerability to EP.

Figure 8 presents the relationship between annual income and total energy expenditure. A moderate positive correlation was observed ($R^2 = 0.38$), indicating that higher-income households generally spend more on energy. However, the blue dots representing households in EP consistently lie above the regression line, meaning that their energy expenditure represents a disproportionate burden relative to their available income. This result confirms that absolute expenditure levels alone are not sufficient to identify vulnerable households.

The comparison between apartment size and primary energy consumption, shown in Figure 9, does not reveal a strong linear relationship ($R^2 = 0.28$). Apartments of similar size exhibited markedly different levels of specific primary energy demand. Since the building envelope and construction quality are common to all dwellings in the private multi-family building, these differences are more likely to result from factors such as the number of occupants, operational patterns of heating and hot water systems, and household energy behaviors. This indicates that apartment size, while an easily accessible parameter, is not sufficient to explain household vulnerability, which instead derives from the interaction between socio-economic conditions and technical characteristics shared across the building.

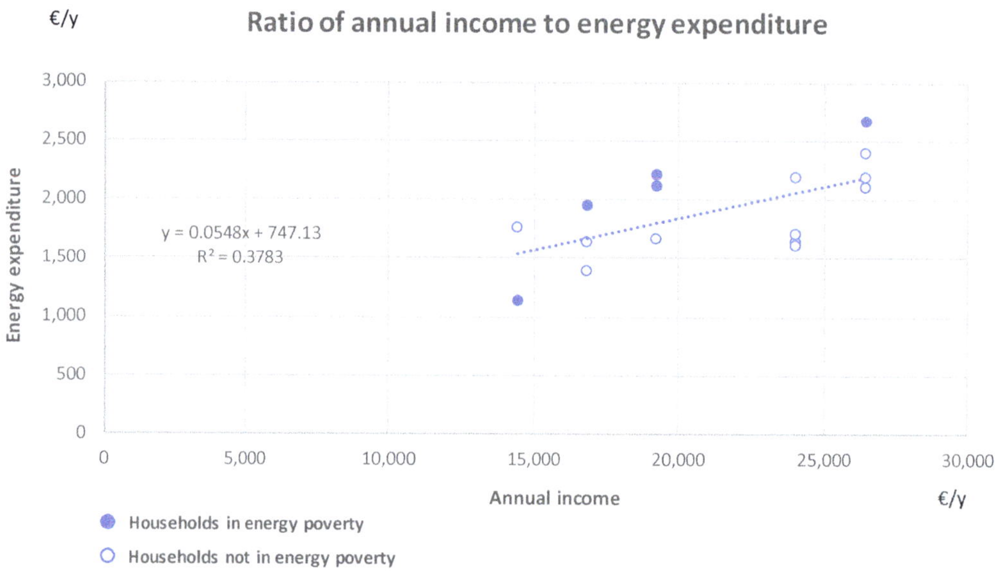

Figure 8. Ratio of annual income to energy expenditure.

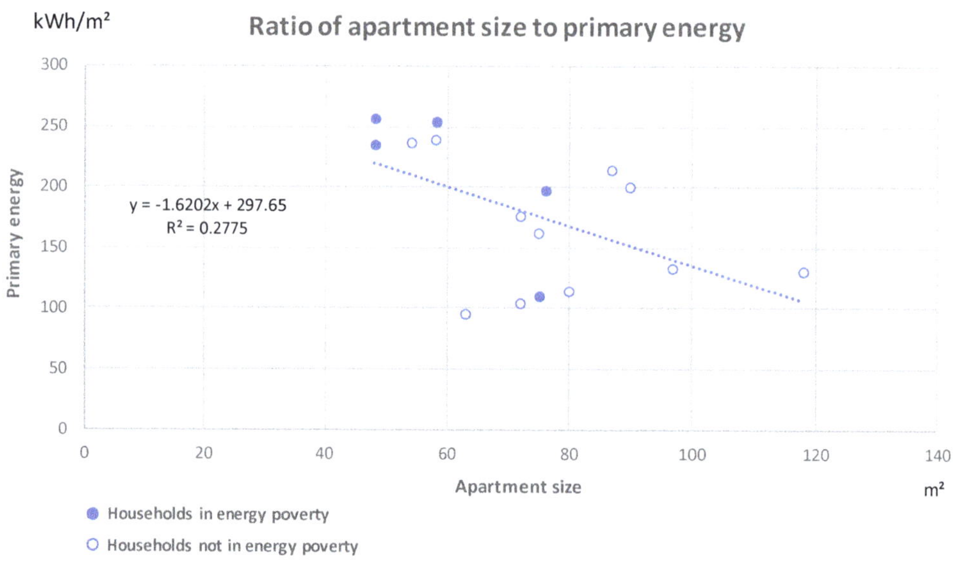

Figure 9. Ratio of apartment size to primary energy.

The relationship between apartment size and total energy expenditure, reported in Figure 10, further illustrates this complexity. The correlation is extremely weak ($R^2 = 0.04$), showing that units of similar size may sustain very different energy expenditure depending on income levels, efficiency of domestic systems, and user behavior. Once again, the blue dots representing households in EP stand out with higher relative burdens than the rest of the sample.

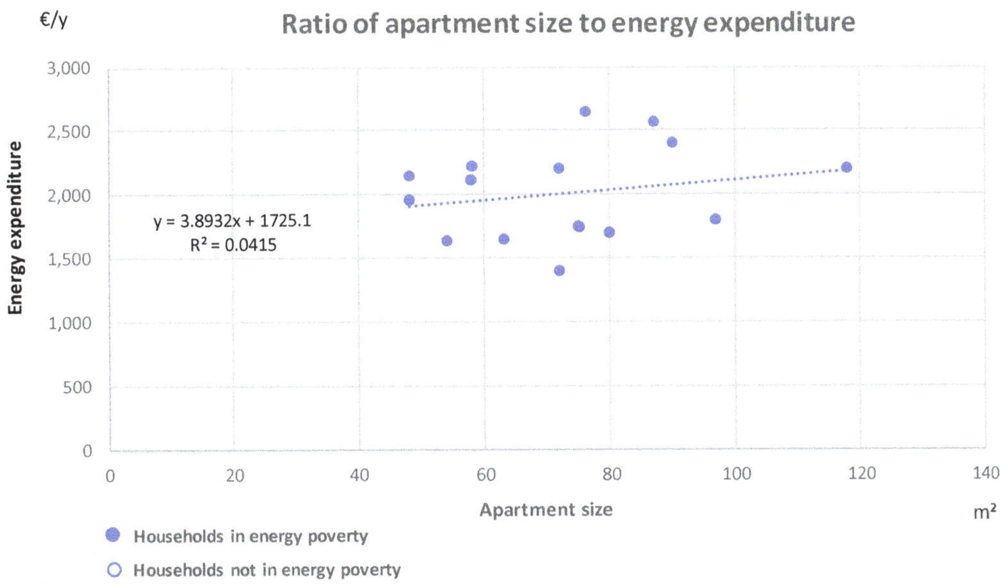

Figure 10. Ratio of apartment size to energy expenditure.

Overall, these analyses confirm that no single parameter can adequately explain household vulnerability. Income, dwelling size, and energy demand interact with technical conditions and socio-demographic characteristics, underscoring the necessity of an integrated approach. This multidimensional perspective provides the basis for linking the diagnosis of EP with the formulation of mitigation strategies that are not only technical, but also responsive to the specific needs of households.

5. Development of Strategies

5.1. Proposed Interventions

The development of intervention strategies was guided by evidence that emerged from the results, both on the socio-demographic and energy side. The survey highlighted that most households are composed of one or two people, often living in dwellings oversized compared to their actual needs. This discrepancy opens up the opportunity to optimize living spaces, for instance, by dividing larger apartments into smaller units such as studios or two-room flats. These transformations can be carried out without altering ownership boundaries or compromising the load-bearing structure of the building, allowing individual owners a certain degree of autonomy in their implementation. This approach is particularly suitable in privately owned multi-family buildings, where residents can act independently on their property.

From an energy perspective, the balance analysis clearly showed that heating represents the most significant component of household expenditure, primarily due to the simple masonry walls and outdated windows. Based on these considerations, four design strategies were proposed:

1. Reconfiguration of apartments, subdividing oversized units to align living spaces with actual household composition.
2. Insulation of the opaque envelope, with the addition of a 12.5 cm layer of insulation, reducing transmittance from $U_{op} = 0.816$ W/(m^{2}K) to $U_{op} = 0.208$ W/(m^{2}K).

3. Improvement of the transparent envelope, by replacing windows, lowering transmittance from U_{tra} = 5.70 W/(m^{2}K) to U_{tra} = 0.88 W/(m^{2}K).
4. Installation of a photovoltaic system on the roof, reducing dependence on the electricity grid from 100% to 59% and producing 10.1 kWh/m^{2}a.

The choice of photovoltaics instead of solar thermal was motivated by the high heterogeneity of domestic hot water systems in the building, which made a centralized solar solution less effective.

It should be noted that these interventions are not necessarily intended to be implemented simultaneously as a single package. Rather, they can be carried out progressively, depending on the specific technical, regulatory, and economic constraints that may arise in real renovation processes.

This flexibility is particularly important in the context of existing private multi-family buildings, where the feasibility of interventions is often conditioned by ownership structures, financial resources, and the capacity of residents to invest over time. Framing the strategies in this way allows them to serve both as an integrated long-term vision and as a set of modular actions that can be adopted step by step, according to priorities and available resources.

Table 6 summarizes the four proposed strategies, comparing the state of the art with the design assumptions.

Table 6. Intervention strategies.

	Strategy	State of the Art	Design Assumptions
(1)	Change in the size of the apartments	1 five-room flat 2 four-room flats 3 three-room flats	2 three-room flats 2 two-room flats 5 studio flats
(2)	Opaque envelope	U_{op} = 0.816 W/(m^{2}K)	U_{op} = 0.208 W/(m^{2}K)
(3)	Transparent envelope	U_{tra} = 5.70 W/(m^{2}K)	U_{tra} = 0.88 W/(m^{2}K)
(4)	Installation of a photovoltaic system	Dependence on the electricity grid 100%	Dependence on the electricity grid 59% electricity production 10.1 kWh/m^{2}a

Using the updated transmission coefficients, the heating demand was recalculated according to UNI 11300 [45], assuming a heating period of 183 days (October–April). The results, reported in Appendix B, show a reduction from 154.6 kWh/m^2 in the current state to 51.4 kWh/m^2 after the interventions, corresponding to a 67% improvement.

Combined with photovoltaic production, this reduction indicates that households currently identified as energy poor, slightly above the 10% threshold, with excess values ranging between 0.2% and 2.3% could potentially be brought below the critical limit. It is worth stressing that apartment reconfiguration within private ownership boundaries represents an intervention that allows owners to directly adapt their dwelling to personal needs (e.g., creating a studio to rent out). Such transformations thus hold significant potential in private buildings, as they can be undertaken independently by individual households.

5.2. Critical Issues and Limitations

While the proposed interventions demonstrate clear potential in mitigating EP, several critical issues and limitations should be acknowledged.

First, the reconfiguration of apartments poses legal, financial, and organizational challenges. Although technically feasible without altering the building's structural system, subdivision requires agreement within the private multi-family building assembly, administrative approvals, and potentially high upfront costs. Moreover, the social implications of densifying the building through smaller units, such as increased occupancy or changes in the resident profile, should be carefully considered.

Second, the energy retrofitting measures (insulation of the opaque and transparent envelopes) imply substantial investment. Although they provide the highest energy savings, low-income households, precisely those most affected by EP, may lack the financial resources or access to credit to afford these renovations. Without targeted support mechanisms, such as subsidies or third-party financing, the risk of exclusion remains high.

Third, the installation of a photovoltaic system faces both technical and governance constraints. Roof availability, shading, and multi-family building level decision-making may limit implementation. Furthermore, while self-consumption significantly reduces grid dependence, it does not directly address heating demand, which remains the primary driver of vulnerability.

Finally, it should be emphasized that the calculated reduction in heating demand (67%) and the potential lifting of households out of EP are theoretical projections based on standardized models (UNI 11300) [45]. Real-world outcomes may diverge due to behavioral factors, rebound effects, or variations in energy prices.

In summary, the strategies presented here highlight the technical and social potential for reducing household vulnerability but also underline the need for supportive policy instruments, financial tools, and participatory governance to ensure their feasibility and equitable implementation.

6. Discussion

The findings of this study highlight both the potential and the limitations of adopting a direct, field-based methodology to assess EP at the building scale. Compared to approaches that rely exclusively on aggregated statistical data, this method allows for a more detailed understanding of household conditions, as it integrates socio-economic characteristics, building features, and real consumption profiles. This multidimensional perspective not only measures the incidence of EP but also sheds light on its underlying causes, offering a robust basis for targeted interventions.

A first important outcome is the systematic confirmation that space heating represents the most significant component of household energy expenditure, outweighing all other final uses. While this finding is consistent with the literature, the detailed reconstruction of household energy balances carried out here enriches the analysis by showing how heating demand is conditioned by the interplay between building quality, system efficiency, and user practices. Such nuances are often invisible in macro-scale studies, underlining the added value of micro-scale approaches.

However, several limitations must be acknowledged. The methodology is based on the active cooperation of households and their willingness to provide socio-economic and energy information. In this case study, a participation rate of 55% was achieved, supported by a favorable climate of trust and cohesion. Yet, some households, especially elderly residents, declined to participate, while others withheld utility bills after initially filling in the questionnaire. This partial fragmentation of data inevitably reduces completeness and highlights the practical challenges of field-based research.

Another limitation relates to the data sources. No direct meter readings were collected; instead, consumption was reconstructed from utility bills and self-reported information. While this proved sufficiently robust for analytical purposes, it introduces margins of approximation, particularly in the absence of standardized criteria for allocating centralized heating costs.

Furthermore, the study adopted Boardman's 10% indicator as the primary benchmark for identifying households in EP. While this indicator is widely recognized and allows comparability across contexts, it only captures the relationship between energy expenditure and income. Other relevant dimensions, such as thermal comfort, arrears on energy bills, or service disconnections, were not systematically included. It should be emphasized that the chosen field methodology, based on interviews and questionnaires, makes it difficult to collect such sensitive information in a structured and comparable way, thereby limiting the possibility of adopting more complex, multidimensional indicators.

Despite these constraints, the approach has proven effective not only as a diagnostic tool for local interventions but also as a valuable complement to national and regional statistics. The generation of detailed micro-scale datasets provides a ground-truth reference that can be used to validate and refine macro-scale indicators, which often rely on overly aggregated assumptions. In this way, field-based methodologies contribute to bridging the gap between quantitative measurement and lived household experiences, ensuring that EP policies are designed with greater accuracy and fairness.

Finally, the case study illustrates the operational relevance of this methodology for policy and practice. By combining technical data with socio-economic conditions, it becomes possible to move beyond generalized subsidies and to design more targeted measures. The identification of heating as the main expenditure burden, for example, points directly to the need for interventions on building envelopes and heating systems, ideally supported by financial mechanisms calibrated to household economic capacity. Embedding this type of micro-scale evidence into decision-making processes would not only increase the effectiveness of EP alleviation strategies but also strengthen their social legitimacy.

7. Conclusions

Ensuring a fair energy transition remains one of the key challenges for the European Union. The recent revision of the EPBD (March 2024) explicitly emphasizes the need to tackle EP through structural measures, prioritizing the renovation of buildings with poor energy performance. In this perspective, the availability of reliable and disaggregated data is essential to design interventions that are not only technically effective but also socially fair, targeting the households most at risk of vulnerability.

This study contributes to this debate by presenting a direct, field-based methodology that integrates socio-economic and technical information at the building scale. Applied to a private multi-family building, the approach enabled the collection of detailed data on household composition, income, and energy consumption, complemented by architectural reconstruction of the building and its systems. On this basis, household energy balances were elaborated, and the incidence of EP was measured using Boardman's 10% indicator. The results showed that 31% of households fell into EP, confirming the vulnerability of pre-regulation housing stock.

Beyond diagnosis, the methodology supported the design of a redevelopment scenario including spatial reconfiguration of dwellings, insulation of opaque and transparent envelopes, and the installation of a photovoltaic system. These measures, while requiring significant upfront investment, have the potential to substantially reduce heating demand (−67%) and to promote partial energy self-sufficiency, thereby lifting households slightly above the EP threshold out of this condition.

The study highlights two broader contributions. First, it demonstrates how micro-scale, field-based evidence can provide insights into the socio-technical roots of EP, which are often hidden in aggregated statistical analyses. Second, it shows the operational relevance of linking household-level data with targeted mitigation strategies, thus offering policymakers concrete tools to prioritize structural interventions over temporary income support.

At national and European level, this type of micro-scale evidence can be aligned with existing policy frameworks that aim to identify and prioritize worst-performing buildings and vulnerable households. The methodology proposed in this paper can provide an operational complement to instruments such as the recast EPBD, national long-term renovation strategies and dedicated energy poverty observatories, by supplying building-level data that help to target renovation programs, define eligibility criteria for financial support, and assess whether public incentives effectively reach those most in need.

Nevertheless, the research also acknowledges some limitations. The reliance on household participation implies partial representativeness and occasional data gaps. The use of the Boardman indicator, while ensuring comparability, does not capture non-monetary dimensions of EP such as arrears or comfort levels, which could not be systematically collected through interviews. These constraints underline the need for future studies to experiment with integrated indicators and to broaden the scale of field-based research.

Looking ahead, the methodology can be extended from the scale of a single private multi-family building to larger building complexes, particularly in the context of public housing, where the presence of a single managing body facilitates large-scale renovations. Such an expansion would allow the formulation of incentive schemes that combine technical efficiency with social equity, ensuring that the energy transition is not only environmentally sustainable but also socially inclusive.

Although the empirical application is rooted in the Italian context, the stepwise process, combining household socio-economic data, technical building information and detailed energy balances, is not country-specific and can, in principle, be adapted to other regulatory environments and building stocks.

In conclusion, the proposed approach provides a replicable framework for assessing and addressing EP at the micro scale, while also offering valuable inputs for policy design at the macro level. Embedding such methodologies in policy-making processes would strengthen the fairness, effectiveness, and legitimacy of EP alleviation strategies across Europe.

Author Contributions: Conceptualization, M.F. and E.F.; Methodology, A.L.G., M.F. and E.F.; Software, A.L.G. and G.D.; Validation, A.L.G. and M.F.; Formal analysis, A.L.G.; Investigation, A.L.G. and G.D.; Resources, A.L.G.; Data curation, A.L.G. and G.D.; Writing—original draft, A.L.G. and G.D.; Writing—review & editing, A.L.G., M.F. and E.F.; Visualization, A.L.G. and G.D.; Supervision, M.F. and E.F.; Project administration, E.F.; Funding acquisition, E.F. All authors have read and agreed to the published version of the manuscript.

Funding: This research received no external funding.

Data Availability Statement: The data presented in this study are available on request from the corresponding author due to privacy restrictions and the sensitive nature of household-level information.

Acknowledgments: The present work has been developed within the activities inside the FULL project "Poveri noi: la povertà energetica delle famiglie italiane".

Conflicts of Interest: The authors declare no conflict of interest.

Abbreviations

I_{EPR}	limit income EP risk [€/year]
I	annual family income [€/year]
t_{ep}	EP risk threshold, set to 10% from the Boardman's definition
C_{ENE}	total Energy expenditures for electrical and thermal uses [€]
$C_{ENE,gas}$	sum of consumption of natural gas [€]
Q_h	heating consumption [kWh_g]

Q_{cook}	cooktop consumption [kWh_g]
Q_{DHW}	domestic hot water consumption [kWh_g]
C_{gas}	natural gas cost factor [€/kWh_g]
$C_{ENE,ele}$	sum of consumption dedicated to electricity [€]
E_{app}	electricity consumption dedicated to appliances [kWh_e]
E_{Light}	electricity consumption dedicated to lighting [kWh_e]
E_{DHW}	electricity consumption dedicated to domestic hot water [kWh_e]
C_{ele}	energy cost factor [€/kWh_e]
$E_{p,gl,tot}$	total primary energy [$kWhg/m^2a$]
$E_{P,gl,nren}$	non-renewable component [$kWhg/m^2a$]
$E_{ENE,gas}$	natural gas input energy carrier [kWh_g/m^2a]
$f_{P,nren,gas}$	natural gas non-renewable conversion factor [-]
$f_{P,ren,gas}$	natural gas renewable conversion factor [-]
$f_{P,tot,gas}$	natural gas total conversion factor [-]
$E_{ENE,ele}$	electricity input energy carrier [kWh_e/m^2a]
$f_{P,nren,ele}$	electricity non-renewable conversion factor [-]
$f_{P,ren,ele}$	electricity renewable conversion factor [-]
$f_{P,tot,ele}$	electricity total conversion factor [-]
$U_{op,sa}$	transmittance value of the opaque envelope according to the state of the art [$W/(m^2k)$]
$U_{op,de}$	transmittance value of the opaque envelope according to the state of design [$W/(m^2k)$]
$U_{tra,sa}$	transmittance value of the transparent envelope according to the state of the art [$W/(m^2k)$]
$U_{tra,de}$	transmittance value of the transparent research envelope according to the state of design [$W/(m^2k)$]

Appendix A

This appendix reports all consumption, costs and value of primary energy of households from the case study.

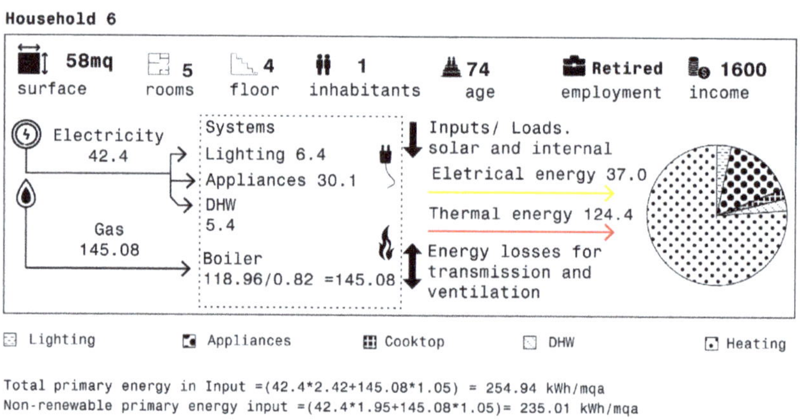

Figure A1. *Cont.*

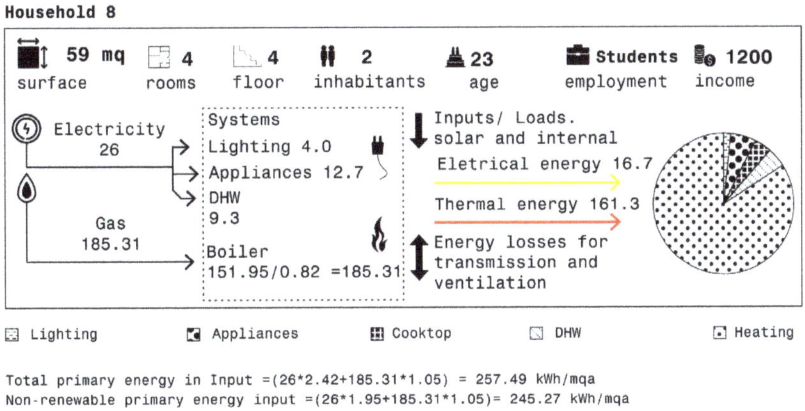

Figure A1. The energy balance of the household 6 and 8.

Table A1. Consumption of electricity, natural gas, and heating is expressed monthly and annually for all households.

		Annual Consumption Average kWh/m²a						
		Electrical Energy			Gas			
Households	Size [m²]	Appliances	Lighting	Total	Cooktop	DHW	Heating Space	Total
1	58	14.9	5.8	20.7 *	16.6 *	0	163.79	180.49
2	72	18.5	1.2	19.7	7.4 *	0	46.82	54.22
3	63	7.3	1.3	8.6	11.9	5.4	53.86	71.16
4	59	19.5	2.9	22.4	9.1 *	0	164.04	173.14
5	45	23.5	5.7	29.24	11.1	28.04	105.20	144.34
6	58	36.0	6.4	42.4	0	0	145.08	145.1
7	54	25.5	2.5	28.0	0	0	161.08	161.1
8	59	22.0	4.0	26.0 *	0	0	185.31	185.31
9	72	5.6	1.9	7.5	6.8	3.0	98.79	107.3
10	118	19.9	6.1	26.0 *	6.3 *	0	102.28	108.6
11	75	5.0	6.0	6.4 *	8.5	6.0	82.74	97.24
12	76	10.8	4.7	15.5	7.1	3.1	81.19	91.39
13	75	11.6	6.2	17.8	12.8	28.2	23.65	64.65
14	80	11.1	3.0	14.1	4.0	2.2	69.72	75.92
15	76	12.7	1.9	14.6	10.0	5.7	141.00	156.70
16	75	13.3	7.2	20.5	9.9 *	2.8 *	128.73	141.43

The data presented with * indicate the absence of electricity or gas bills for that household and the following value estimate.

Table A2. Costs of electricity, gas, and heating expressed monthly and annually for all households.

	Annual Expenditure €/Year			
	Bill Costs			
Household	Electrical Energy	Gas	Heating	Total
1	420.0	337.0	1352.0	2109.0
2	640.0	187.0	571.0	1398.0

Table A2. *Cont.*

	Annual Expenditure €/Year			
	Bill Costs			
Household	Electrical Energy	Gas	Heating	Total
3	550.0	471.0	620.0	1640
4	570.0	187.0	1374.0	2130
5	705.0	180	726.0	1611
6	1167.0	0	1064.0	2231
7	395.0	0	1245.0	1639
8	468.0	0	1492.0	1960
9	389.0	458.0	1355.0	2201
10	749.0	262.0	1191.0	2202
11	168.0	600.0	946.0	1714
12	449.0	377.0	946.0	1772
13	388.0	389.46	365.0	1142
14	633.0	180.0	860.0	1674
15	460.0	357.0	1580.0	2398
16	889.0	337.0	1475.0	2683

Table A3. Average consumption of primary energy.

Average Consumption kWh/m^2	
	Primary Energy
Household 1	239.50
Household 2	104.60
Household 3	95.53
Household 4	236.01
Household 5	222.32
Household 6	254.94
Household 7	236.90
Household 8	257.49
Household 9	130.78
Household 10	176.93
Household 11	162.61
Household 12	133.46
Household 13	110.96
Household 14	113.84
Household 15	199.86
Household 16	198.11

Appendix B

This appendix reports the thermal requirement $Q_{H,nd}$ dedicated to the heating of the case study before and after the application of the intervention.

Heating demand was calculated according to UNI/TS 11300 [45] using a steady-state procedure. A heating set-point temperature of 20 °C was assumed for all residential units, in accordance with national regulatory standards. A constant natural air change rate of

0.5 h^{-1} was applied, and sensible internal heat gains were computed using the normative relationship between useful floor area and internal gains reported in UNI/TS 11300 [45]. The heating period was fixed at 183 days (October–April).

Table A4. State of the art.

Months	$Q_{H,Tr}$ [kWh]	$Q_{H,ve}$ [kWh]	$Q_{H,los}$ [kWh]	Q_{int} [kWh]	$Q_{sol,w}$ [kWh]	Q_{gain} [kWh]	$\eta_{H\,gn}$ [-]	$Q_{H,nd}$ [-]
January	4393.8	8966.4	52,160.2	334.8	13,736.26	14,071.06	0.776	41,239.32
February	34,919.6	7275.8	42,195.5	302.4	15,616.63	15,919.03	0.714	30,824.91
March	26,476.2	5562.0	32,038.2	334.8	18,902.71	18,902.71	0.614	20,235.57
April	8511.4	1856.1	10,368.5	162	8415.50	8577.50	0.537	5761.48
May	-	-	-	-	-	-	-	-
June	-	-	-	-	-	-	-	-
July	-	-	-	-	-	-	-	-
August	-	-	-	-	-	-	-	-
September	-	-	-	-	-	-	-	-
October	14,330.3	3056.3	17,386.6	280.8	14,354.93	14,635.73	0.533	9589.19
November	29,220.1	6078.6	35,298.7	324	11,221.30	11,545.30	0.742	26,734.10
December	39,952.5	8295.1	48,247.6	334.8	12,929.88	13,264.68	0.773	37,995.34
Total	196,603.9	41,091.4	237,695.1	2074.6	95,177.2	97,251.0	-	172,379.90
								154.6 kWh/m^2

Table A5. State of design.

Months	$Q_{H,Tr}$ [kWh]	$Q_{H,ve}$ [kWh]	$Q_{H,los}$ [kWh]	Q_{int} [kWh]	$Q_{sol,w}$ [kWh]	Q_{gain} [kWh]	$\eta_{H\,gn}$ [-]	$Q_{H,nd}$ [-]
January	22,373.5	8966.4	31,339.4	334.8	13,736.26	14,071.06	0.991	17,394.64
February	18,085.4	7275.8	25,361.2	302.4	15,616.63	15,919.03	0.954	10,011.70
March	13,708.6	5562.0	19,270.7	334.8	18,902.71	19,237.51	0.838	3144.44
April	4401.3	1856.1	6257.3	162	8415.50	8577.50	0.684	388.72
May	-	-	-	-	-	-	-	-
June	-	-	-	-	-	-	-	-
July	-	-	-	-	-	-	-	-
August	-	-	-	-	-	-	-	-
September	-	-	-	-	-	-	-	-
October	7416.0	3056.3	10,472.3	280.8	14,354.93	14,635.73	0.674	607.80
November	15,134.3	6078.6	21,213.0	324	11,221.30	11,545.30	0.980	9901.78
December	20,684.5	8295.1	28,989.6	334.8	12,929.88	13,264.68	0.990	15,853.70
Total	10,813.6	41,090.4	142,904.0	2074.6	95,177.2	97,251.0	-	57,302.77
								51.4 kWh/m^2

References

1. Fragkos, P.; Kanellou, E.; Konstantopoulos, G.; Nikas, A.; Fragkiadakis, K.; Filipidou, F.; Fotiou, T.; Doukas, H. Energy Poverty and Just Transformation in Greece. In *Vulnerable Households in the Energy Transition*; Springer: Berlin/Heidelberg, Germany, 2023; pp. 235–267. [CrossRef]
2. Priesmann, J.; Spiegelburg, S.; Madlener, R.; Praktiknjo, A. Does renewable electricity hurt the poor? Exploring levy programs to reduce income inequality and energy poverty across German households. *Energy Res. Soc. Sci.* **2022**, *93*, 102812. [CrossRef]

3. Carfora, A.; Scandurra, G. Boosting green energy transition to tackle energy poverty in Europe. *Energy Res. Soc. Sci.* **2024**, *110*, 103451. [CrossRef]
4. Widuto, A. Energy poverty in the EU. *Eur. Parliam. PE* **2022**, *733*, 583.
5. Simcock, N.; Walker, G.; Day, R. Fuel poverty in the UK: Beyond heating? *People Place Policy Online* **2016**, *10*, 25–41. [CrossRef]
6. Bouzarovski, S.; Thomson, H. Energy Vulnerability in the Grain of the City: Toward Neighborhood Typologies of Material Deprivation. *Ann. Am. Assoc. Geogr.* **2017**, *108*, 695–717. [CrossRef]
7. Karpinska, L.; Śmiech, S. Multiple faces of poverty. Exploring housing-costs-induced energy poverty in Central and Eastern Europe. *Energy Res. Soc. Sci.* **2023**, *105*, 103273. [CrossRef]
8. Oliveras, L.; Artazcoz, L.; Borrell, C.; Palència, L.; López, M.J.; Gotsens, M.; Peralta, A.; Marí-Dell'oLmo, M. The association of energy poverty with health, health care utilisation and medication use in southern Europe. *SSM-Popul. Health* **2020**, *12*, 100665. [CrossRef]
9. Parliament, E. *Directive 2012/27/EU on Energy Efficiency, Amending Directives 2009/125/EC and 2010/30/EU and Repealing Directives 2004/8/EC and 2006/32/EC*; European Union: Brussels, Belgium, 2012.
10. Parliament, E. *Directive (EU) 2018/844 Amending Directive 2010/31/EU on the Energy Performance of Buildings and Directive 2012/27/EU on Energy Efficiency*; European Union: Brussels, Belgium, 2018.
11. European Parliament. Directive 2009/72/EC of the European Parliament and of the Council of 13 July 2009 Concerning Common Rules for the Internal Market in Electricity and Repealing Directive 2003/54/EC. Available online: https://eur-lex.europa.eu/eli/dir/2009/72/oj/eng (accessed on 12 August 2025).
12. Commission, E. *Recommendation (UE) 2023/2407 Commission of 20 October 2023. On Energy Poverty*; European Union: Brussels, Belgium, 2023.
13. Robić, S.; Ančić, B. Exploring health impacts of living in energy poverty: Case study Sisak-Moslavina County, Croatia. *Energy Build.* **2018**, *169*, 379–387. [CrossRef]
14. Grazini, C. Energy poverty as capacity deprivation: A study of social housing using the partially ordered set. *Socio-Econ. Plan. Sci.* **2024**, *92*, 101843. [CrossRef]
15. Simionescu, M.; Radulescu, M.; Cifuentes-Faura, J.; Balsalobre-Lorente, D. The role of renewable energy policies in TACKLING energy poverty in the European UNION. *Energy Policy* **2023**, *183*, 113826. [CrossRef]
16. Osservatorio Italiano sulla Povertà Energetica (OIPE). *La Povertà Energetica in Italia Nel 2021*; OIPE: Padova, Italy, 2022.
17. Hwang, I.C.; Baek, J.-R. Energy poverty in Seoul: Current status and the effect of government supports. *World Dev. Sustain.* **2025**, *7*, 100253. [CrossRef]
18. Marcoje, B.M.; Hallack, M.; Schiavon, L.d.C.; Machado, D.C.; de Abreu, M.W. Energy Poverty in Brazil: A look at access to energy services. *Dev. Sustain. Econ. Financ.* 2025; *in press*. [CrossRef]
19. Bouzarovski, S.; Petrova, S.; Sarlamanov, R. Energy poverty policies in the EU: A critical perspective. *Energy Policy* **2012**, *49*, 76–82. [CrossRef]
20. Bollino, C.A.; Botti, F. Energy Poverty in Europe: A Multidimensional Approach. *PSL Q. Rev.* **2017**, *70*, 283.
21. Faiella, I.; Lavecchia, L. Energy poverty. How can you fight it, if you can't measure it? *Energy Build.* **2021**, *233*, 110692. [CrossRef]
22. Grant Agreement No. 889385. ENPOR—Actions to Mitigate Energy Poverty in the Private Rented Sector, (n.d.). 2020–2023. Available online: https://cordis.europa.eu/project/id/889385 (accessed on 8 August 2025).
23. Moore, R. Definitions of fuel poverty: Implications for policy. *Energy Policy* **2012**, *49*, 19–26. [CrossRef]
24. Ntaintasis, E.; Mirasgedis, S.; Tourkolias, C. Comparing different methodological approaches for measuring energy poverty: Evidence from a survey in the region of Attika, Greece. *Energy Policy* **2019**, *125*, 160–169. [CrossRef]
25. Boardman, B. *Fuel Poverty: From Cold Homes to Affordable Warmth*; Belhaven Press: London, UK, 1991.
26. Boardman, B. *Fixing Fuel Poverty: Challenges; Solutions; Earthscan*; Belhaven Press: London, UK, 2013. [CrossRef]
27. Papada, L.; Kaliampakos, D. Measuring energy poverty in Greece. *Energy Policy* **2016**, *94*, 157–165. [CrossRef]
28. Guevara, Z.; Mendoza-Tinoco, D.; Silva, D. The theoretical peculiarities of energy poverty research: A systematic literature review. *Energy Res. Soc. Sci.* **2023**, *105*, 103274. [CrossRef]
29. Simoes, S.G.; Gregório, V.; Seixas, J. Mapping Fuel Poverty in Portugal. *Energy Procedia* **2016**, *106*, 155–165. [CrossRef]
30. Camboni, R.; Corsini, A.; Miniaci, R.; Valbonesi, P. Mapping fuel poverty risk at the municipal level. A small-scale analysis of Italian Energy Performance Certificate, census and survey data. *Energy Policy* **2021**, *155*, 112324. [CrossRef]
31. Fabbri, K. *Urban Fuel Poverty*; Elsevier: Amsterdam, The Netherlands, 2019. [CrossRef]
32. Fabbri, K.; Gaspari, J. Mapping the energy poverty: A case study based on the energy performance certificates in the city of Bologna. *Energy Build.* **2021**, *234*, 110718. [CrossRef]
33. Desvallées, L. Low-carbon retrofits in social housing: Energy efficiency, multidimensional energy poverty, and domestic comfort strategies in southern Europe. *Energy Res. Soc. Sci.* **2022**, *85*, 102413. [CrossRef]
34. Castaño-Rosa, R.; Solís-Guzmán, J.; Marrero, M. Energy poverty goes south? Understanding the costs of energy poverty with the index of vulnerable homes in Spain. *Energy Res. Soc. Sci.* **2020**, *60*, 101325. [CrossRef]

35. Ruiz-Rivas, U.; Tirado-Herrero, S.; Castaño-Rosa, R.; Martínez-Crespo, J. Disconnected, yet in the spotlight: Emergency research on extreme energy poverty in the Cañada Real informal settlement, Spain. *Energy Res. Soc. Sci.* **2023**, *102*, 103182. [CrossRef]
36. Ballesteros-Arjona, V.; Oliveras, L.; Muñoz, J.B.; Lima, A.O.d.L.; Carrere, J.; Ruiz, E.M.; Peralta, A.; León, A.C.; Rodríguez, I.M.; Daponte-Codina, A.; et al. What are the effects of energy poverty and interventions to ameliorate it on people's health and well-being?: A scoping review with an equity lens. *Energy Res. Soc. Sci.* **2022**, *87*, 102456. [CrossRef]
37. Grdenić, G.; Delimar, M.; Robić, S. Framing the context of energy poverty in Croatia: A case-study from Zagreb. *Energy Policy* **2020**, *147*, 111869. [CrossRef]
38. Menyhért, B. Energy poverty in the European Union. The art of kaleidoscopic measurement. *Energy Policy* **2024**, *190*, 114160. [CrossRef]
39. Briot-Arthur, S.; Fournier, V.; Lee, B. Quantifying energy poverty vulnerability with minimal data–A building energy simulation approach. *Energy Build.* **2024**, *309*, 114062. [CrossRef]
40. Matos, A.M.; Delgado, J.M.P.Q.; Guimarães, A.S. Linking Energy Poverty with Thermal Building Regulations and Energy Efficiency Policies in Portugal. *Energies* **2022**, *15*, 329. [CrossRef]
41. Fabbri, K.; Marchi, L.; Antonini, E.; Gaspari, J. Exploring the Role of Building Envelope in Reducing Energy Poverty Risk: A Case Study on Italian Social Housing. *Energies* **2023**, *16*, 8093. [CrossRef]
42. Palladino, D.; Pagliaro, F.; Del Fatto, V.; Lavinia, C.; Margiotta, F.; Colasuonno, L. *Stato Dell'arte del Patrimonio Edilizio Nazionale e Analisi Degli Attestati di Prestazione Energetica (APE)*; Ministero dello Sviluppo Economico: Rome, Italy, 2021.
43. Horta, A.; Gouveia, J.P.; Schmidt, L.; Sousa, J.C.; Palma, P.; Simões, S. Energy poverty in Portugal: Combining vulnerability mapping with household interviews. *Energy Build.* **2019**, *203*, 109423. [CrossRef]
44. Ministro dello Sviluppo Economico. Ministro dell'Ambiente e della Tutela del Territorio e del Mare, Ministro delle Infrastrutture e dei Trasporti, DM 26-05-2015 Requisiti Minimi. 2015. Available online: http://www.ing-russo.it/norme/energia/DM26_6_2015_requisiti_minimi.pdf (accessed on 12 August 2025).
45. *UNI/TS 11300-1*; Energy Performance of Buildings—Part 1: Evaluation of Energy Need for Space Heating and Cooling. Ente Italiano di Normazione (UNI): Milan, Italy, 2014.

energies

MDPI

Article

Analysis of Useful Energy Demand for Heating Purposes in a Building with a Self-Supporting Polystyrene Structure in a Temperate Climate

Krzysztof Wąs [1,*], Grzegorz Nawalany [1] and Miroslav Žitňák [2]

[1] Department of Rural Building, Faculty of Environmental Engineering, University of Agriculture in Krakow, al. Mickiewicza 24/28, 30-059 Krakow, Poland; grzegorz.nawalany@urk.edu.pl

[2] Faculty of Engineering, Institute of Agricultural Engineering, Transport and Bioenergetics, Slovak University of Agriculture in Nitra, Trieda Andreja Hlinku 2, 94976 Nitra, Slovakia; miroslav.zitnak@uniag.sk

* Correspondence: krzysztof.was@urk.edu.pl

Abstract

This article presents an analysis of the useful energy demand for heating purposes in a single-family, single-storey building with a self-supporting polystyrene structure, which is a relatively niche solution, in relation to a traditional masonry structure with similar partition thickness. The structures considered met the requirements for passive buildings. The analysis was performed on three locations in Europe with a temperate climate, i.e., Kołobrzeg in Poland, Vienna in Austria, and Essen in Germany. The research showed significant savings in the energy demand of the polystyrene structure compared to the masonry structure for each location, ranging from 38% to 52%. Similarly, the heating period was 21% to 38% shorter in individual locations. This shows that polystyrene construction allows for a significant reduction in building energy demand, leading to lower operating costs.

Keywords: polystyrene construction; useful energy demand; energy efficient construction; passive building

1. Introduction

The construction industry is facing a series of challenges, with the primary concern being the reduction in energy demand in response to increasing energy costs. According to the IEA (International Energy Agency), the construction sector is responsible for 30% of global energy consumption [1]. The observed increase in energy prices is associated with economic development, a process that requires substantial energy consumption, as well as with the ongoing energy transition, characterized by a gradual shift away from fossil fuels, whose reserves are constantly diminishing. Consequently, it is necessary to implement solutions that will reduce energy demand across a variety of sectors, including construction. The present age is witnessing an intensive effort to identify novel technologies and materials that exhibit high energy efficiency, a development that is of supreme importance to the field of construction.

Energy-efficient construction focuses on minimizing energy consumption through the implementation of various strategies. These include the utilization of partitions that restrict heat losses, the employment of materials with superior thermal insulation properties, and the incorporation of active systems such as mechanical ventilation with heat recovery. The efficacy of these systems in reducing ventilation loss is substantial. The implementation of

https://doi.org/10.3390/en18246514

renewable energy systems is a common practice in such contexts. An example of energy-efficient construction is the so-called passive house (PH) developed by the Passive House Institute in Darmstadt. The foundation of this approach is rooted in five fundamental principles: high-quality insulation of partitions with a heat transfer coefficient (U-value) below 0.15 W/(m^2K), windows with a heat transfer coefficient of less than 0.8 W/(m^2K), ventilation with heat recovery values greater than 75%, a building's air tightness being no greater than 0.6 air changes per hour, and the elimination of thermal bridges [2]. A comprehensive review of the principles and technologies employed in passive construction is provided in [3], while studies delineating energy consumption in this category of buildings are presented in [4–7].

It is evident that thermal insulation of partitions constitutes a key element in the design and construction of low-energy buildings. An analysis of the insulation properties of inorganic materials, such as concrete and brick, and organic materials, such as wood and straw, is presented in [8]. The available technologies easily enable the achievement of the required heat transfer coefficient values, as stipulated by the passive building standard, in the context of residential construction. Masonry structures, when employed in conjunction with the optimal insulation thickness, have been demonstrated to meet stringent requirements. Analyses of such a structure can be found in [9,10]. However, due to the total thickness of the partition, it may be necessary to abandon ceramic or silicate materials in favor of aerated concrete, which has better insulating properties, or to use insulating materials that have higher parameters yet are more expensive. In the context of single-family houses characterized by elevated energy standards, it is more common to find frame structures that allow for better insulation parameters with a smaller partition thickness, due to the fact that the thermal insulation fills the structural element of the partition. The insulation properties of various types of frame partitions are described in the article by J. Kosny et al. [11]. This research indicates that such structures can achieve U values in the range of 0.30–0.11 W/(m^2K). A comprehensive review of the latest technological advancements in wooden partition wall systems, with a focus on enhancing energy efficiency, is presented in [12]. Research on passive buildings with different structures is presented in [13–15].

Another technology that can be used in the construction of low-energy buildings is Insulated Concrete Forms (ICFs), where polystyrene forms serve both as an insulating layer and as a framework for the reinforced concrete layer, which is the load-bearing element of the structure. Such structures are distinguished by their enhanced thermal capacity in comparison to frame structures. Research on insulated concrete form (ICF) partitions in two variants—with solid concrete filling and with a frame structure, as described in publication [16]—demonstrated a minor difference with regard to thermal insulation properties. At the same time, the superior capacity of solid partitions to stabilize internal conditions was highlighted. An analysis of the use of this type of wall for energy storage is presented in [17,18], while the use of ICF foundations for energy storage is described in [19]. An analysis of different types of ICF partition cores and their impact on energy consumption reduction and strength is described in [20].

Another significant aspect pertains to the comparison of diverse technologies for the construction of energy-efficient buildings. In [21], a multi-criteria comparison of technologies was carried out. As demonstrated in [22], a comparison of energy consumption in ICF and timber frame buildings reveals that ICF buildings exhibit a reduced energy demand. However, it is imperative to note that the utilization of materials characterized by specific thermal parameters exerts a substantial influence on this outcome. In turn, Ref. [15] underscores the significance of environmental impacts in addition to the insulation parameters of a given technology. A comprehensive comparison of timber and concrete technologies, with consideration for the building lifecycle, is presented in [23].

Another type of construction that can use polystyrene in partition design is a Structural Insulated Panel (SIP) system. A comprehensive overview of this technology is provided in [24], while its utilization in the construction of nZEB (nearly zero energy building) and buildings of PH standard is elaborated in [25].

The technologies described above enable the construction of buildings that are highly energy-efficient. However, there is a clear gap in research on the subject of self-supporting polystyrene structures. Despite the growing interest in technologies that improve energy efficiency of buildings through the use of new materials or alternative uses of existing materials, there is currently a lack of systematic analyses of the energy requirements of this niche structure. Hence, the authors proposed to conduct such analyses in order to initiate a broader discussion covering other aspects related to microclimate, comfort, and durability prospectively. The ultimate aim of this discussion is to highlight the possibilities and limitations of polystyrene structures.

Contemporary analyses of the energy demand of buildings are based on various computer tools that have been validated to enable effective simulations of heat and moisture transport in both building envelopes and entire buildings. One such software product is WUFI®Plus 3.5, developed by the Fraunhofer Institute for Building Physics [26]. The tool's validation is documented in the following publications: [27–29]. The WUFI®Plus tool has been employed in a variety of residential, agricultural and historical building types [30–34].

2. Materials and Methods

2.1. Research Object

The aim of the research is to analyze useful energy demand for heating in a single-family, single-storey building with a polystyrene structure, which is a relatively niche construction solution. The building's partitioning is composed of self-supporting blocks of compressed polystyrene, which adopt an arch shape, seamlessly integrating with the external walls and the roof surface. The constituent modules are interconnected, as illustrated in Figure 1. This cross-section is notable for the absence of a wooden or steel frame, a feature common to frame structures, and the lack of a concrete layer, which is characteristic of ICF technology. Consequently, the polystyrene foam functions both as a structural and insulating layer. This innovative solution facilitates the comprehensive utilization of the partition cross-section for the purpose of enhancing the thermal insulation of the building. It also eliminates the occurrence of thermal bridges, which are characteristic of other structural types, due to the seamless integration of external walls with the roof slope.

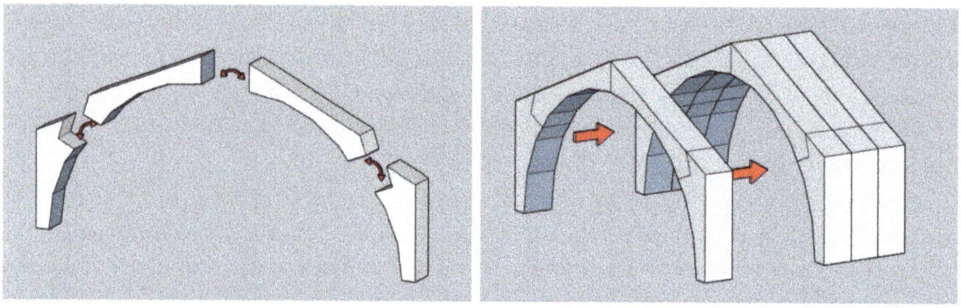

Figure 1. Self-supporting polystyrene elements.

In the technology under discussion, masonry structures are utilized exclusively in the construction of gable walls, with the primary function of these walls being to enhance the structural integrity and rigidity of the building as a whole. Furthermore, masonry walls have been shown to possess an accumulative function, thereby buffering thermal energy within themselves. This property stands in contrast to that of polystyrene walls, which have been demonstrated to possess low thermal capacity. As a result, such a structure is being considered a hybrid structure. The diagram is shown in the graphic (Figure 2).

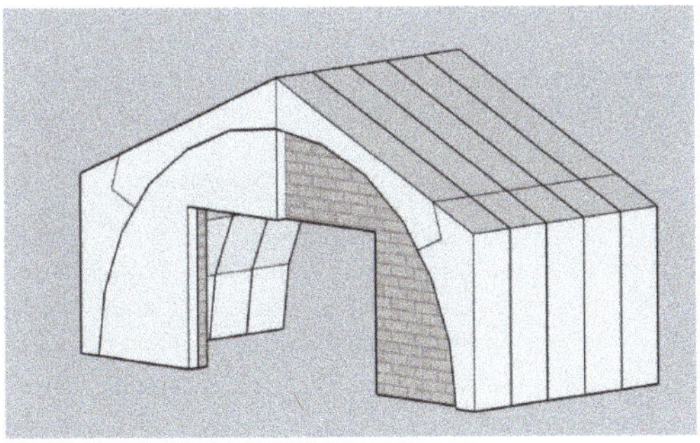

Figure 2. Gable wall of the building.

The polystyrene partitions of the building are reinforced on both sides with fiberglass mesh. Aluminum profiles are utilized internally for the purpose of securing plasterboard panels. Externally, the roof surface is reinforced with waterproof plywood strips to support the installation of steel sheet roofing. The building is founded on a conventional foundation slab, where a layer of lean concrete is initially applied, followed by a layer of thermal insulation composed of compacted polystyrene, construction foil, and subsequently, a reinforced concrete slab and a leveling layer are poured. The entire building is founded on this prepared base Figure 3. Due to its construction, the floor on the ground performs an accumulation function, similar to the gable walls. The construction of the building on a foundation slab insulated from below and the continuity of insulation serve to eliminate the formation of typical thermal bridges at partition joints.

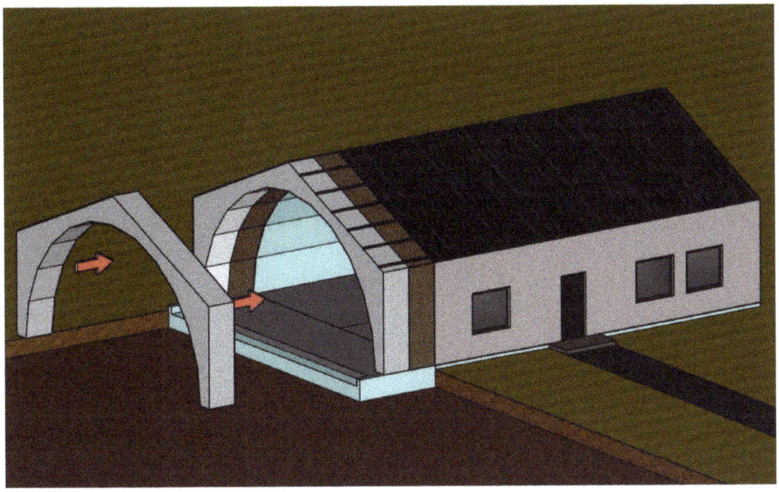

Figure 3. Diagram of the polystyrene building structure.

2.2. Assumptions for Calculations

The calculations are based on a generic single-storey detached house with a usable area of 106 m^2 (Figure 4). The floor plan of the building (Figure 5) is based on the letter 'L'.

The total dimensions of the building are 12.96 m × 12.96 m, and its height is 4.5 m. The net volume of the building is 177 m^3. The total window area measures 21.61 m^2. The roof pitch measures 27°. The building partitions and window frames included in the analysis meet the requirements for passive buildings; i.e., the heat transfer coefficient for walls is $U < 0.15$ W/m^2K, and that for windows is $U < 0.8$ W/m^2K.

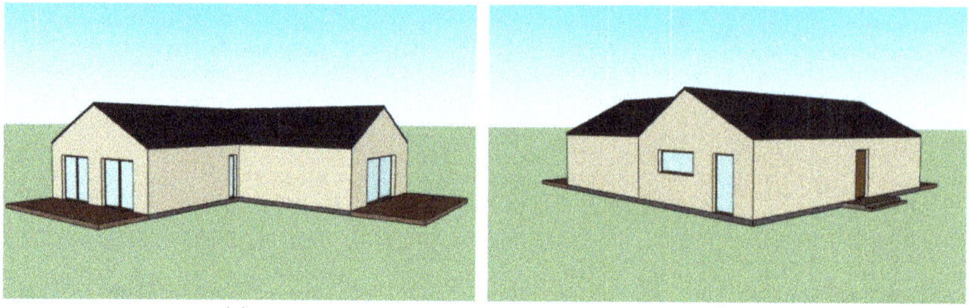

(a) (b)

Figure 4. Building model: (**a**) South-west elevation, (**b**) North-east elevation.

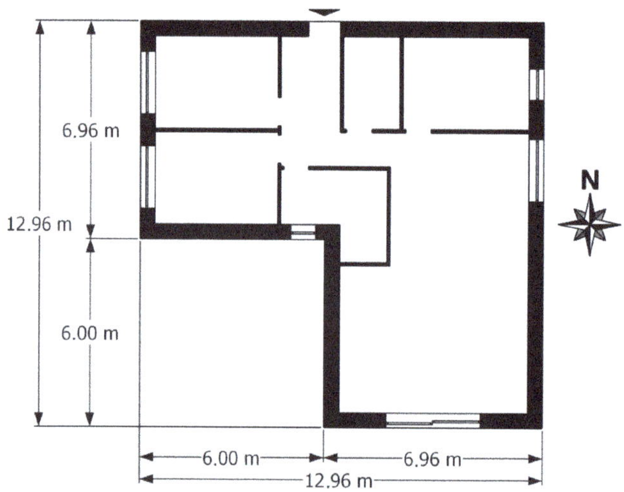

Figure 5. Plan of the analyzed polystyrene building.

The calculations of the useful energy demand for heating purposes were performed using the WUFI®Plus 3.5 calculation platform based on a coupled heat and moisture exchange model for non-stationary building analyses. The simulations were conducted on an annual basis, with a 1-h time step. In order to facilitate the process of simulation and comparison with traditional masonry partition walls, the arched shape of the partitions was simplified to a flat form. The thickness of the polystyrene structure ranges from 0.30 to 0.85 m, and thus its cross-section was divided into a total of 180 segments. The mean thickness of the polystyrene cross-section employed for subsequent calculations was determined to be 0.48 m. A schematic representation of the simplified arched cross-section is presented in Figure 6. Simulations frequently require simplification of the building geometry due to limitations of the calculation tool or the requirement to combine different structures. It has been demonstrated that this may result in disparities in the outcomes when compared to the original geometry. Consequently, prior to conducting additional simulations, it is imperative to estimate the potential impact of the simplifications employed on the calculations. In this analysis, the arched polystyrene wall was simplified to a flat form, characteristic of most buildings. The non-uniform thickness of the material may result in the formation of local thermal bridges in cross-section constructions when compared to the average variant. In order to estimate potential discrepancies in the relevant simulations, both variants were analyzed. It was demonstrated that the simplification of the cross-section generates an error of 2.45%, thereby limiting heat exchange in comparison to the original partition. This low discrepancy confirmed the high consistency between the analyzed cross-sections and allowed for effective simulations to be carried out.

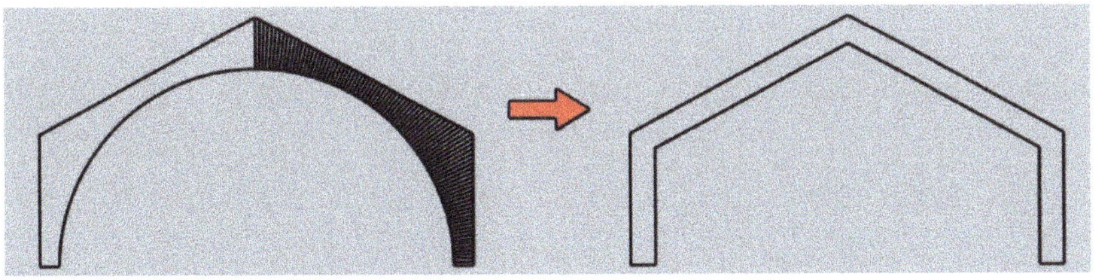

Figure 6. Simplification of the cross-section for simulation.

The analysis incorporated a total of six calculation variants. A comparative study was conducted for two structures: polystyrene and masonry structure. The study was conducted for three locations characterized by a temperate climate.

In order to compare the construction variants analyzed, the thickness of the external walls and roof slopes was set at 0.5 m. The layers of individual partitions are presented in the tables below. The layer arrangement for the polystyrene wall is shown in Table 1, while the heat transfer coefficient U for this structure is 0.073 W/m^{2}K. The layers of the masonry structure are shown in Table 2, and the heat transfer coefficient was 0.12 W/m^{2}K.

Table 1. Polystyrene wall—layers.

Layers	Thickness [m]	Thermal Conductivity [W/m·K]
Plasterboard	0.0125	0.2
Fibreglass mesh	0.0005	0.035
Compacted polystyrene	0.48	0.036
Fibreglass mesh	0.0005	0.035
External plaster	0.01	0.87

Table 2. Masonry wall—layers.

Layers	Thickness [m]	Thermal Conductivity [W/m·K]
Interior plaster	0.01	0.2
Aerated concrete	0.25	0.14
Compacted polystyrene	0.23	0.036
External plaster	0.01	0.87

With regards to roof construction, the layer arrangement for a polystyrene building is presented in Table 3, where the thermal conductivity coefficient U was 0.074 W/m^{2}K. In the case of the masonry structure, a rafter roof construction was adopted (Table 4), where U was 0.095 W/m^{2}K.

Table 3. Roof of polystyrene structure—layers.

Layers	Thickness [m]	Thermal Conductivity [W/m·K]
Sheet metal	0.0005	60
Fibreglass mesh	0.001	0.035
Compacted polystyrene	0.48	0.036
Fibreglass mesh	0.001	0.035
Plasterboard	0.0125	0.2

Table 4. Rafter roof structure—layers.

Layers	Thickness [m]	Thermal Conductivity [W/m·K]
Sheet metal	0.0005	60
Air layer	0.12	0.035
Vapour barrier foil	0.001	2.3
Mineral wool/rafter	0.16	0.036
Mineral wool	0.2	0.036
Vapour barrier foil	0.001	2.3
Plasterboard	0.0125	0.2

The foundation slab on which the building was constructed was the same for both structures. The layers are delineated in Table 5, and the heat transfer coefficient U for this partition was determined to be 0.087 W/m^2K.

Table 5. Foundation slab—layers.

Layers	Thickness [m]	Thermal Conductivity [W/m·K]
Concrete screed	0.0005	1.6
Reinforced concrete slab	0.2	1.6
Compacted polystyrene	0.4	0.036
Construction film	0.0001	2.3
Lean concrete	0.1	1.6

2.3. Indoor Climate

In the course of the simulations, the internal temperature was assumed to be 20 °C. This value denotes the minimum temperature for rooms. Once exceeded, the instantaneous demand necessary to achieve the above-mentioned value within a given time step is calculated. It was further hypothesized that the building is inhabited by a family of four. On this basis, a scenario of internal heat gains was adopted according to the WUFI®Plus database for a typical weekday for a family (two adults + two children) including gains from people and appliances. The daily pattern that has been assumed in the calculations is demonstrated in Figure 7. Furthermore, solar radiation gains are calculated on an hourly basis. This calculation is based on the shading of partitions, which takes into account climate data. The ventilation flow in the building was assumed in the calculations at a level of 1 air change per hour.

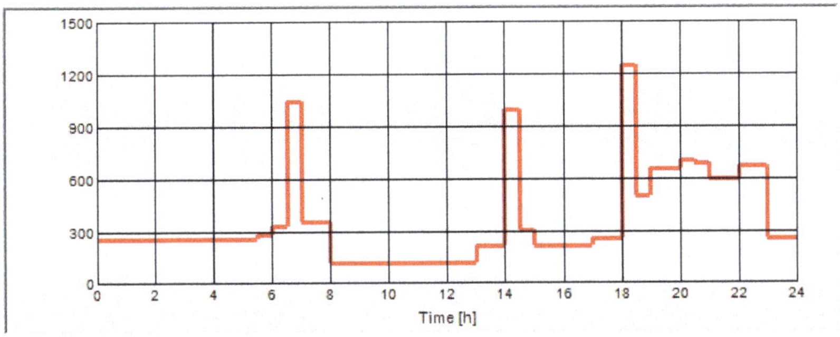

Figure 7. Diagram of internal heat gains [W]—WUFI®Plus database.

2.4. Outdoor Climate

The calculations were performed for a period of one year in three locations with a temperate climate: Kołobrzeg in Poland, Vienna in Austria and Essen in Germany. The assumed locations are illustrated in Figure 8. The climate data employed in the simulations is drawn from the WUFI® Plus database, comprising long-term data that has undergone statistical processing. This data set has been used to derive a reference year for each location, thereby ensuring the consistency and reliability of the data.

Figure 8. Locations assumed in the analysis.

2.5. Kołobrzeg

The climate of Kołobrzeg is classified as temperate maritime. The range of temperatures is from −9.8 to 29.7 °C, with an average temperature of 8.6 °C, which is also the lowest among the locations analyzed. The mean relative humidity is 81.3%. The total radiation is 2253.2 kWh/m^{2}a. The annual precipitation is 560.9 mm/a and the prevailing winds are south-westerly winds, with an average speed of 2.3 m/s. The fundamental climate parameters are illustrated in Figures 9 and 10.

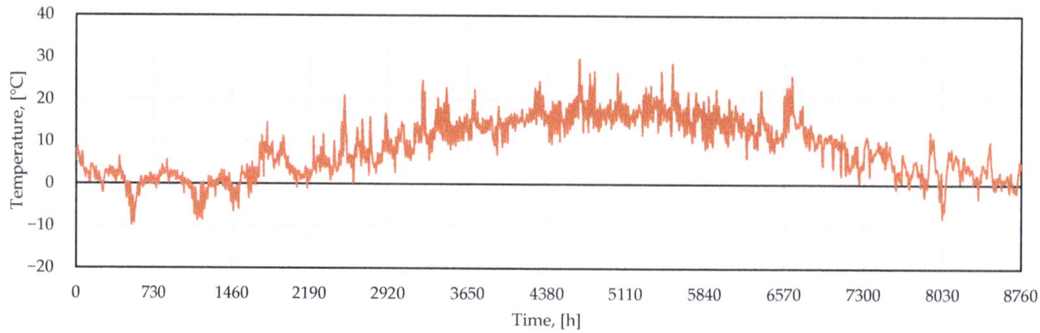

Figure 9. Air temperature—Kołobrzeg.

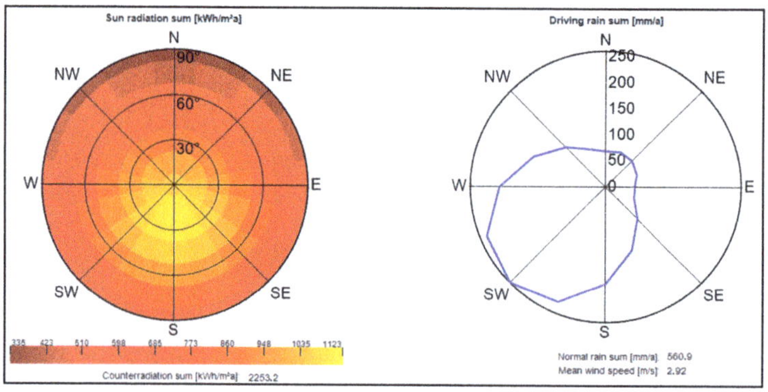

Figure 10. Sun radiation sum and driving rain sum with wind direction—Kołobrzeg.

2.6. Vienna

The climate of Vienna is characterized as temperate transitional with continental influences. The annual temperature range extends from −10.9 to 32.4 °C, with an average temperature of 10.4 °C, representing the highest of the locations analyzed. The mean relative humidity is 73.0%. The total radiation is 2317.2 kWh/m^{2}a. The annual precipitation is 625.1 mm/a. Vienna is subject to westerly and south-easterly winds, with an average speed of 3.6 m/s. The fundamental climate parameters for Vienna are illustrated in Figures 11 and 12.

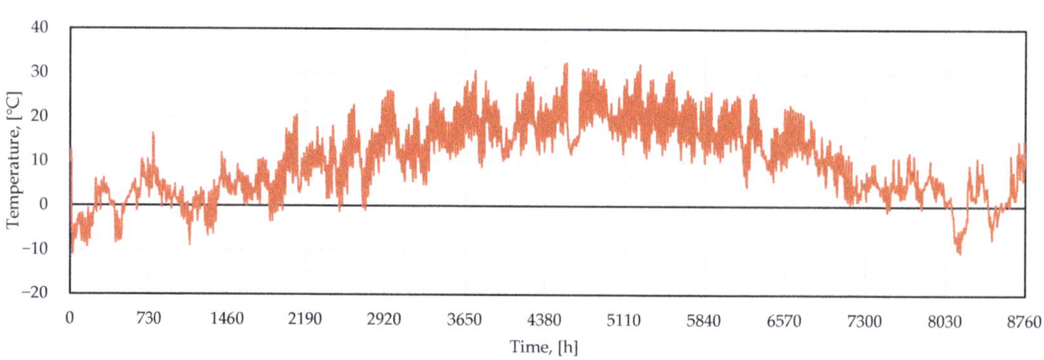

Figure 11. Air temperature—Vienna.

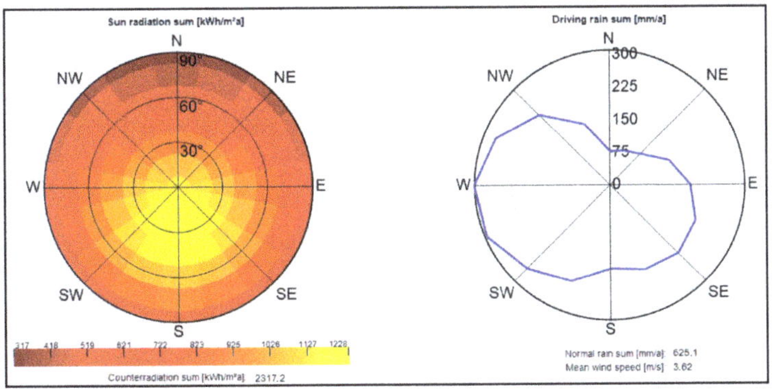

Figure 12. Sun radiation sum and driving rain sum with wind direction—Vienna.

2.7. Essen

The climate in Essen is described as temperate. The annual temperature range is from −9.7 to 30.2 °C, with an average temperature of 9.5 °C. The mean relative humidity is 78.1%. The total radiation is 2434 kWh/m^{2}a. The annual precipitation is 1094.6 mm. Essen is subject to westerly and south-westerly winds, with an average speed of 2.74 m/s. The fundamental climate parameters for Essen are demonstrated in Figures 13 and 14.

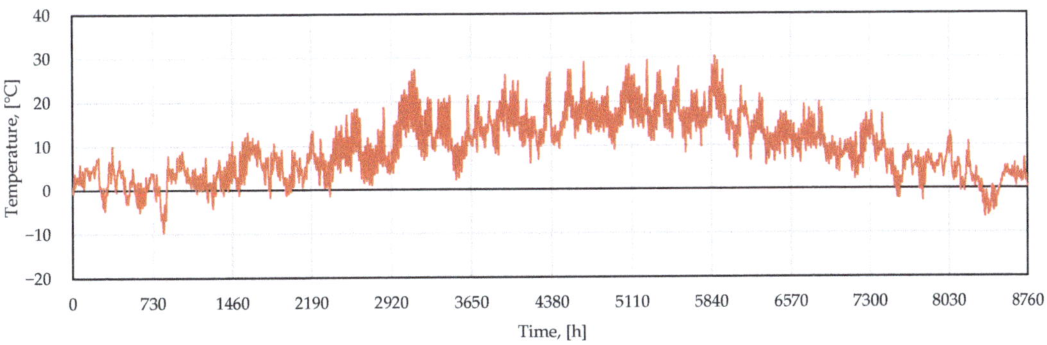

Figure 13. Air temperature—Essen.

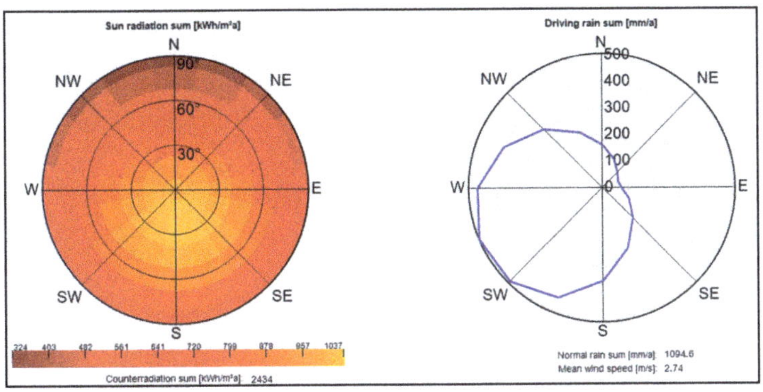

Figure 14. Sun radiation sum and driving rain sum with wind direction—Essen.

3. Results

3.1. Energy Demand—Kołobrzeg

In the first of the analyzed locations, the annual energy demand for heating in the polystyrene building was 1173.7 kWh, while in the masonry structure, the demand was 1851.1 kWh. Hourly data for heating power is presented in Figure 15.

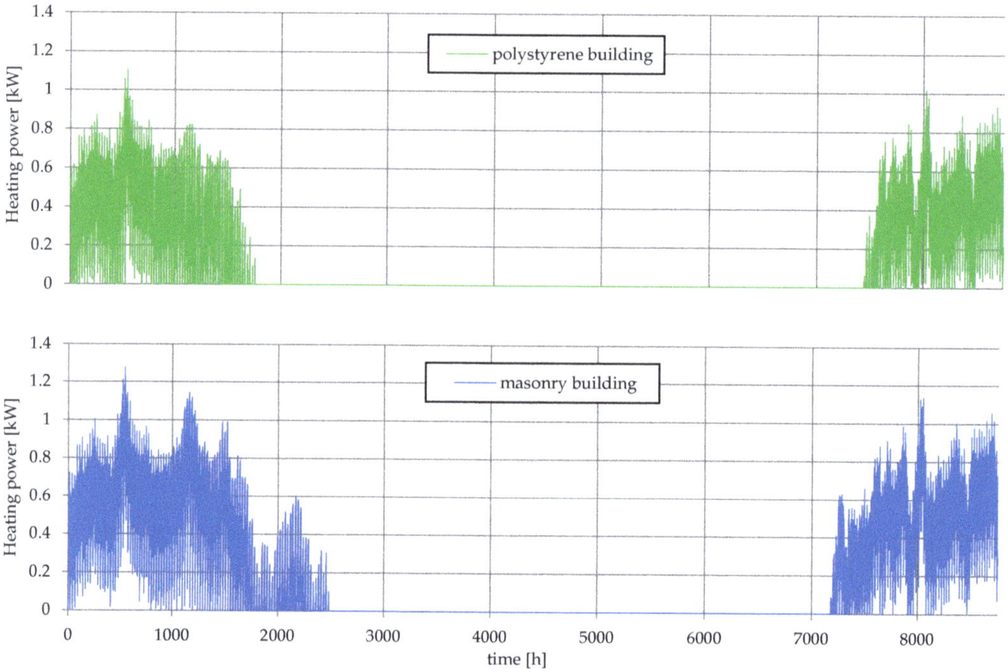

Figure 15. Hourly data for heating power—Kołobrzeg.

The heating season in Kołobrzeg was found to be the longest among the locations analyzed. For the polystyrene structure, the mean duration of heating season was 128 days, and for the masonry structure, it was 169 days. The heating season values presented for Kołobrzeg were calculated based on the hourly energy demand when the temperature inside the building falls below the set limit of 20 °C. It was determined that, in the event of inadequate energy from external sources and internal heat gains, the heating system should be activated in order to supply the building with the requisite heat flow to achieve the set temperature. This approach was replicated in the case of the other locations. Figure 16 illustrates a monthly summary of the useful energy demand for heating purposes for both structures.

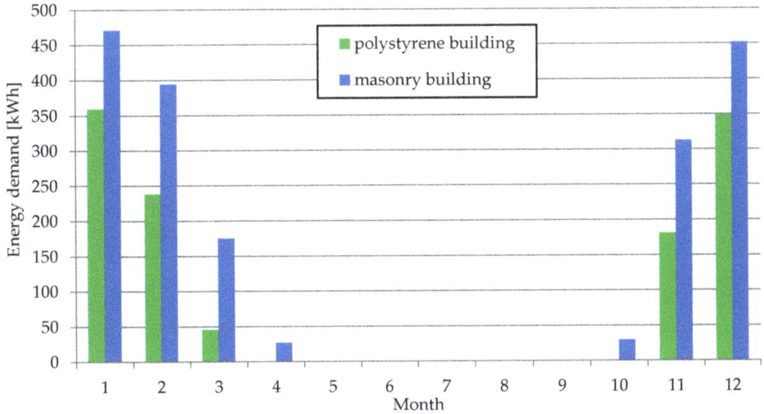

Figure 16. Monthly energy demand in the analyzed structures for Kołobrzeg.

3.2. Energy Demand—Vienna

In the second location analyzed, Vienna, the annual energy demand for heating in the polystyrene building was 803.3 kWh, while in the masonry structure it was 1681.1 kWh. Hourly data for heating power is presented in Figure 17.

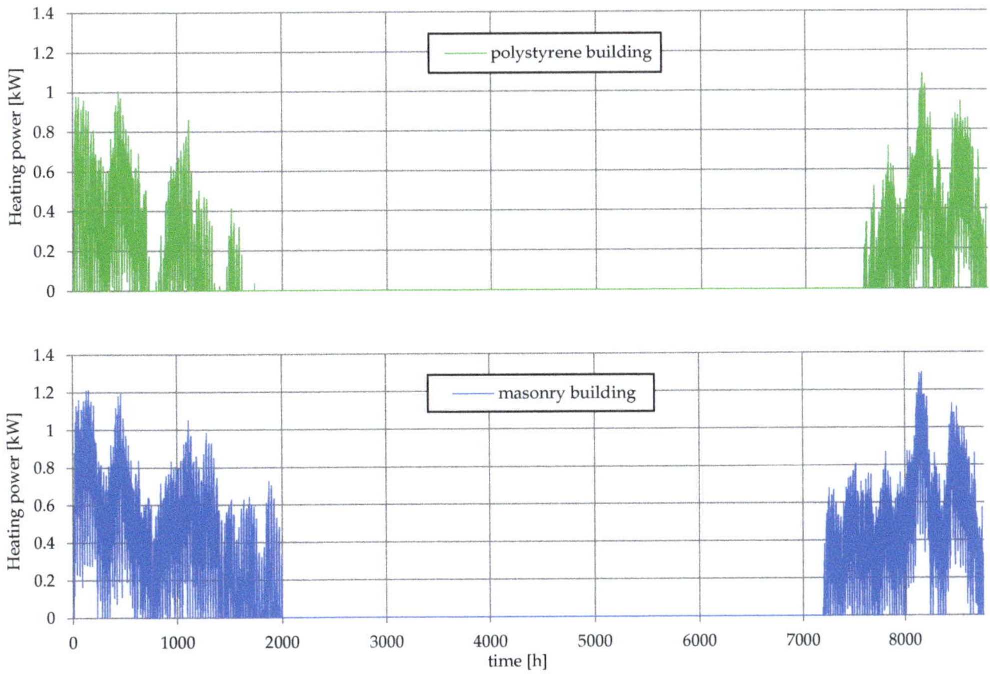

Figure 17. Hourly data for heating power—Vienna.

For Vienna, the heating season lasted 117 days for the polystyrene construction and 149 days for the masonry construction. Figure 18 illustrates the monthly summaries for both constructions.

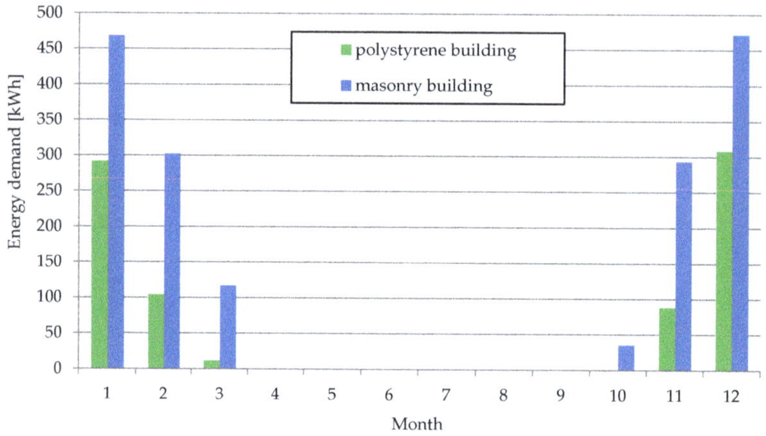

Figure 18. Monthly energy consumption summary for the analyzed structures for Vienna.

3.3. Energy Demand—Essen

In the third location analyzed, in Essen, the annual energy demand for heating in the polystyrene building was 854.9 kWh, while in the masonry structure the demand was 1622.3 kWh. Figure 19 illustrates hourly data for heating power.

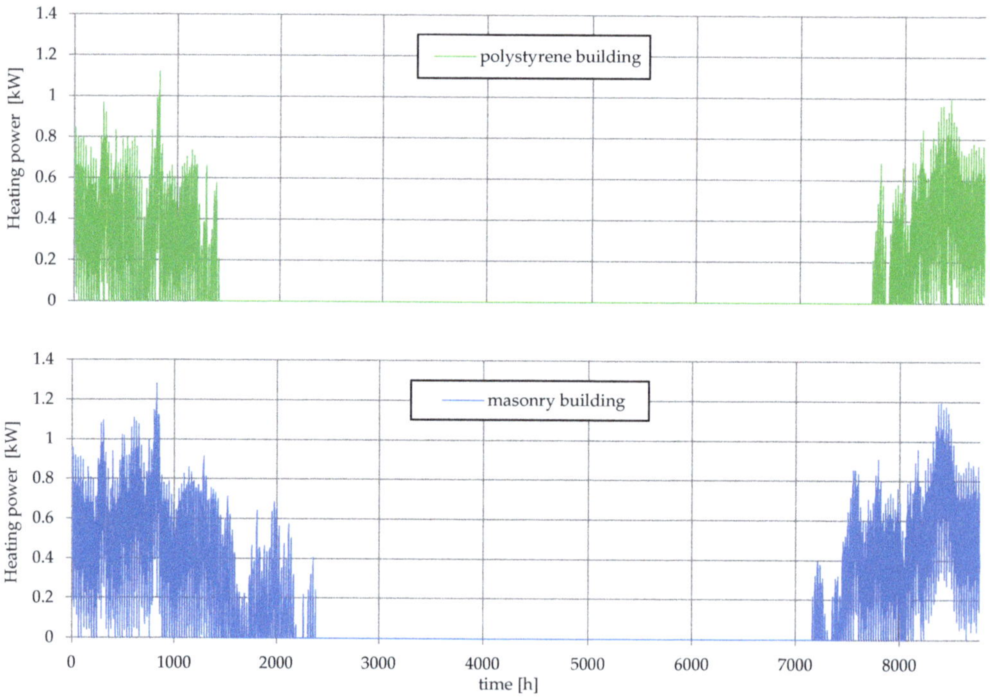

Figure 19. Hourly data for heating power—Essen.

In Essen, the heating season for the polystyrene structure lasted 103 days, whereas for the masonry structure it lasted 166 days. Figure 20 illustrates a monthly summary for both structures.

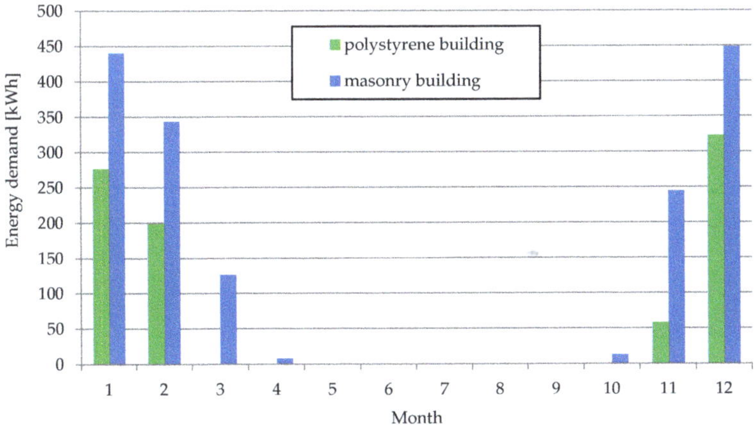

Figure 20. Monthly energy demand summary for the analyzed structures in Essen.

4. Discussion

The study compared the annual energy demand for heating for two types of buildings meeting passive building standards: polystyrene and masonry, in three locations with a temperate climate: Kołobrzeg in Poland, Vienna in Austria and Essen in Germany. The two structures possessed comparable thicknesses in their external walls and roof surfaces. Both the floor on the ground and the window and door frames exhibited identical characteristics. The polystyrene structure was found to exhibit a substantially lower energy demand across all calculation variants. The most significant disparity in energy consumption was observed in Vienna, with a recorded difference of 878 kWh per year, representing a 52% saving in energy consumption. In Essen, the discrepancy was 767 kWh, which corresponds to a 47% variance. The lowest recorded difference was observed in Kołobrzeg, with a total of 683 kWh, representing an energy saving of 37%.

With regard to the duration of the heating season, the highest advantage of polystyrene construction was observed in Essen. In this case, the heating season was shorter by 63 days, which represents a 38% reduction compared to masonry construction. In Kołobrzeg, the season was shorter by 41 days, which corresponds to a 24% difference. The discrepancy in the length of the heating season was smallest in Vienna, where the recorded value was less than 32 days, representing a 21% difference.

The disparity in useful energy demand in the studied structures is predominantly ascribed to the marked difference in their heat transfer coefficient: with polystyrene insulation exhibiting a coefficient of 0.073 W/m^2K, in comparison to masonry coefficient of 0.12 W/m^2K. This represents a 31.2% difference.

Another factor that significantly affects energy efficiency and differences in useful energy demand is thermal capacity of the building. In the case of low thermal mass of a building, full utilisation of heat gains is not possible. This is a characteristic feature of lightweight structures. In order to confirm this, additional calculations were made for one of the locations (Essen). In the simulation of energy demand for a polystyrene structure, the reinforced concrete slab and thermal insulation of the floor on the ground were switched. As a result, thermal mass of the concrete slab was excluded outside the building's insulation shell. Consequently, the annual energy demand of the building grew from 854.9 kWh to 918 kWh, representing an increase of 6.9%. Analogous calculations were conducted for the masonry version. In this case, upon excluding the reinforced concrete slab situated outside the thermal envelope of the building, the energy demand exhibited a comparatively minor increase, from 1622 kWh to 1659 kWh, representing an increment of 2.2%. The lower increase in energy demand can be attributed to the fact that the reinforced concrete slab, excluded from the insulation, constitutes a smaller proportion of the total storage mass of the masonry building.

Well-insulated buildings with a lightweight structure and low mass are characterised by rapid thermal response, in contrast to heavy masonry structures, which in turn enable effective suppression of short-term temperature fluctuations and better storage of heat gains. The polystyrene structure analysed can be regarded as a hybrid solution, combining the advantages of both light and heavy structures. The employment of lightweight external walls and roofing materials has been demonstrated to facilitate superior thermal insulation properties. In contrast, the integration of brick gable walls and a reinforced concrete floor, insulated from below, has been shown to enable enhanced temperature stabilisation when compared to conventional lightweight structures.

Such a solution allows for increased utilisation of internal gains and solar radiation gains. As a result, the energy demand for heating the building is reduced. Therefore, it is believed that the savings generated in the building in question, compared to the masonry structure, result mainly from the insulation parameters of the polystyrene partitions and the possibility of heat buffering in the building's capacitive elements.

The simplification of the polystyrene wall cross-section used in the simulations resulted in a 2.45% reduction in heat exchange in this element. For a more accurate analysis, it is necessary to conduct experimental tests to validate and refine the calculation model.

The findings of the analyses evidently demonstrate a substantial discrepancy in energy demand between polystyrene and masonry structures, while the equivalent thickness of the analyzed partitions is maintained. The polystyrene structure is composed predominantly of insulating material, which also functions as a load-bearing element. In contrast, the secondary structure incorporates a load-bearing layer characterized by reduced thermal parameters. A similar situation will occur when we compare, for example, frame structures with masonry structures or even ICF structures with similar partition thicknesses. A review of construction practice indicates that there are a number of objective reasons why the thickness of external partitions or the thermal insulation layer itself is limited. These limitations are the result of technical, aesthetic and economic considerations. This is confirmed, among others, by publication [35], where the use of thermal insulation above 0.24 m on a masonry wall made of aerated concrete is rarely economically justified. Similar conclusions can be drawn from article [36], where a layer of 0.18 m to 0.24 m is indicated as the optimal thermal insulation layer for a masonry structure. Therefore, further increasing the thickness of the partitions is not justified. Hence the attempts at using increasingly better materials or optimizing of the structure of the partitions themselves by making full use of their cross-sections. This is exemplified by the analyzed polystyrene structure. The structure described creates a specific margin, limiting the energy demand compared to a masonry structure, which may allow for greater freedom in the selection of a heat source and internal installations. This leads to measurable savings related to the operation of the building. Therefore, further research on this type of structure is important in order to learn about other aspects such as such as detailed microclimate analysis and the thermal comfort of residents, which the authors intend to undertake in the future. Nevertheless, further studies are required in the domain of structural strength, fire resistance and sound insulation. This approach will facilitate an objective evaluation of the technology's potential and constraints.

5. Conclusions

The analyses conducted clearly demonstrated that, while maintaining a comparable thickness of external partitions, the polystyrene structure exhibits a substantially diminished annual energy demand for heating in comparison to masonry structures across all studied locations with a temperate climate. The key factor contributing to this advantage is the significantly lower heat transfer coefficient and the favourable combination of light and heavy construction properties, allowing both heat loss to be reduced and internal gains and solar radiation gains to be used effectively. Conventional increases in insulation thickness in masonry structures are limited by technological, economic and functional considerations, reinforcing the rationale for seeking alternative material and structural solutions. The polystyrene structure analysed may provide a solution to these challenges, offering significant potential for reducing energy demand. In view of the findings, further research is required. This should include experimental measurements in real conditions. Such measurements will allow the calculation model to be validated. They will also allow the analysis to be extended to include aspects such as internal microclimate and thermal comfort of users. A comprehensive verification of these and other parameters is an essential step towards a complete and objective assessment of the applicability of polystyrene structures in modern buildings with high energy standards.

Author Contributions: Conceptualization, K.W.; methodology, K.W.; software, K.W.; validation, K.W.; formal analysis, K.W.; investigation, K.W.; resources, K.W. and G.N. and M.Ž.; data curation, K.W.; writing—original draft preparation, K.W.; writing—review and editing, G.N. and M.Ž.; visualization, K.W.; supervision, K.W. All authors have read and agreed to the published version of the manuscript.

Funding: This research was funded by the Faculty of Environmental Engineering, University of Agriculture in Krakow, through the project "Subvention 030001-D014 Environmental Engineering, Mining, and Energy".

Data Availability Statement: The original contributions presented in this study are included in the article. Further inquiries can be directed to the corresponding author.

Conflicts of Interest: The authors declare no conflicts of interest.

Abbreviations

The following abbreviations are used in this manuscript:

IEA	International Energy Agency
PH	Passive House
ICF	Insulated Concrete Forms
SIP	Structural Insulated Panels

References

1. IEA. *Energy Efficiency 2023*; IEA: Paris, France, 2023. Available online: https://www.iea.org/reports/energy-efficiency-2023 (accessed on 6 October 2025).
2. Passive House Institute (PHI). Available online: https://passivehouse.com (accessed on 6 October 2025).
3. Anand, V.; Kadiri, V.L.; Putcha, C. Passive buildings: A state-of-the-art review. *J. Infrastruct. Preserv. Resil.* **2023**, *4*, 3. [CrossRef]
4. Firląg, S.; Baig, A.S.; Koc, D. Historical Analysis of Real Energy Consumption and Indoor Conditions in Single-Family Passive Building. *Sustainability* **2025**, *17*, 717. [CrossRef]
5. Mihai, M.; Tanasiev, V.; Dinca, C.; Badea, A.; Vidu, R. Passive house analysis in terms of energy performance. *Energy Build.* **2017**, *144*, 74–86. [CrossRef]
6. Wąs, K.; Radoń, J.; Sadłowska-Sałęga, A. Maintenance of Passive House Standard in the Light of Long-Term Study on Energy Use in a Prefabricated Lightweight Passive House in Central Europe. *Energies* **2020**, *13*, 2801. [CrossRef]
7. Johnston, D.; Siddall, M.; Ottinger, O.; Soeren, P.; Feist, W. Are the energy savings of the passive house standard reliable? A review of the as-built thermal and space heating performance of passive house dwellings from 1990 to 2018. *Energy Effic.* **2020**, *13*, 1605–1631. [CrossRef]
8. Yu, J.; Dong, Y.; Wang, T.-H.; Chang, W.-S.; Park, J. U-Values for Building Envelopes of Different Materials: A Review. *Buildings* **2024**, *14*, 2434. [CrossRef]

9. Kočí, V.; Bažantová, Z.; Černý, R. Computational analysis of thermal performance of a passive family house built of hollow clay bricks. *Energy Build.* **2014**, *76*, 211–218. [CrossRef]
10. Liang, X.; Wang, Y.; Rouapoor, M.; Wu, Q.; Roskilly, T. Comparison of building performance between Conventional House and Passive House in the UK. *Energy Procedia* **2017**, *142*, 1823–1828. [CrossRef]
11. Kosny, J.; Asiz, A.; Smith, I.; Shresta, S.; Fallahi, A. A review of high R-value wood framed and composite wood wall technologies using advanced insulation techniques. *Energy Build.* **2014**, *72*, 441–456. [CrossRef]
12. Cabral, M.R.; Blanchet, P. A State of the Art of the Overall Energy Efficiency of Wood Buildings—An Overview and Future Possibilities. *Materials* **2021**, *14*, 1848. [CrossRef]
13. Feist, W.; Schnieders, J.; Dorer, V.; Haas, A. Re-inventing air heating: Convenient and comfortable within the frame of the Passive House concept. *Energy Build.* **2005**, *37*, 1186–1203. [CrossRef]
14. Mahapatra, K.; Olsson, S. Energy Performance of Two Multi-Story Wood-Frame Passive Houses in Sweden. *Buildings* **2015**, *5*, 1207–1220. [CrossRef]
15. Mazur, Ł.; Szlachetka, O.; Jeleniewicz, K.; Piotrowski, M. External Wall Systems in Passive House Standard: Material, Thermal and Environmental LCA Analysis. *Buildings* **2024**, *14*, 742. [CrossRef]
16. Maref, W.; Armstrong, M.M.; Saber, H.H.; Rousseau, M.; Ganapathy, G.; Nicholls, M.; Swinton, M.C. *Field Energy Performance of an Insulating Concrete form (ICF) Wall*; National Research Council of Canada: Ottawa, ON, Canada, 2012. [CrossRef]
17. Ekrami, N.; Garat, A.; Fung, A.S. Thermal Analysis of Insulated Concrete Form (ICF) Walls. *Energy Procedia* **2015**, *75*, 2150–2156. [CrossRef]
18. Mantesi, E.; Hopfe, C.J.; Mourkos, K.; Glass, J.; Cook, M. Empirical and computational evidence for thermal mass assessment: The example of insulating concrete formwork. *Energy Build.* **2019**, *188–189*, 314–332. [CrossRef]
19. Kashan, M.E.; Fung, A.S.; Eisapour, A.H. Insulated concrete form foundation wall as solar thermal energy storage for Cold-Climate building heating system. *Energy Convers. Manag. X* **2023**, *19*, 100391. [CrossRef]
20. El-Maghraby, Y.; Tarabieh, K.; Sharkass, M.; Mashaly, I.; Fahmy, E. Thermal and Structural Performances of Screen Grid Insulated Concrete Forms (SGICFs) Using Experimental Testing. *Buildings* **2024**, *14*, 2599. [CrossRef]
21. Fard, F.A.; Jafarpour, A.; Nasiri, F. Comparative assessment of insulated concrete wall technologies and wood-frame walls in residential buildings: A multi-criteria analysis of hygrothermal performance, cost, and environmental footprints. *Adv. Build. Energy Res.* **2019**, *15*, 466–498. [CrossRef]
22. Ghosh, S.; Bigelow, B. Comparative Analysis of Energy Efficiency: Insulated Concrete Form vs. Wood-Framed Residential Construction. *Buildings* **2025**, *15*, 804. [CrossRef]
23. Nässén, J.; Hedenus, F.; Karlsson, S.; Holmberg, J. Concrete vs. wood in buildings—An energy system approach. *Build. Environ.* **2012**, *51*, 361–369. [CrossRef]
24. Amran, Y.H.M.; El-Zeadani, M.; Lee, Y.H.; Lee, Y.Y.; Murali, G.; Feduik, R. Design innovation, efficiency and applications of structural insulated panels: A review. *Structures* **2020**, *27*, 1358–1379. [CrossRef]
25. Finnegan, S.; Edwards, R.; Al-Derbi, B.; Campbell, I.; Fulton, M. The Potential of Structurally Insulated Panels (SIPs) to Supply Net Zero Carbon Housing. *Buildings* **2022**, *12*, 2081. [CrossRef]
26. WUFI-Wärme Und Feuchte Instationär. Available online: https://wufi.de/en (accessed on 15 October 2025).
27. Lengsfeld, K.; Holm, A. Entwicklung und Validierung einer hygrothermischen Raumklima-Simulationssoftware WUFI®-Plus. *Bauphysik* **2007**, *29*, 178–186. [CrossRef]
28. Sadłowska-Sałęga, A.; Radoń, J. Feasibility and limitation of calculative determination of hygrothermal conditions in historical buildings: Case study of st. Martin church in Wiśniowa. *Build. Environ.* **2020**, *186*, 107361. [CrossRef]
29. Radon, J.; Was, K.; Flaga-Maryanczyk, A.; Schnotale, J. Experimental and theoretical study on hygrothermal long-term performance of outer assemblies in lightweight passive house. *J. Build. Phys.* **2017**, *41*, 299–320. [CrossRef]
30. Nawalany, G.; Sokołowski, P. Numerical Analysis of the Effect of Ground Dampness on Heat Transfer between Greenhouse and Ground. *Sustainability* **2021**, *13*, 3084. [CrossRef]
31. Nawalany, G.; Lendelova, J.; Sokołowski, P.; Zitnak, M. Numerical Analysis of the Impact of the Location of a Commercial Broiler House on Its Energy Management and Heat Exchange with the Ground. *Energies* **2021**, *14*, 8565. [CrossRef]
32. Sadłowska-Sałęga, A.; Wąs, K. Moisture Risk Analysis for Three Construction Variants of a Wooden Inverted Flat Roof. *Energies* **2021**, *14*, 7898. [CrossRef]
33. Coelho, G.B.A.; Silva, H.E.; Henriques, F.M.A. Calibrated hygrothermal simulation models for historical buildings. *Build. Environ.* **2018**, *142*, 439–450. [CrossRef]
34. Laska, M.; Reclik, K. Analysis of Internal Conditions and Energy Consumption during Winter in an Apartment Located in a Tenement Building in Poland. *Sustainability* **2024**, *16*, 3958. [CrossRef]

35. Dylewski, R. Optimal Thermal Insulation Thicknesses of External Walls Based on Economic and Ecological Heating Cost. *Energies* **2019**, *12*, 3415. [CrossRef]
36. Aviža, D. Investigation of the Determination of the Cost-Optimal Thickness of the Thermal Insulation Layer of the Walls of a Modernized Public Building. *Taikom. Tyrim. Stud. Ir. Prakt. Appl. Res. Stud. Pract.* **2021**, *17*, 79–84. Available online: https://ojs.panko.lt/index.php/ARSP/article/view/148 (accessed on 6 October 2025).

energies

MDPI

Article

Intelligent Optimization of Ground-Source Heat Pump Systems Based on Gray-Box Modeling

Kui Wang [1,2], Zijian Shuai [3] and Ye Yao [3,*]

[1] Guangxi Guiwu Jin'an Refrigeration and Air-Conditioning Technology Co., Ltd., Nanning 530200, China; drwkui@163.com
[2] Guangxi Guiwu Energy Saving Co., Ltd., Nanning 530200, China
[3] Institute of Refrigeration and Cryogenics, Shanghai Jiao Tong University, Shanghai 200240, China; hbri818@sjtu.edu.cn
* Correspondence: yeyao10000@sjtu.edu.cn

Abstract

Ground-source heat pump (GSHP) systems are widely regarded as an energy-efficient solution for building heating and cooling. However, their actual performance in large commercial buildings is often limited by rigid control strategies, insufficient equipment coordination, and suboptimal load matching. In the Liuzhou Fengqing Port commercial complex, the seasonal coefficient of performance (SCOP) of the GSHP system remains at a relatively low level of 3.0–3.5 under conventional operation. To address these challenges, this study proposes a gray-box-model-based cooperative optimization and group control strategy for GSHP systems. A hybrid gray-box modeling approach (YFU model), integrating physical-mechanism modeling with data-driven parameter identification, is developed to characterize the energy consumption behavior of GSHP units and variable-frequency pumps. On this basis, a multi-equipment cooperative optimization framework is established to coordinate GSHP unit on/off scheduling, load allocation, and pump staging. In addition, continuous operational variables (e.g., chilled-water supply temperature and circulation flow rate) are globally optimized within a hierarchical control structure. The proposed strategy is validated through both simulation analysis and on-site field implementation, demonstrating significant improvements in system energy efficiency, with annual electricity savings of no less than 3.6×10^5 kWh and an increase in SCOP from approximately 3.2 to above 4.0. The results indicate that the proposed framework offers strong interpretability, robustness, and engineering applicability. It also provides a reusable technical paradigm for intelligent energy-saving retrofits of GSHP systems in large commercial buildings.

Keywords: ground-source heat pump (GSHP); gray-box modeling; AI-based control; group control optimization; energy efficiency; commercial building HVAC

1. Introduction

With the rapid economic development and rising living standards in China, building energy consumption has increased continuously, posing significant challenges to energy conservation and carbon emission reduction. According to the China Building Energy Consumption and Carbon Emission Research Report (2023) [1,2], the building sector accounted for 36.3% of national energy consumption in 2021, with more than 60% occurring during the operational stage. In large public buildings, centralized heating, ventilation, and air-conditioning (HVAC) systems typically account for over 40% of total building energy use [3]. Therefore, improving the operational efficiency of HVAC systems is regarded as a critical pathway for achieving energy conservation and low-carbon building operation under China's dual-carbon targets.

Among various low-carbon HVAC technologies, ground-source heat pump (GSHP) systems have attracted increasing attention due to their high theoretical efficiency and environmental advantages [4,5]. By utilizing the relatively stable thermal conditions of shallow geothermal resources, GSHP systems can operate with smaller temperature lifts than air-source systems, resulting in higher coefficients of performance and improved seasonal efficiency [5]. Owing to these advantages, GSHP technology has gradually matured in terms of system configurations and engineering applications and has been widely applied in building heating and cooling, particularly in large commercial and public buildings [4,5]. Long-term field measurements and performance assessments further confirm the feasibility and energy-saving potential of GSHP systems in real engineering applications [6,7].

However, the actual operational performance of GSHP systems often deviates from their theoretical potential. Unlike conventional chiller systems, GSHP systems involve strong coupling among multiple subsystems, including heat pump units, ground heat exchangers, circulation pumps, and terminal systems. Previous studies have reported that building load imbalance may lead to soil thermal imbalance in ground heat exchangers, which is recognized as a common long-term operational challenge in GSHP systems [8–11]. Long-term monitoring and numerical investigations have shown that such thermal imbalance can cause gradual degradation of ground temperature conditions and system coefficient of performance [10,12–14]. These strong coupling effects and long-term source–load interactions significantly increase the complexity of GSHP system operation and performance prediction [15]. In engineering practice, GSHP systems are commonly operated using rule-based control strategies with fixed setpoints or simple load-following logic [16], which lack adaptability to varying operating conditions and limit the potential for system-level energy efficiency improvement.

To address these challenges, increasing research efforts have been devoted to the modeling and optimization-based control of GSHP systems. Physics-based models have been widely used to describe system dynamics and heat-transfer processes, particularly for long-term performance assessment and thermal imbalance analysis [9,10,12–14]. Meanwhile, numerical simulation platforms and detailed component models have been applied to investigate hybrid GSHP configurations and operational strategies [17,18]. In addition, variable-flow control methods have been explored to improve the thermal and economic performance of GSHP systems under partial-load operation [19]. However, purely physics-based models usually require detailed system knowledge and extensive calibration effort, which limit their applicability in existing buildings. Conversely, purely data-driven approaches, although capable of capturing complex nonlinear relationships, often suffer from limited interpretability and unreliable extrapolation under unseen operating conditions [20]. These limitations have constrained the practical deployment of advanced optimization and control strategies in real GSHP installations.

In recent years, gray-box modeling approaches, which integrate simplified physical structures with data-driven parameter identification, have attracted growing interest in building energy modeling and control [20,21]. Gray-box models are capable of balancing modeling accuracy, robustness, and engineering feasibility, making them particularly suitable for HVAC systems characterized by strong coupling and slow thermal dynamics [20]. Reduced-order gray-box models, typically formulated using state-space or resistance–capacitance (RC) structures, have demonstrated reliable performance in HVAC energy prediction and model predictive control while maintaining physical interpretability [21]. For GSHP systems, such approaches are particularly promising, as they are well suited to

https://doi.org/10.3390/en19030608

capturing essential thermal behaviors of ground loops and equipment interactions while remaining computationally tractable for online optimization and group control [20,21].

More recently, advanced artificial intelligence-based control strategies have been increasingly explored for HVAC and heat pump systems to overcome the limitations of conventional rule-based operation. Reinforcement learning (RL), in particular, has attracted growing attention due to its model-free and adaptive learning capability. For example, Fu et al. [22] proposed a multi-agent deep reinforcement learning framework for coordinated control of building cooling water systems, demonstrating notable energy-saving performance through parallel learning and load distribution optimization. These studies indicate that multi-agent RL has strong potential for handling high-dimensional control problems in complex HVAC systems.

In parallel, model predictive control (MPC) has been widely applied to heat pump-based energy systems by explicitly incorporating system dynamics and operational constraints into the optimization process. Zhao et al. [23] developed an adaptive MPC strategy for a heat pump-assisted solar water heating system, in which time-varying heat pump efficiency and adaptive control boundaries were considered, resulting in improved energy and cost performance compared with conventional MPC and PID control.

Despite their promising performance, RL- and MPC-based control methods often face challenges when applied to existing large-scale commercial ground-source heat pump (GSHP) systems. These challenges include high data requirements for RL training, substantial computational burden for real-time MPC optimization, and limited physical interpretability for engineering deployment and long-term maintenance. In this context, gray-box-model-based optimization approaches offer a favorable compromise between physical interpretability, modeling accuracy, and computational efficiency, making them particularly suitable for coordinated control and energy optimization of GSHP systems with strong source–load coupling and multiple interacting components.

Motivated by the above considerations, this study proposes a gray-box-model-based group control optimization strategy for existing large-scale commercial ground-source heat pump (GSHP) air-conditioning systems. Energy consumption models are developed for GSHP units and variable-frequency pumps, and a multi-parameter cooperative optimization framework is established by jointly coordinating supply water temperature, circulation flow rate, and pump operating frequency. The proposed strategy is validated through both simulation analysis and on-site field testing in a real commercial GSHP installation, demonstrating its effectiveness, robustness, and strong potential for practical energy retrofits and intelligent operation of GSHP systems.

2. Optimization Model Development

2.1. System and Model

To implement the proposed gray-box-model-based group-control optimization strategy, a simplified yet physically meaningful representation of the GSHP system is first required. This representation captures the system structure and key operating characteristics for optimization. In this study, a typical GSHP air-conditioning system is considered as the modeling object, and its configuration is abstracted to capture the essential thermal and hydraulic interactions relevant to operational optimization, without reference to any specific engineering project.

A typical GSHP system comprises buried ground heat exchanger loops, ground-side circulation pumps, heat pump units, chilled-water circulation pumps, and terminal air-conditioning equipment. These components form a closed-loop structure of 'ground heat exchanger–heat pump units–building load', as schematically illustrated in Figure 1. This

https://doi.org/10.3390/en19030608

configuration represents the fundamental energy transfer path and coupling relationships that govern GSHP system performance during cooling operation.

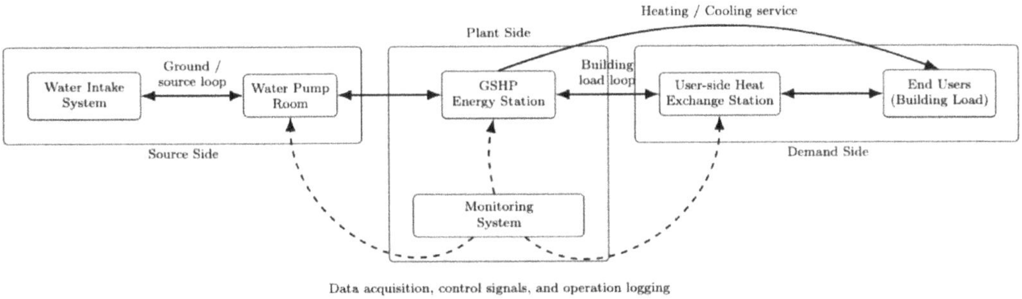

Figure 1. Schematic diagram of a typical ground-source heat pump (GSHP) system, illustrating the source-side loop, plant-side components, and demand-side heat exchange. Bidirectional arrows indicate hydraulic circulation loops, the curved arrow represents the direction of heating/cooling service delivery, and dashed arrows denote data acquisition, control signals, and operational logging.

In this section, a gray-box model of GSHP units (the GSHP-YFU model) and an energy-consumption model of variable-frequency pumps are developed based on the physical mechanisms of GSHP operation, forming the foundation for a multi-parameter cooperative optimization framework. Unlike conventional water-cooled chiller systems, GSHP systems rely on ground heat exchanger loops, and their performance is primarily affected by ground-side inlet and outlet water temperatures, load ratio, and circulation flow rate. These characteristics result in relatively small temperature fluctuations but complex coupling relationships.

To balance modeling accuracy and computational efficiency for optimization-oriented applications, recent advances in HVAC and GSHP modeling are incorporated to construct an engineering-applicable YFU model, which is further integrated with a particle swarm optimization (PSO) algorithm to achieve system-level group-control optimization.

2.2. *Gray-Box Model of GSHP Units (GSHP-YFU Model)*

2.2.1. Modeling Requirements

Ground-source heat pump (GSHP) systems utilize soil as a heat source during heating operation and as a heat sink during cooling operation. Ground temperature varies only mildly with seasonal changes, but it can exhibit long-term cumulative effects. Therefore, GSHP performance is influenced not only by instantaneous load but also by the evolution of the underground thermal field. To accurately describe unit performance, this study adopts a gray-box modeling approach combining physical mechanisms with data-driven parameter identification. This hybrid method incorporates fundamental thermodynamic laws together with measured engineering data, forming a model framework that is interpretable and suitable for real-time optimization.

The gray-box model for GSHP units must satisfy the following requirements: (1) The coefficient of performance (COP) must reflect simultaneous influences from chilled-water temperature, ground-loop water temperature, and unit part-load ratio (PLR); (2) internal parameters such as evaporating and condensing temperatures cannot be directly measured under practical engineering conditions; therefore, the model must be formulated using measurable quantities; (3) the model must provide sufficiently high prediction accuracy to support real-time optimization and control.

Based on these requirements, the classical YAO universal model [24] is extended and adapted to GSHP operation characteristics, forming the GSHP-YFU model.

https://doi.org/10.3390/en19030608

2.2.2. Evaporator and Condenser Heat-Transfer Models

The modeling approach follows the principle of using "physically meaningful yet practically measurable parameters," avoiding unobservable refrigerant-state variables.

- Evaporator Model: The evaporator absorbs heat from the chilled-water loop. Its evaporating temperature depends on the chilled-water return temperature and the water flow rate. The evaporating temperature model is expressed as follows:

$$T_e = T_{w,e,E} - \frac{Q_e}{c_w G_{w,e}\left[1 - \exp\left(-\frac{K_e A_e}{c_w G_{w,e}}\right)\right]}, \tag{1}$$

where T_e is the evaporating temperature (K); $T_{w,e,E}$ is the chilled-water return temperature (K); Q_e is the evaporator cooling load (kW); c_w is the specific heat capacity of water (kJ/(kg·K)); $G_{w,e}$ is the chilled-water mass flow rate (kg/s); and $K_e A_e$ represents the effective overall heat-transfer coefficient multiplied by heat-transfer area (W/K·m^2). These parameters collectively characterize evaporator heat-transfer performance.

- Condenser Model: The GSHP condenser rejects heat into the underground heat-exchanger loop. The condensing temperature depends on the ground-side inlet water temperature and the ground-loop flow rate. The condensing temperature model is expressed as follows:

$$T_c = T_{w,c,E} + \frac{Q_c}{c_w G_{w,c}\left[1 - \exp\left(-\frac{K_c A_c}{c_w G_{w,c}}\right)\right]}, \tag{2}$$

where T_c is the condensing temperature (K); $T_{w,c,E}$ is the ground-side inlet temperature (K); Q_c is the condenser load (kW); $G_{w,c}$ is the ground-loop mass flow rate (kg/s); and K_c A_c represents the condenser heat-transfer capacity. These parameters describe the condenser thermal behavior during GSHP operation.

2.2.3. Gray-Box COP Model for GSHP Units

There are various mathematical models for the energy efficiency of GSHP units, including white-box models, gray-box models, and black-box models. In practical engineering applications, gray-box and black-box models are commonly used, and a gray-box model is adopted in this subsection. The traditional YAO model constructs the COP based on the evaporation temperature, condensation temperature, and load ratio, as shown in Equation (3):

$$COP = \frac{r}{\frac{T_c - T_e}{T_c} + a_1 \frac{T_c}{T_e} + a_2}, \tag{3}$$

where r is the loading ratio of the chiller and a_1 and a_2 are empirical factors determined by experimental data.

The generalized YAO model is derived from the first and second laws of thermodynamics, under the assumption that the entropy of the refrigerant remains constant throughout the refrigeration cycle. The model further accounts for the influence of the unit load ratio on the COP. It employs three independent variables—load ratio, evaporating temperature, and condensing temperature—and includes two regression coefficients.

In commonly used gray-box models, the effects of load ratio, chilled-water flow rate, and cooling-water flow rate on the COP of GSHP units are typically not considered. The YAO model incorporates these influential parameters. However, in practical applications, quantities such as evaporating temperature, condensing temperature, heat-transfer coefficients, and heat-exchange areas are often difficult to measure accurately. Moreover, the heat-transfer coefficient is sensitive to variations in water flow rate. For these reasons, the

https://doi.org/10.3390/en19030608

original YAO formulation is revised and simplified in this study, and a new COP model for GSHP units is proposed.

The load ratio reflects the relative magnitude of the cooling demand on the chiller, and the water flow rate is generally positively correlated with the cooling load. Therefore, in the reformulated model, the product of the heat-transfer coefficient and heat-exchange area appearing in Equations (4) and (5) is replaced with a constant that is assumed to vary linearly with the load ratio within a bounded range, as follows:

$$K_e A_e = (m_1 - m_0)\,r + m_0, \tag{4}$$

$$K_c A_c = (n_1 - n_0)\,r + n_0, \tag{5}$$

In the revised formulation, m_1 and m_0 denote the minimum and maximum values of the empirical constant for the evaporator, respectively; similarly, n_1 and n_0 denote the minimum and maximum values for the condenser.

Accordingly, Equations (1) and (2) can be rewritten as follows:

$$T_e' = T_{w,e,E} - \frac{Q_e}{c_w G_{w,e}\left[1 - \exp\left(-\frac{(m_1 - m_0)r + m_0}{c_w G_{w,e}}\right)\right]}, \tag{6}$$

$$T_c' = T_{w,c,E} + \frac{Q_c}{c_w G_{w,c}\left[1 - \exp\left(-\frac{(n_1 - n_0)r + n_0}{c_w G_{w,c}}\right)\right]}, \tag{7}$$

Equations (3)–(7) establish a COP model that incorporates the supply and return temperature of chilled water, the inlet and outlet temperature of cooling water, and the chilled and cooling water flow rates. This model is hence named the YFU model.

2.2.4. Power Model of the GSHP Unit

The power consumption of the GSHP unit is determined by the cooling load and the coefficient of performance (COP), as expressed in Equation (8):

$$P_{PGHSP} = \frac{Q_e}{COP}, \tag{8}$$

where P_{PGHSP} is the heat pump unit power (kW).

2.3. Energy Consumption Model of Variable-Frequency Pumps

Pumps provide the circulation power for chilled water and ground-loop water in GSHP systems and constitute one of the most energy-intensive components aside from the heat pump units themselves. Because building loads fluctuate with weather and internal heat gains, constant-speed pumps often operate under high-flow, low-temperature-difference conditions in which the actual temperature difference falls below the design value, resulting in significant pumping energy waste. To support an optimal control scheme, this subsection develops a mathematical energy-consumption model for variable-frequency pumps.

2.3.1. Single Variable-Frequency Pump

The hydraulic output power of a single pump is expressed as follows:

$$P_{\text{pump},O} = \frac{\rho_w g H G_w}{3600 \times 1000}, \tag{9}$$

where $P_{\text{pump},O}$ is the hydraulic output power of the pump (kW), ρ_w is water density (kg/m^3), g is gravitational acceleration (m/s^2), H is the pump head (mH_2O), and G_w is the volume flow rate (m^3/h). These parameters characterize the pump's energy conversion capability.

The head–flow characteristic of the water pump can be approximated by a quadratic expression:

$$H = a_0 + a_1 G_w + a_2 G_w^2, \tag{10}$$

The electrical input power of the pump should account for pump efficiency, motor efficiency, and variable-frequency drive (VFD) efficiency:

$$P_{\text{pump}} = \frac{P_{\text{pump},O}}{\eta_p \eta_m \eta_{\text{vfd}}}, \tag{11}$$

where η_p, η_m , and η_{vfd} denote the pump, motor, and VFD efficiencies, respectively.

For motors rated above 18.375 kW, the motor efficiency is expressed as a function of the speed ratio k, as follows:

$$\eta_m = 0.94187(1 - \exp(-9.04k)), \tag{12}$$

where k is the ratio of actual to rated motor speed.

Similarly, the VFD efficiency is given by the following equation:

$$\eta_{\text{vfd}} = 0.5087 + 1.283k - 1.42k^2 + 0.5842k^3, \tag{13}$$

According to pump affinity laws, the relationships between flow rate, head, and rotational speed n under two operating conditions can be expressed by Equation (14):

$$\frac{G_1}{G_2} = \frac{n_1}{n_2}, \ \frac{H_1}{H_2} = \left(\frac{n_1}{n_2}\right)^2, \tag{14}$$

The key parameters H, η_p, η_m , and η_{vfd} can be expressed as functions of flow rate. The pump input power is reformulated as a polynomial of real-time flow rate:

$$P_{\text{pump}} = b_0 + b_1 G_w + b_2 G_w^2, \tag{15}$$

where b_{i} is a coefficient identified by field data.

2.3.2. Parallel Variable-Frequency Pumps

Parallel pump operation is adopted when the flow requirement exceeds the capacity of a single pump or when redundancy is needed for reliability. For n set of pumps operating in parallel, the total system flow rate equals the sum of the individual flow rates, while all pumps share the same head:

$$G_{w,t} = \sum_{i=1}^{n} G_{w,i}, \tag{16}$$

$$H_t = H_1 = H_2 = \cdots = H_n, \tag{17}$$

Under temperature-difference-based control, flow-regulating valves remain fixed, allowing system hydraulic resistance to be treated as constant. The relationship between head and flow rate is as follows:

$$H_t = S G_{w,t}^2, \tag{18}$$

where S is the system hydraulic resistance coefficient.

https://doi.org/10.3390/en19030608

The head–flow characteristic of each pump in parallel can be rewritten as follows:

$$H_t = a_{0,i} + a_{1,i}\frac{G_{w,i}}{n} + a_{2,i}\left(\frac{G_{w,i}}{n}\right)^2, \tag{19}$$

which provides the basis for determining flow distribution under parallel operation.

Although the total input power of n set of pumps can be theoretically derived by solving Equations (9)–(19) simultaneously, this approach is computationally intensive and sensitive to uncertainties in system hydraulic resistance. Therefore, for engineering applicability, the total power consumption is modeled as the summation of single-pump models:

$$P_{\text{pump},t} = \sum_{i=1}^{n}\left(b_{0,i} + b_{1,i}G_{w,i} + b_{2,i}\frac{G_{w,i}}{n}\right), \tag{20}$$

where $b_{j,i}$ are identified from measured data under different pump staging and operating conditions. This formulation enables real-time optimization and integration into system-level cooperative control strategies.

2.4. Optimization Strategies for GSHP System

2.4.1. Overall Optimization Problem Formulation of the GSHP System

The ground-source heat pump (GSHP) system considered in this study consists of GSHP units, chilled-water circulation pumps, ground-side circulation pumps, and terminal heat exchangers. The overall system energy performance is determined by the combined effects of GSHP unit staging, load allocation, supply water temperature, and circulation flow rates. These decision variables are strongly coupled through thermodynamic and hydraulic relationships, resulting in a highly nonlinear and constrained optimization problem.

The primary optimization objective is to minimize the total system energy consumption while satisfying terminal cooling or heating demand and ensuring safe and stable system operation. The total system power consumption is composed of the power consumption of GSHP units, chilled-water pumps, and ground-side circulation pumps, and can be expressed as

$$\min J = P_{\text{GSHP}} + P_{\text{cwp}} + P_{\text{gsp}}, \tag{21}$$

where P_{GSHP}, P_{cwp}, and P_{gsp} denote the power consumption of the GSHP units, chilled-water pumps, and ground-side circulation pumps, respectively.

From an optimization perspective, the GSHP system involves both discrete and continuous decision variables, as follows.

Discrete variables:

- On/off states of GSHP units;
- On/off states (or operating numbers) of chilled-water pumps;
- On/off states (or operating numbers) of ground-side circulation pumps.

Continuous variables:

- Chilled-water supply temperature;
- Chilled-water circulation flow rate;
- Ground-side circulation flow rate.

The coexistence of discrete and continuous variables renders the GSHP system optimization problem a mixed-integer nonlinear programming (MINLP) problem.

https://doi.org/10.3390/en19030608

The optimization is subject to multiple physical and operational constraints, which are summarized as follows.

- Cooling-load balance constraint:

$$Q_{\text{load}} = \rho_w c_w G_{\text{chw}} (T_{\text{chw,r}} - T_{\text{chw,s}}) \tag{22}$$

where G_{chw} is the chilled-water flow rate, and $T_{\text{chw,r}}$ and $T_{\text{chw,s}}$ are the chilled-water supply and return temperatures, respectively.

- Chilled-water temperature constraints:

$$T_{s,min} \leq T_{\text{chw,s}} \leq T_{s,max}, \tag{23}$$

$$T_{\text{chw,r}} - T_{\text{chw,s}} \leq \Delta T_{\text{chw,max}}, \tag{24}$$

- Ground-side temperature constraints:

$$T_{g,min} \leq T_{g,\text{in}} \leq T_{g,max}, \tag{25}$$

$$T_{g,\text{out}} - T_{g,\text{in}} \leq \Delta T_{g,max}, \tag{26}$$

- Flow-rate constraints:

$$G_{\text{chw,min}} \leq G_{\text{chw}} \leq G_{\text{chw,max}}, \tag{27}$$

$$G_{g,min} \leq G_g \leq G_{g,max}, \tag{28}$$

- GSHP unit and pump operational constraints:

GSHP units and pumps must satisfy minimum on/off time requirements, minimum part-load ratio constraints, and rated capacity limits to ensure operational safety and equipment protection.

Optimization decomposition strategy: Due to the high dimensionality and mixed-variable nature of the problem, directly solving the full MINLP formulation would lead to excessive computational complexity. To ensure computational feasibility while preserving optimization accuracy, a hierarchical decomposition strategy is adopted.

The optimization problem is decomposed into two coordinated stages:

- Discrete optimization stage, in which the on/off combinations of GSHP units and circulation pumps as well as the load allocation among the running GSHP units, are determined;
- Continuous optimization stage, in which the chilled-water supply temperature and circulation flow rates are optimized under fixed unit and pump configurations.

Due to the strong coupling among GSHP units with respect to on/off control, load allocation, supply-water temperature, and circulation flow rates, the overall optimization problem constitutes a high-dimensional mixed-integer nonlinear programming (MINLP) problem with complex constraints. Directly solving the complete MINLP formulation would impose excessive computational burden and limit its applicability in real-time engineering practice.

To address this challenge, a hierarchical two-stage cooperative optimization strategy is adopted in this study. In the first stage, discrete decision variables—including the on/off states of GSHP units and circulation pumps, as well as load allocation among operating units—are optimized using combinatorial logic and exhaustive search. In the second stage, continuous operational variables, such as the chilled-water supply temperature and ground-side circulation flow rate, are globally optimized under fixed on/off configurations using particle swarm optimization (PSO).

https://doi.org/10.3390/en19030608

This two-stage strategy effectively decouples combinatorial on/off control decisions from continuous nonlinear optimization, significantly reducing computational complexity while preserving solution quality. The overall framework of the proposed optimization strategy is illustrated in Figure 2.

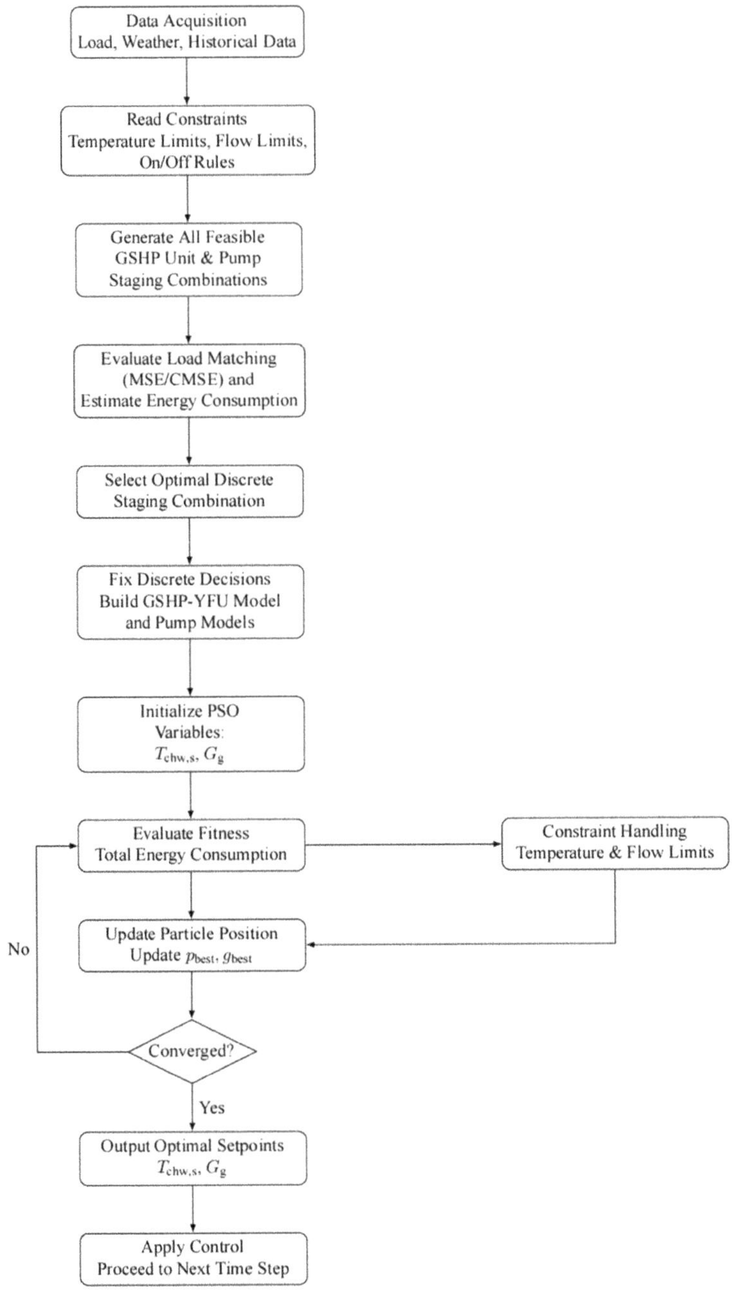

Figure 2. Two-stage cooperative optimization framework for the GSHP system.

https://doi.org/10.3390/en19030608

2.4.2. GSHP Unit On/Off Control and Load Allocation

As a key component of the discrete optimization stage described in Section 2.4.1, this subsection discusses the coupled optimization of GSHP unit on/off control and load allocation. Conventional on/off control strategies for ground-source heat pump (GSHP) units usually rely on fixed load thresholds or predefined staging sequences, while load allocation among operating units is treated independently, typically following equal-load or equal-load-ratio principles. Such decoupled control logic neglects the fact that GSHP units exhibit different efficiency characteristics under varying load levels, which often leads to suboptimal operating conditions and limits the achievable system efficiency.

To address this issue, this study proposes a coupled optimization strategy for GSHP unit on/off control and load allocation, aiming to maximize the system-level coefficient of performance (COP) while satisfying the terminal cooling or heating demand. The core idea is to coordinate unit staging and load distribution simultaneously, allowing different GSHP units to operate at different load levels so that as many units as possible remain close to their high-efficiency regions.

Based on the performance curves of individual GSHP units, the optimal load ratio corresponding to the maximum COP of each unit is identified. A Combined Mean Squared Error (CMSE) index is then defined to quantify the deviation between the actual load ratios of operating units and their respective optimal load ratios. A smaller CMSE indicates that the overall operating state of the GSHP system is closer to its optimal efficiency envelope.

The overall coupled optimization procedure is illustrated in Figure 3.

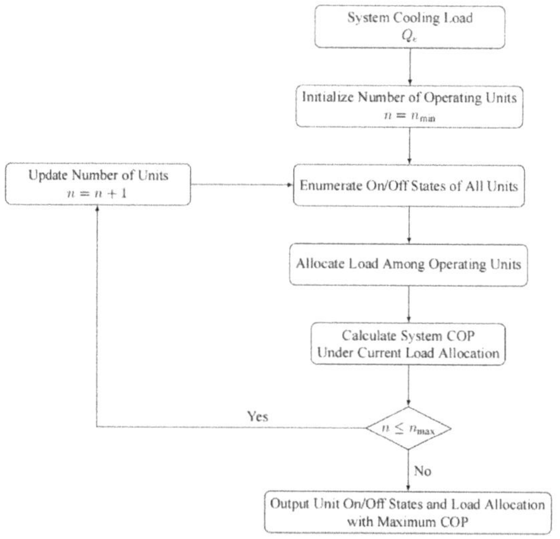

Figure 3. COP-based unit staging and load allocation optimization procedure.

First, according to the real-time system cooling or heating load, the minimum number of GSHP units required to meet the demand is determined. For a given number of operating units, all possible unit on/off combinations are enumerated using an exhaustive search. For each candidate combination, a load allocation procedure is performed, and the corresponding system COP is calculated. The unit combination and load allocation scheme yielding the maximum system COP are retained as the optimal solution. The number of operating units is then increased incrementally, and the above process is repeated until all feasible configurations have been evaluated.

https://doi.org/10.3390/en19030608

Within each candidate on/off configuration, the detailed load allocation procedure is conducted within the discrete optimization stage shown in Figure 4. A subset of GSHP units is allowed to operate at their optimal load ratios, while the remaining units share the residual load. Load allocation schemes that result in negative unit loads are discarded. When no unit is assigned to operate at its optimal load ratio, an equal-load-ratio allocation is adopted to avoid excessively low load levels that would cause severe COP degradation. Through this coupled optimization of unit staging and load allocation, inefficient operating conditions such as "large units serving small loads" or "small units overloaded by large demands" are effectively avoided, leading to a higher average system COP and reduced energy consumption associated with frequent on–off cycling.

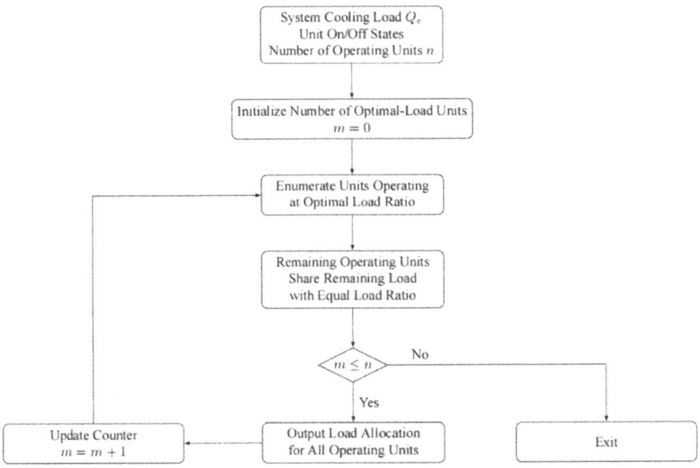

Figure 4. Inner-loop load allocation procedure under fixed unit staging.

2.4.3. Optimization of Chilled-Water and Ground-Side Pump On/Off Control

The objective of the optimized on/off control strategy for chilled-water pumps and ground-side circulation pumps is to minimize the total pump power consumption while satisfying the system flow-rate requirements. By optimally determining the operating status of each pump, unnecessary pump operation can be avoided and overall pump-system energy efficiency can be improved. The optimization problem is formulated with the total pump power consumption as the objective function, expressed as follows:

$$P_{\text{total}} = \sum_{i=1}^{N} u_i P_i(Q_i), \tag{29}$$

where u_i is a binary variable indicating the operating status of pump$_i$, Q_i is the flow rate of pump$_i$, and $P_i(Q_i)$ denotes the power consumption of pump$_i$ as a function of its flow rate.

In this formulation, the total system flow rate must satisfy the required demand, which is determined by the cooling or heating load of the GSHP system. The flow rate of each pump is constrained within its allowable operating range, defined by the minimum and maximum flow-rate limits of the pump.

The optimization procedure for pump on/off control is implemented as follows. First, based on the required system flow rate, the minimum number of pumps necessary to meet the demand is determined and set as the initial number of operating pumps. For a given number of operating pumps, all possible pump on/off combinations are enumerated using an exhaustive search method. Since the pumps operate in parallel and are of identical or

https://doi.org/10.3390/en19030608

similar types, equal load ratio (equal flow-rate ratio) allocation is adopted for the operating pumps to ensure consistent head conditions and to prevent adverse hydraulic phenomena such as reverse flow caused by head mismatch.

For each candidate on/off configuration, the total pump power consumption is calculated based on the assigned flow rates. If the total pump power under the current configuration is lower than the previously recorded minimum value, the current pump on/off configuration is recorded as the optimal solution. When the number of operating pumps is smaller than the total number of installed pumps, the number of operating pumps is increased by one, and the above enumeration and evaluation process is repeated. Finally, the pump on/off configuration and corresponding flow-rate distribution that yield the minimum total pump power consumption are obtained, which are then used to determine the operating frequency of each pump. The detailed optimization procedure of the pump on/off control strategy is illustrated in Figure 5.

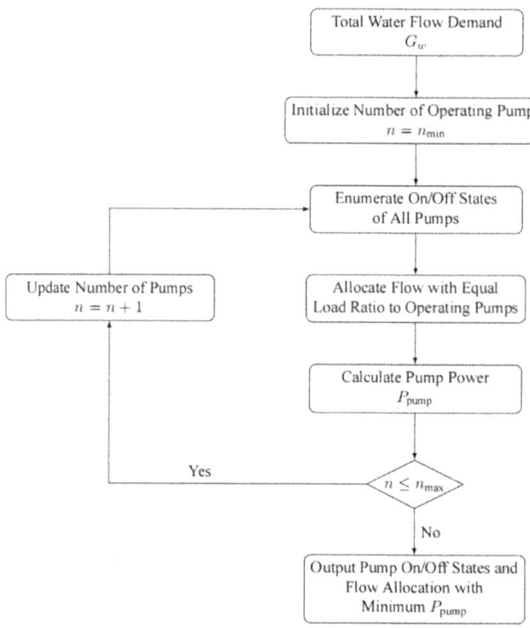

Figure 5. Pump staging and flow allocation optimization procedure based on minimum power consumption.

2.4.4. Determination of the Optimal Staging Combination Based on Start–Stop Strategies

Following the optimization logic described in Sections 2.4.1 and 2.4.2, the first stage of the proposed framework identifies feasible staging combinations for GSHP units and circulation pumps. The objective is to determine combinations that satisfy cooling-load requirements while minimizing total system energy consumption. The process consists of three steps.

- Construction of the feasible on/offsets: Considering minimum on/off time constraints, minimum part-load ratios, motor capacity limits, and operational safety requirements, a complete set of feasible unit start–stop combinations is generated.
- Load-matching-based combination screening: For each candidate combination, the mean squared error (MSE) between the operating point and the optimal efficiency region of each GSHP unit is calculated. This metric is used to select one or several candidate staging combinations that best match the prevailing load.

https://doi.org/10.3390/en19030608

- Optimal pump-staging selection: Based on the head–flow characteristics of parallel pumps, the power consumption under different numbers of operating pumps is estimated. The staging combination that satisfies the target flow rate with the minimum energy consumption is selected.

  The output of this first stage includes the following:

1. The optimal on/off combination of GSHP units;
2. The optimal number of operating chilled-water pumps;
3. The optimal number of operating ground-side circulation pumps. These discrete decision variables serve as fixed boundary conditions for the second-stage PSO optimization.

2.4.5. PSO-Based Continuous Variable Optimization Using the Gray-Box Model

Once the unit and pump staging strategies have been determined, particle swarm optimization (PSO) is applied to perform global optimization of continuous operational variables. The optimization objective is to minimize the total system energy consumption while satisfying terminal cooling-load requirements and maintaining stable system operation.

The total system energy consumption is defined as the sum of the power consumption of GSHP units, chilled-water pumps, and ground-side circulation pumps. Operational constraints include limits on chilled-water supply temperature, ground-loop inlet temperature, temperature differences across heat exchangers, and allowable flow-rate ranges. To reduce problem dimensionality and improve computational efficiency, a hierarchical variable-selection strategy is adopted.

In the chilled-water loop, the supply chilled-water temperature is selected as the optimization variable, while the return temperature is treated as an uncontrollable variable determined by load conditions. In the ground-side loop, the circulation flow rate is selected as the optimization variable, whereas inlet and outlet temperatures are constrained by soil thermal characteristics. Continuous variables are optimized using PSO, while discrete variables are fixed based on the results of Stage 1. This hybrid optimization strategy ensures both computational efficiency and high-quality solutions for GSHP system operation.

3. Case Study

3.1. System Equipment Overview

3.1.1. Configuration of Cooling Source Equipment

The chilled-water plant of the Liuzhou Fengqing Port commercial complex is located in an independent area on the first floor of the building, fully isolated from other functional zones. The cooling-source subsystem is equipped with three DunAn ground-source heat pump (GSHP) units. Among these, one large-capacity unit provides a cooling capacity of 2627 kW and a rated power of 507 kW, while the other two smaller units each provide a cooling capacity of 2388 kW and a rated power of 364 kW.

The associated hydronic network includes four chilled-water pumps and four ground-loop circulation pumps, all operating under variable-frequency control. Each pump has a rated power of 75 kW, a design flow rate of 460 m^3/h, and a head of 37 m. The chilled-water supply temperature is normally fixed as 9 °C.

The GSHP system operates year-round, and the annual operating cost is estimated using an average electricity tariff of 0.8 RMB/kWh. The annual load profile indicates approximately 90 days of high-load operation and 120 days of low-load operation. Based on on-site investigations and operational records, the GSHP System Equipment Parameter Table (Table 1) is compiled.

https://doi.org/10.3390/en19030608

Table 1. Equipment parameters of the GSHP system.

Category	Quantity	Cooling Capacity (kW)	Power (kW)	Flow Rate (m^3/h)	Head (m)
GSHP Unit	1	2627	507	–	–
GSHP Unit	2	2388	364	–	–
Chilled-water Pump	4	–	75	460	37
Ground-loop Pump	4	–	75	460	37

3.1.2. Current Status of Variable-Frequency Pump Control

All chilled-water and ground-loop circulation pumps are equipped with ABB variable-frequency drives (VFDs), enabling continuous modulation of pump speed. The control architecture consists of a centralized group-control panel and an electrical power cabinet, which together facilitate integrated control of GSHP units, pumps, and plate heat exchangers. This control infrastructure provides a reliable foundation for the subsequent implementation of AI-based group-control optimization.

3.1.3. System Operation and Data Acquisition

The GSHP system undergoes frequent start–stop cycles and wide variations in cooling load throughout the year. The system's critical operational variables—including chilled-water supply and return temperatures, ground-side inlet and outlet temperatures, flow rates, pump and unit power consumption, and ambient temperature—are continuously monitored and recorded through on-site sensors. These measurements form a full-spectrum dataset that captures the system's behavior across all operating conditions and serves as the basis for model training and parameter identification.

3.2. Model Development and Parameter Identification

Based on the GSHP-YFU model and the gray-box pump models introduced in Section 2, parameter identification is performed using measured operational data from the Liuzhou Fengqing Port GSHP system. Prior to model calibration, the dataset undergoes preprocessing, including outlier removal, segmentation according to load level, and imputation of missing values, ensuring data quality and improving the effectiveness of model fitting.

Subsequently, the least-squares method is applied to identify the model parameters. The resulting parameter sets allow the GSHP unit and pump models to accurately reflect equipment energy efficiency characteristics under varying operating conditions, and the results are presented in Table 2.

Table 2. Model parameters of the GSHP-YFU units.

Unit	a_1	a_2
1 GSHP-YFU Unit	0.3485	0.3662
2 GSHP-YFU Unit	0.3814	0.3917

For the chilled-water pumps operating in parallel, separate parameter sets are identified for one, two, three, and four-pump operating scenarios, resulting in four corresponding gray-box pump models. Each model follows the cubic polynomial form described in Section 2, and the fitted coefficients are summarized in Table 3.

https://doi.org/10.3390/en19030608

Table 3. Model parameters of chilled-water pumps under different operating modes.

Operating Mode	b_0	b_1	b_2
Single Pump	8.2563	−0.2257	0.0022
Two Pumps in Parallel	11.5840	−0.3655	0.0039
Three Pumps in Parallel	12.4910	−0.4600	0.0056
Four Pumps in Parallel	13.2159	−0.5545	0.0073

The ground-loop circulation pumps operate in series with the GSHP units and do not interfere with one another. Therefore, parameter identification is conducted individually for each pump, and the results are presented in Table 4.

Table 4. Model parameters of ground-loop circulation pumps.

Pump No.	b_0	b_1	b_2
1 Ground-loop Pump	16.4840	−0.2895	0.0017
2 Ground-loop Pump	14.9624	−0.2701	0.0016
3 Ground-loop Pump	10.0260	−0.1985	0.0014
4 Ground-loop Pump	17.5911	−0.3055	0.0017

After parameter identification, the prediction accuracy of the proposed GSHP-YFU model was further evaluated using measured operational data. The modeled coefficient of performance (COP) was compared with field measurements under different load conditions. Considering the strong load dependency of GSHP system performance, a load-ratio-based grouping strategy was adopted for model evaluation. The quantitative prediction accuracy of the GSHP-YFU model is summarized in Table 5. The results show that the GSHP-YFU model achieves an RMSE of approximately 0.22 and a coefficient of variation (CV) of about 3.5%, indicating satisfactory prediction accuracy and robustness for optimization-oriented control applications.

Table 5. Prediction accuracy of the GSHP-YFU model.

Metric	Value	Description
RMSE	0.22	Root mean square error of COP prediction
CV (%)	3.5	Coefficient of variation under load-ratio-based grouping

3.3. Simulation Platform and Performance Validation

To evaluate the effectiveness of the proposed multi-parameter cooperative optimization strategy, an integrated simulation platform is developed for the GSHP cooling-source system of the Liuzhou Fengqing Port commercial complex. The platform is constructed based on measured operational data and the identified gray-box models of GSHP units and pumps, enabling dynamic simulation of system operation, component-level power calculation, real-time cooperative optimization, and quantitative assessment of energy-saving performance.

Two representative operating scenarios are designed for comparative analysis during the cooling season:

- a conventional non-optimized operation mode, in which the chilled-water supply temperature is fixed at 9 °C, pump frequency is maintained at 50 Hz, and no load-adaptive control is applied; and

https://doi.org/10.3390/en19030608

- an AI-based group-control optimization mode, in which chilled-water supply temperature and circulation flow rates are dynamically adjusted using the proposed optimization framework.

3.3.1. Performance Under Low-Load Operating Conditions

Figure 6a illustrates the system cooling-load profile under typical low-load operating conditions during the cooling season. Based on this load characteristic, Figure 6b compares the system energy consumption before and after the application of the AI-based optimization strategy.

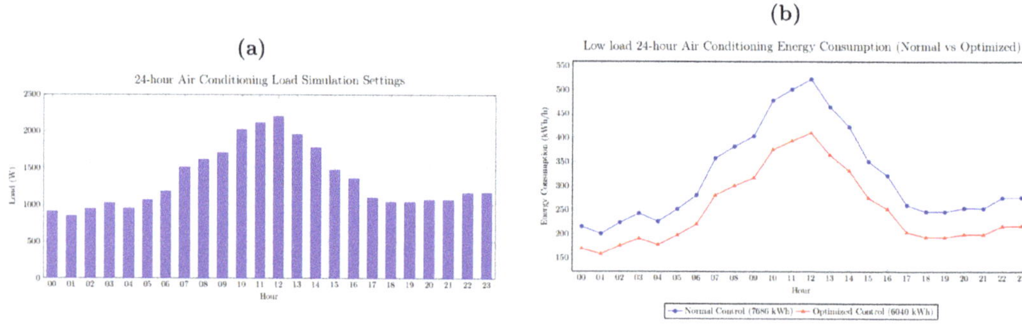

Figure 6. System performance under low-load operating conditions during the cooling season: (**a**) cooling-load variation; (**b**) comparison of energy consumption before and after AI-based optimization.

Simulation results indicate that under low-load conditions, the proposed control strategy effectively coordinates GSHP unit staging and pump operation, thereby maintaining system operation close to the high-efficiency region. As a result, unnecessary unit cycling and redundant pump operation are significantly reduced. Quantitative analysis shows that the optimized control achieves an average daily energy-saving rate of 21.4%, corresponding to a daily electricity saving of 1646 kWh. Assuming an electricity tariff of 0.8 RMB/kWh, the resulting daily cost reduction is approximately RMB 1316.

3.3.2. Performance Under High-Load Operating Conditions

Figure 7a presents the cooling-load variation of the GSHP system under typical high-load operating conditions. Figure 7b shows the corresponding comparison of system energy consumption between conventional control and AI-based optimized control.

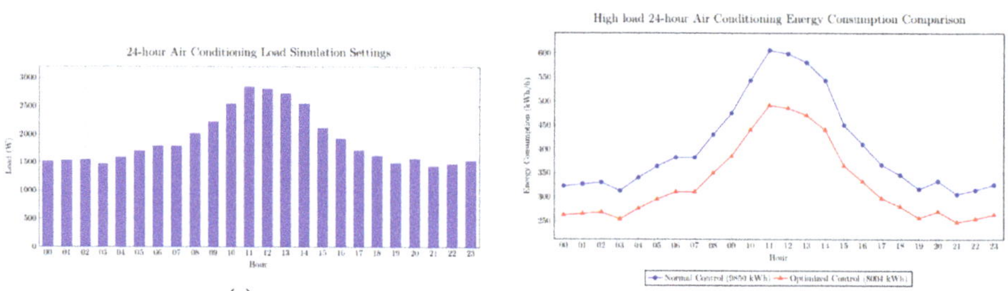

(a)

24-hour Air Conditioning Load Simulation Settings

(b)

High load 24-hour Air Conditioning Energy Consumption Comparison

Figure 7. System performance under high-load operating conditions during the cooling season: (**a**) cooling-load variation; (**b**) comparison of energy consumption before and after AI-based optimization.

https://doi.org/10.3390/en19030608

Despite the significantly increased cooling demand under high-load conditions, the proposed optimization strategy remains effective in reducing total system energy consumption through real-time adjustment of chilled-water supply temperature and ground-loop circulation flow rate. The simulation results indicate that the system achieves an average daily energy-saving rate of 18.7% under high-load conditions, with a daily electricity saving of 1841 kWh. At an electricity price of 0.8 RMB/kWh, this corresponds to a daily cost saving of approximately RMB 1472.

3.3.3. Annual Energy-Saving Potential

Based on the annual operating profile of the Liuzhou Fengqing Port GSHP system, approximately 120 days of low-load operation and 90 days of high-load operation are assumed. Using the simulated daily energy-saving results, the total annual electricity saving of the cooling-source subsystem can be estimated as follows:

$$E_{\text{annual}} = 1646 \times 120 + 1841 \times 90 = 3.63 \times 10^5 \text{ kWh}, \quad (30)$$

With an average electricity tariff of 0.8 RMB/kWh, the corresponding annual cost saving is approximately RMB 2.91×10^5. These results demonstrate the significant energy-saving potential and practical applicability of the proposed AI-based cooperative optimization strategy for large-scale GSHP systems.

3.4. Field Implementation and Measured Energy-Saving Performance of AI-Based Optimization Control

After completing the modeling and simulation-based validation of the proposed optimization strategy, the AI-based cooperative optimization control was deployed to the cooling and heating source system of the Liuzhou Fengqing Port project for on-site verification. A continuous field test lasting approximately two months was conducted to evaluate the actual energy-saving performance under real operating conditions.

The AI control platform adopts a programmable logic controller (PLC) as the execution layer.

Operating data from ground-source heat pump (GSHP) units and pumps are acquired through standard industrial communication protocols, while optimized control setpoints—including unit start–stop scheduling, variable-frequency pump control, and supply-water temperature adjustment—are issued in real time. The overall control architecture forms a closed-loop structure consisting of data acquisition, model prediction, optimization decision, and command execution, enabling the optimization algorithm to continuously adapt to changing operating conditions in practice.

3.4.1. Test Methodology and Baseline Configuration

Energy-saving performance was evaluated using an alternating-day experimental method, in which one day of AI-based control was followed by one day of conventional (non-AI) control. The test period spanned from 19 September to 17 November 2024. Daily electricity consumption was used as the primary evaluation metric.

Under the non-AI baseline condition, system settings were configured as follows. In September, the chilled-water supply temperature was fixed at 10 °C, with the ground-side circulation pumps operating at 44 Hz and the chilled-water pumps at 50 Hz. During October and November, the chilled-water supply temperature remained at 10 °C, both pump types operated at a fixed frequency of 50 Hz, and the cooling tower ran at constant speed. These baseline settings reflect typical conventional control strategies without load-adaptive optimization.

https://doi.org/10.3390/en19030608

3.4.2. Energy-Saving Performance of the Cooling System

Figure 8 presents the comparative results of daily electricity consumption for the cooling system under AI-based control and conventional operation during the test period. In September, the AI-based strategy achieved energy-saving rates of 17.4% for the ground-side circulation pumps, 12.5% for the GSHP units, and 43.7% for the chilled-water pumps, resulting in an overall system energy-saving rate of 24.1%.

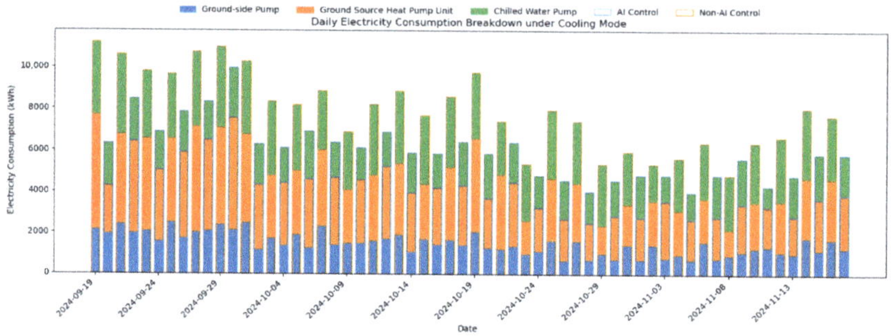

Figure 8. Daily electricity consumption breakdown of the cooling system under different control strategies. Each stacked bar represents the total daily electricity consumption, consisting of ground-side circulation pumps, ground-source heat pump units and chilled-water pumps. Bars outlined in blue indicate AI-based control, while those outlined in orange indicate conventional (non-AI) operation.

In October, the energy-saving performance further improved, with the ground-side pump energy-saving rate increasing to 29.5%, while the chilled-water pumps maintained a stable saving rate of 42.2%. The overall system energy-saving rate reached 25.8%. As the system entered a low-load operating regime in November, the total system energy-saving rate remained at 21.8%, and the chilled-water pumps continued to exhibit a substantial energy-saving rate of 33.3%. These results indicate that the AI-based control strategy maintains robust performance across varying load conditions.

3.4.3. Energy-Saving Performance of the Heating System

The measured energy-saving performance of the heating system during the same test period is shown in Figure 9. For the No. 1 hot-water heat pump unit, the AI-based control strategy achieved system-level energy-saving rates of 27.2%, 23.7%, and 21.8% in September, October, and November, respectively.

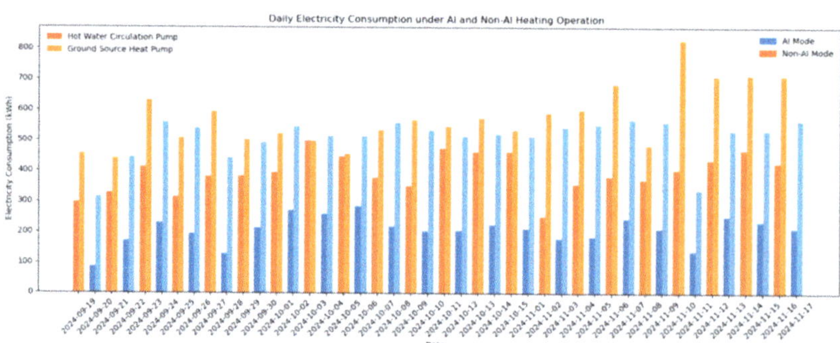

Figure 9. Measured energy-saving performance of the heating system under AI-based control during the test period from September to November.

https://doi.org/10.3390/en19030608

Throughout the entire heating test period, the chilled-water pumps consistently exhibited high energy-saving rates ranging from 39% to 51%. This sustained performance demonstrates the strong capability of the AI-based optimization strategy in continuously optimizing variable-flow operation, even under changing seasonal and load conditions.

3.4.4. Field Test Results

Based on the comprehensive multi-condition field test results, the AI-based control system achieved an average energy-saving rate of 24.1% during the cooling season and 24.2% during the heating season, with the chilled-water pumps identified as the primary contributors to overall energy reduction. Notably, even during low-load and transitional seasons, the system maintained an energy-saving rate exceeding 21.8%, demonstrating the stability, robustness, and strong potential for large-scale application of the proposed AI-based cooperative optimization strategy.

4. Conclusions

This study investigates the intelligent optimization of ground-source heat pump (GSHP) systems for large commercial buildings, taking the Liuzhou Fengqing Port commercial complex as a representative case. A comprehensive AI-based group-control optimization framework founded on gray-box modeling is proposed to address the long-standing issues of rigid operational strategies, insufficient equipment coordination, and suboptimal energy efficiency in conventional GSHP system operation. The main conclusions are summarized as follows.

- A gray-box performance modeling method for GSHP units is developed based on thermodynamic principles and measured operational data. By extending and reformulating the classical YAO model, the proposed GSHP-YFU model accurately captures the coupled effects of load ratio, chilled-water temperature, ground-loop temperature, and flow rate on the coefficient of performance (COP), while maintaining strong interpretability and suitability for real-time engineering applications.
- Energy consumption models for both single and parallel variable-frequency pumps are established using data-driven identification techniques. These models effectively characterize the nonlinear relationship between pump power consumption and real-time flow rate under different staging conditions, providing a reliable foundation for system-level hydraulic optimization.
- A multi-parameter cooperative optimization strategy is proposed, integrating GSHP unit start–stop scheduling, load allocation, pump-staging optimization, and continuous–variable regulation. To efficiently handle the coexistence of discrete and continuous decision variables, a two-stage hierarchical optimization framework is designed, in which discrete staging decisions are determined first, followed by particle swarm optimization (PSO)for continuous operational variables. This structure significantly reduces computational complexity while ensuring global optimization performance.
- An integrated simulation platform is constructed based on measured data and the developed gray-box models to validate the proposed optimization strategy under different operating conditions. Simulation results indicate that the AI-based optimization strategy achieves average daily energy-saving rates of 21.4% under low-load conditions and 18.7% under high-load conditions during the cooling season, demonstrating strong robustness across varying load regimes.
- Field implementation and alternating-day on-site testing further confirm the practical effectiveness of the proposed strategy. The AI-based control system achieves average energy-saving rates of 24.1% in the cooling season and 24.2% in the heating season,

https://doi.org/10.3390/en19030608

with chilled-water pumps identified as the primary contributors to energy reduction. Even during low-load and transitional periods, the system consistently maintains energy-saving rates exceeding 21.8%. The seasonal coefficient of performance (SCOP) of the GSHP system is improved from approximately 3.2 to above 4.0, corresponding to an annual electricity saving of more than 3.6×10^5 kWh.

- Overall, the proposed AI-based cooperative optimization framework demonstrates strong engineering applicability, stability, and scalability. It provides a reusable technical paradigm for intelligent energy-saving retrofits of GSHP systems in large commercial buildings and contributes to the advancement of low-carbon building operation under the dual-carbon targets.

Despite the promising results, several aspects warrant further investigation. Future work will focus on improving the scalability of the proposed optimization framework by replacing the current exhaustive search strategy in the discrete decision stage with more efficient solution methods, such as genetic algorithms or mixed-integer particle swarm optimization, to enable application to larger-scale GSHP systems. In addition, longer-term field tests covering multiple seasons, together with soil temperature profiling, will be conducted to further evaluate the long-term thermal behavior of ground heat exchangers and the effectiveness of thermal imbalance mitigation under practical operating conditions.

Author Contributions: Conceptualization, K.W.; Methodology, Z.S. and Y.Y.; Software, Z.S. and Y.Y.; Formal analysis, K.W. and Y.Y.; Investigation, K.W., Z.S. and Y.Y.; Resources, K.W. and Y.Y.; Data curation, Z.S.; Writing—original draft, Z.S.; Writing—review & editing, Y.Y.; Supervision, Y.Y.; Project administration, K.W. All authors have read and agreed to the published version of the manuscript.

Funding: This research was funded by the Key R&D Program of Guangxi Province, Grant No. 2024AB02026.

Data Availability Statement: The original contributions presented in the study are included in the article, further inquiries can be directed to the corresponding author.

Conflicts of Interest: Author Kui Wang was employed by the company Guangxi Guiwu Jin'an Refrigeration and Air-Conditioning Technology Co., Ltd. and Guangxi Guiwu Energy Saving Co., Ltd. The remaining authors declare that the research was conducted in the absence of any commercial or financial relationships that could be construed as a potential conflict of interest.

References

1. China Academy of Building Energy Efficiency. *China Building Energy Consumption and Carbon Emission Research Report (2023)*; China Academy of Building Energy Efficiency: Beijing, China, 2024.
2. Research Institute of Urban Construction, Department of Chongqing University. *China Building Energy Consumption and Carbon Emission Research Report (2023)*; Chongqing University: Chongqing, China, 2024.
3. Guo, S.Y.; Jiang, Y.; Hu, S. The pathways toward carbon peak and carbon neutrality in China's building sector. *Chin. J. Urban Environ. Stud.* **2022**, *10*, 2250011. [CrossRef]
4. Maghrabie, H.M.; Abdeltwab, M.M.; Tawfik, M.H.M. Ground-source heat pumps (GSHPs): Materials, models, applications, and sustainability. *Energy Build.* **2023**, *299*, 113560. [CrossRef]
5. Adebayo, P.; Jathunge, C.B.; Darbandi, A.; Fry, N.; Shor, R.; Mohamad, A.; Wemhöner, C.; Mwesigye, A. Development, modeling, and optimization of ground source heat pump systems for cold climates: A comprehensive review. *Energy Build.* **2024**, *320*, 114646. [CrossRef]
6. Spitler, J.D.; Gehlin, S. Measured performance of a mixed-use commercial-building ground source heat pump system in Sweden. *Energies* **2019**, *12*, 2020. [CrossRef]
7. Zhi, C.Q.; Zhou, X.; Hong, X.Y.; Wang, A.X.; Ye, W.; Zhang, X.; Chen, H.X. Field study of the long-term performance and reliability of large-scale ground source heat pump systems applied to a mixed-use complex of buildings. *Renew. Energy* **2025**, *241*, 122344. [CrossRef]
8. Bisengimana, E.; Zhou, J.Z.; Binama, M.; Zhao, K.M.; Abbas, S.; Yuan, Y.P. The frosting and soil imbalance performance issues of building heat pumps: An overview. *Energy Build.* **2022**, *273*, 112387. [CrossRef]

https://doi.org/10.3390/en19030608

9. Li, W.X.; Li, X.D.; Wang, Y.; Tu, J.Y. An integrated predictive model of the long-term performance of ground source heat pump (GSHP) systems. *Energy Build.* **2018**, *159*, 309–318. [CrossRef]
10. Bina, S.M.; Fujii, H.; Kosukegawa, H.; Inagaki, F. A predictive model of long-term performance assessment of ground source heat pump (GSHP) systems in Japanese regions. *Geothermics* **2024**, *119*, 102955. [CrossRef]
11. Liu, J.; Wang, F.H.; Gao, Y.; Zhang, Y.P.; Cai, W.L.; Wang, M.; Wang, Z.H. Influencing factors analysis and operation optimization for the long-term performance of medium-deep borehole heat exchanger coupled ground source heat pump systems. *Energy Build.* **2020**, *226*, 110385. [CrossRef]
12. Li, J.L.; Qi, X.P.; Li, X.L.; Huang, H.J.; Gao, J. Coupled study on the building load dynamics and thermal response of ground sources in shallow geothermal heat pump systems under severe cold climate conditions. *Modelling* **2025**, *6*, 63. [CrossRef]
13. Alavy, M.; Shirazi, P.; Rosen, M.A. Long-term energy performance of thermal caisson geothermal systems. *Energy Build.* **2023**, *292*, 113152. [CrossRef]
14. Ma, W.T.; Kim, M.K.; Hao, J.L. Numerical simulation modeling of a GSHP and WSHP system for an office building in the hot summer and cold winter region of China: A case study in Suzhou. *Sustainability* **2019**, *11*, 3282. [CrossRef]
15. Wang, J.L.; Yan, T.; Tang, X.; Pan, W.G. Design and operation of hybrid ground source heat pump systems: A review. *Energy* **2025**, *316*, 134537. [CrossRef]
16. Noye, S.; Martinez, R.M.; Carnieletto, L.; De Carli, M.; Aguirre, A.C. A review of advanced ground source heat pump control: Artificial intelligence for autonomous and adaptive control. *Renew. Sustain. Energy Rev.* **2022**, *153*, 111685. [CrossRef]
17. Cupeiro Figueroa, I.; Picard, D.; Helsen, L. Short-term modeling of hybrid geothermal systems for model predictive control. *Energy Build.* **2020**, *215*, 109884. [CrossRef]
18. Hou, G.Y.; Taherian, H.; Li, L.J.; Fuse, J.; Moradi, L. System performance analysis of a hybrid ground source heat pump with optimal control strategies based on numerical simulations. *Geothermics* **2020**, *86*, 101849. [CrossRef]
19. Zhou, K.; Mao, J.F.; Li, Y.; Zhang, H.; Chen, S.Y.; Chen, F. Thermal and economic performance of horizontal ground source heat pump systems with different flowrate control methods. *J. Build. Eng.* **2022**, *53*, 104554. [CrossRef]
20. Li, Y.F.; O'Neill, Z.; Zhang, L.; Chen, J.L.; Im, P.; DeGraw, J. Grey-box modeling and application for building energy simulations: A critical review. *Renew. Sustain. Energy Rev.* **2021**, *146*, 111174. [CrossRef]
21. Afram, A.; Janabi-Sharifi, F. Gray-box modeling and validation of residential HVAC system for control system design. *Appl. Energy* **2015**, *137*, 134–150. [CrossRef]
22. Fu, Q.M.; Chen, X.Y.; Ma, S.A.; Fang, N.W.; Xing, B.; Chen, J.P. Optimal control method of HVAC based on multi-agent deep reinforcement learning. *Energy Build.* **2022**, *270*, 112284. [CrossRef]
23. Zhao, Z.H.; Wang, C.L.; Wang, B.L. Adaptive model predictive control of a heat pump-assisted solar water heating system. *Energy Build.* **2023**, *300*, 113682. [CrossRef]
24. Yao, Y.; Yu, Y. *Modeling and Control in Air-Conditioning Systems*; Springer: Berlin/Heidelberg, Germany, 2017.

https://doi.org/10.3390/en19030608

MDPI AG
Grosspeteranlage 5
4052 Basel
Switzerland
Tel.: +41 61 683 77 34

Energies Editorial Office
E-mail: energies@mdpi.com
www.mdpi.com/journal/energies

www.ingramcontent.com/pod-product-compliance
Lightning Source LLC
Chambersburg PA
CBHW061848210926
51850CB00006B/1001